Centrifugal Microfluidics for Nanobiotechnology

Tae Seok Seo
Professor
Integrative NanoBioMEMS Laboratory
Kyung Hee University
South Korea

""

Preface

Since the inception of lab-on-a-chip (LOC) and micro-total-analysis-systems (μTAS) concepts, a diverse array of microsystems founded on microfluidics has been introduced. Among these, centrifugal microfluidics emerges as particularly noteworthy, showcasing several distinctive advantages over alternative forms. Its most pronounced merit lies in its divergence from the necessity for intricate tubes or pumps in controlling solution flow, a departure from the complexities inherent in other microfluidic devices. This characteristic streamlines device design and fabrication, thereby amplifying data reproducibility through the elimination of manual interventions.

For several years, my research has focused on microdevices for genetic diagnostics grounded in various driving forces such as electrokinetic and hydrodynamic forces. The challenges encountered in establishing consistent conditions for the same chip, stemming from complications in creating tube connection part on a chip, managing multiple tubes for sample in-and-out, and addressing irregular pump operations, have prompted a search for more efficient solutions. In this context, centrifugal microfluidics has proven invaluable, effortlessly executing fundamental unit operations such as volume splitting, solution merging, passive valving, and flow switching solely through rotational force and direction. This capability, coupled with straightforward microfluidic design and surface treatment, underscores the practicality of centrifugal microfluidics.

Within our specialized genetic diagnostics research lab, centrifugal microfluidics was initially introduced for genomic extraction in 2008. Subsequent developments encompassed various forms of genetic diagnostic centrifugal microfluidics, including polymerase chain reaction, isothermal amplification, a lateral flow strip assay, and a solution-loading cartridge facilitating the entire process from sample loading to diagnosis. Additionally, our lab has expanded its focus to high-throughput immuno-diagnostics and nanoparticle synthesis using centrifugal microfluidics, culminating in a high-throughput concept capable of concurrently processing up to 40 samples in a single device, offering a distinct advantage over other microfluidic forms.

Collating our laboratory's published papers, I organized them into ten chapters. The central theme of this book revolves around two primary applications of centrifugal microfluidic technologies: genetic analysis and nanoparticle synthesis. These sections distinguish themselves through their depth and detail, providing a comprehensive examination of the processes from both theoretical and practical perspectives. Each chapter intricately dissects device design, operational principles, and the unparalleled efficiency of centrifugal microfluidics in molecular diagnostics and nanoparticle synthesis.

It is my hope that this work stimulates contemplation among readers regarding the potential impacts of these technologies across various sectors, particularly in the field of nanobiotechnology. With its resonance within both the scientific community and industry practitioners, the book can offer insights into how centrifugal microfluidics is shaping the future of rapid and accurate biomedical diagnostics, contributing significantly to global health strategies, as well as exploring novel smart nanomaterials for the future.

Acknowledgements

In the process of compiling this manuscript, I find myself reflecting upon my academic career and acknowledging the individuals who played pivotal roles in bringing this book to fruition. Foremost among them are the numerous students who, under my mentorship during my tenure as a professor, conducted research in the realm of centrifugal microfluidics. Their unwavering passion, diligence, and innovative approaches were indispensable in achieving commendable research outcomes, and this book's compilation would have been insurmountable without their contributions. Special recognition is extended to Dr. Jae Hwan Jung, Dr. Byung Hyun Park, Dr. Seung Jun Oh, Dr. Hau Van Nguyen, Dr. Goro Choi, Ji Hyun Seo, Hyun Young Heo, Vu Minh Phan, and Hiep Van Nguyen.

Particularly noteworthy are the substantial contributions of Vu Minh Phan, Hiep Van Nguyen, and Thanh Hoai Nguyen in tasks such as paper editing, figure refinement, formatting standardization, and bibliography compilation.

Above all, my endless love and gratitude is extended to Daniel, Clara, John, Joseph, and Elizabeth, who have served as steadfast companions and sources of intellectual fortitude throughout my academic journey. Finally, but most importantly, I express my thanks to God for allowing all of these endeavors to come to fruition.

Tae Seok Seo
Kyung Hee University, South Korea
February 24, 2024

Contents

Preface   1

Acknowledgements   3

Chapter 1. Centrifugal Microfluidic Device for Sample Pretreatment   8

1. Background   8
2. Chip Design   9
   2.1. A low-throughput centrifugal microfluidic chip for sample pretreatment   10
   2.2. A high-throughput centrifugal microfluidic chip for sample pretreatment   13
3. Chip fabrication   14
4. Chip operation   16
   4.1. Principle of a centrifugal microfluidics   16
   4.2. Operation of a low-throughput centrifugal microfluidic chip for sample pretreatment   17
   4.3. Operation of a high-throughput centrifugal microfluidic chip for sample pretreatment   19
5. Performance of nucleic acid extraction   22
   5.1. RNA extraction on a low-throughput centrifugal microfluidic chip   22
   5.2. DNA extraction on a high-throughput centrifugal microfluidic chip   24
6. Conclusion   26
References   27

Chapter 2. Centrifugal Microfluidic Device for Polymerase Chain Reaction   30

1. Background   30
2. Centrifugal PCR chip   32
2.1. Material for a microfluidic chip   32
2.2. Chip design   33
2.3. Chip fabrication   34
3. Centrifugal PCR system   35
   3.1. Rotational motor   35
   3.2. Heater   36
   3.3. Fluorescence detection   38
4. Chip operation   39
5. Performance of PCR on a centrifugal microdevice   40
6. Conclusion   43
References   43

Chapter 3. Centrifugal Microfluidic Device for Isothermal Amplification 45

1. Background   45
   1.1. A LAMP reaction   46
   1.2. An RCA reaction   46
   1.3. An RPA reaction   47
2. Centrifugal chip for isothermal amplification   48
   2.1. Chip design for LAMP   48
   2.2. Chip design for RCA   49
   2.3. Chip design for RPA   50
3. Chip fabrication process   52
4. Chip operation   55
5. Performance of isothermal amplification on a centrifugal microdevice   60
6. Conclusion   66
References   67

Chapter 4. Colorimetric Loop-mediated Isothermal Amplification Reaction on a Centrifugal Microfluidic Device   71

1. Background   71
   1.1. Colorimetric assay for genetic analysis   71
   1.2. Colorimetric PCR and colorimetric LAMP   72
   1.3. Principle of EBT-based colorimetric detection   74
2. Centrifugal microdevice for colorimetric LAMP reactions   75
   2.1. Chip design   75
   2.2. Chip fabrication   78
3. Chip operation   79
4. Application for food-borne pathogen detection   83
   4.1. Monoplex detection of pathogen   83
   4.2. Multiplex detection of pathogen   85
   4.3. Limit-of-detection test   88
5. Conclusion   90
References   90

Chapter 5. Lateral Flow Strip Assay-Incorporated Centrifugal Microfluidic Device for Genetic Analysis   93

1. Background   93
   1.1. Lateral flow strip assay   93
   1.2. Principle of amplicon detection on a lateral flow strip   95
2. Lateral flow assay-incorporated centrifugal microfluidic chip   97
   2.1. Design of a LAMP-lateral flow strip chip   97
   2.2. Design of a sample pretreatment-LAMP-lateral flow strip chip   99
   2.3. Chip fabrication   100
3. Chip operation   102
4. Application of pathogen detection   104
   4.1. Virus detection on a LAMP-lateral flow strip chip   105
   4.2. Bacterial detection on a sample pretreatment-LAMP-lateral flow strip chip   108
5. Conclusion   111

References 112

Chapter 6. Combination of a Solution-loading Cartridge with a Centrifugal Microfluidic Device — 115

1. Background 115
2. Chip design 117
    2.1. Design and fabrication of a solution-loading cartridge 117
    2.2. Design and fabrication of an integrated centrifugal microdevice 121
3. Development of a portable genetic analyzer 125
4. Chip operation 129
5. Application of multiplex pathogen detection 132
6. Conclusion 142
References 143

Chapter 7. Centrifugal Microfluidic Device for High-throughput Genetic Analysis — 147

1. Background 147
2. Centrifugal microfluidic chip for HTP genetic analysis 148
    2.1. Overall design of a centrifugal HTP microfluidic chip 148
    2.2. A single unit-centrifugal microfluidics chip with multiple reaction chambers 150
    2.3. A centrifugal HTP microfluidics chip with 10 units for COVID-19 diagnostics 151
    2.4. A centrifugal HTP microfluidics chip with 30 units for COVID-19 diagnostics 154
3. Fabrication of a centrifugal HTP chip 155
4. Construction of a portable HTP genetic analyzer 157
5. Chip operation 159
6. Application of pathogen detection 163
    6.1. A single unit-centrifugal microfluidics chip with multiple reaction chambers 163
    6.2. A centrifugal HTP microfluidics chip with 10 units for COVID-19 diagnostics 164
    6.3. A centrifugal HTP microfluidics chip with 30 units for COVID-19 diagnostics 166
7. Conclusion 169
References 170

Chapter 8. Centrifugal Microfluidic Device for High-throughput Enzyme-linked Immunosorbent Assay — 173

1. Background 173
2. Design of a centrifugal ELISA HTP chip 175
3. Immobilization of antibody on a microfluidic device 177
4. Fabrication of a centrifugal ELISA HTP chip 178
5. Construction of a portable HTP genetic analyzer 179
6. Chip operation 181
7. Application of COVID-19 detection 183

8. Conclusion 186
References 187

Chapter 9. Centrifugal Microfluidic Device for High-throughput Nanoparticle Synthesis 189

1. Background 189
   1.1. Limitations of traditional nanoparticle synthetic approaches 189
   1.2. Microfluidics based nanoparticle synthesis 190
   1.3. High-throughput nanoparticle synthesis on a microfluidic device 191
2. Design of a centrifugal HTP chip for nanoparticle synthesis 192
   2.1. A zigzag aliquot structure for efficient and rapid solution loading 193
   2.2. Regulating the release of the solution through centrifugal force 194
   2.3. Design for a 30-unit centrifugal HTP chip 195
   2.4. Design for a 60-unit centrifugal HTP chip 196
   2.5. Design for a 60-unit centrifugal HTP chip with a serially diluting structure 197
3. Fabrication of a centrifugal HTP chip for nanoparticle synthesis 199
4. Chip operation 200
5. Application of nanoparticle synthesis 203
   5.1. Quantum dots 203
   5.2. Gold nanoparticles 206
   5.3. Bimetallic catalysts 210
6. Conclusion 211
References 212

Chapter 10. Automatic Centrifugal Microfluidic Device 216

1. Background 216
   1.1. Automation 217
   1.2. Robot Arm 218
2. Design of components 219
   2.1. A robotic solution pipetting device 219
   2.2. A solution loading cartridge 221
   2.3. Design and fabrication of a centrifugal microfluidic disc 223
3. Integration of a robotic solution pipetting device with a centrifugal chip 224
4. Construction of a portable genetic analyzer 225
5. Chip operation 227
6. Respiratory infectious virus detection from nasopharyngeal swab samples 228
7. Conclusion 232
References 233

Author 235

Chapter 1
Centrifugal Microfluidic Device for Sample Pretreatment

1. Background

The integration of various functional units into a single micro-total-analysis-system (µTAS) or lab-on-a-chip (LOC) device has revolutionized the field of genetic analysis, offering fast analysis time, reduced reagent consumption, high-throughput capability, and portability. These microfluidic devices have proven their potential in biomedical diagnostics, food safety testing, and environmental pollutant screening, moving towards the realization of a µTAS that combines miniaturized functional units on a single platform [1].

Despite the significant advancements in LOC technology, the integration of sample pretreatment units with detection units such as optical sensors, electrochemical devices, and polymerase chain reaction (PCR) systems remains a challenging task [2–11]. The handling of macro-to-micro interfaces, complex tubing, and pumping systems required to control sample flow in microfluidics often results in low data reproducibility, high chip fabrication costs, and expert-dependent operation [12, 13]. Traditional nucleic acid (NA) extraction techniques, which are crucial for most molecular diagnostic assays, rely on liquid-liquid phase extraction using toxic chemicals, making them unsuitable for integration into µTAS. The extraction kits, such as the QIAamp kit from Qiagen, the Maxwell kit from Promega, and the PureLink kit from Thermo Fisher, utilize spin columns with silica membranes for nucleic acid purification. Magnetic bead-assisted NA extraction kits like the MagJET from Thermo Fisher, the AxyPrep from Axygen, and the MagListo from Bioneer have also gained popularity due to their automatic capabilities [14]. While these commercial kits provide high-quality genomics for downstream biochemical reactions such as PCR, next-generation sequencing (NGS), and dot-blot hybridization (DBH), they also have limitations [15–17]. The overall process requires a large amount of disposable plastic, and the kits are expensive with tedious manual operation steps needed for a prolonged time.

Centrifugal microfluidic devices, as a subset of these technologies, have garnered attention, particularly in the realm of sample pretreatment, including nucleic acid (NA) extraction and purification [2, 4]. These devices, as expounded by Park et al. (2012), utilize rotational forces for fluid manipulation, offering a streamlined alternative to conventional microfluidic systems [18]. Their design simplicity, reduced power dependency, and lack of need for intricate external controls render them an attractive option for sample pretreatment, which is often beleaguered by the challenges of handling varying sample scales and integrating multiple process steps. Yet, the integration of sample pretreatment within LOC platforms is not devoid of hurdles. The conventional approaches are frequently encumbered by laborious and time-intensive procedures, necessitating complex tubing and control mechanisms [13, 19, 20]. This complexity becomes particularly pronounced when addressing the need for high-throughput processing, a critical demand in various scenarios such as during health crises or in forensic investigations.

In response to these challenges, recent innovations in centrifugal microfluidics have been introduced. Park et al. present, for the first time, a novel approach that combines rotary systems with silica-based solid-phase extraction for efficient RNA purification. Meanwhile, Jung et al. (2013) propose a microdevice that incorporates glass microbeads, aiming to elevate RNA purification efficiency from viral lysates [21]. This approach is indicative of a broader trend towards multifunctional and sensitive µTAS. In a different vein, Nguyen et al. (2021) focus on a high-throughput centrifugal microfluidic disc designed for forensic applications, capable of processing multiple samples simultaneously with enhanced efficiency, thus addressing the pressing needs in forensic DNA testing [22–24].

This chapter seeks to unravel the complexities and intricacies of centrifugal microfluidic devices as elucidated by Park et al., Jung et al., and Nguyen et al [18, 21, 22]. We will explore their design, fabrication processes, operational methodologies, and assess their efficacy in nucleic acid extraction. Through an amalgamation of insights from these leading-edge studies, our goal is to present a holistic view of the current landscape and the burgeoning potential of centrifugal microfluidic devices in sample pretreatment, setting the stage for the development of more advanced, integrated, and effective diagnostic systems. These advancements include the integration of various functional units on a single platform, the utilization of centrifugal pumps to simplify and speed up DNA/RNA extraction, and the use of innovative materials such as surface-treated glass microbeads for RNA purification. These innovations represent significant steps toward overcoming the existing challenges in sample pretreatment and integration of detection units, thereby paving the way for fully realized µTAS and LOC devices with broad applications in biomedical diagnostics, food safety testing, and environmental pollutant screening.

2. Chip Design

The design of centrifugal microfluidic devices, a critical component of µTAS and LOC systems, has seen remarkable innovations in recent years. These designs are pivotal in determining the efficiency, functionality, and applicability of the devices in various diagnostic and analytical settings. In the realm of centrifugal microfluidics, a key differentiator in design lies in the throughput capacity of the chips, broadly categorized into low-throughput and high-throughput systems. Low-throughput designs are typically tailored for specific, often more specialized applications where the volume of samples processed is relatively small, but precision and accuracy are paramount. These designs focus on detailed manipulation of small fluid volumes and often integrate sophisticated functionalities on a single chip. The precision and control offered by these designs make them ideal for applications where individual sample handling is crucial. Conversely, high-throughput designs are engineered to handle a larger number of samples concurrently, emphasizing efficiency and speed. These systems are essential in scenarios requiring rapid processing of multiple samples, such as during epidemic outbreaks or mass screening processes. High-throughput designs often incorporate parallel processing channels or arrays, leveraging the centrifugal platform to process numerous samples in an automated, efficient manner. Both low-throughput and high-throughput designs share common features essential to centrifugal microfluidic devices, such as the need for precise control of fluid dynamics, integration of various functional units (like mixers, separators, and reactors), and the capability to perform multiple steps of an analytical process in an automated sequence.

In the following sections, we delve deeper into the specific designs of low-throughput and high-throughput centrifugal microfluidic chips, as illuminated by the studies of Park et al., Jung et al., and Nguyen et al., to understand their unique features, functionalities, and applications.

2.1. Low-throughput centrifugal microfluidic chip for sample pretreatment

The low-throughput centrifugal microfluidic devices, as explored by Park et al. and Jung et al., epitomize precision and specialization in the realm of microfluidic technology [18, 21]. These designs, while focused on processing a smaller volume of samples, incorporate intricate functionalities and mechanisms tailored for specific, often complex, analytical tasks. Both designs, while distinct in their approach, share common design principles essential to low-throughput centrifugal microfluidic devices. They exhibit meticulous attention to detail in the layout of fluidic pathways, ensuring that each step of the process—from sample introduction to final analysis—is carried out with precision. The integration of various functional units within a compact space without compromising the efficiency of each operation is a hallmark of these designs.

Park et al.'s microdevice design for RNA purification demonstrates an intricate integration of microfluidics and rotary systems [18]. The device comprises a silica sol-gel matrix for RNA capture and three distinct reservoirs for various solutions, connected by microfluidic channels of varying dimensions. Each channel is specifically designed to handle different solutions: 120 μm wide for the RNA sample, 40 μm for the washing solution, and 20 μm for the elution buffer. The microchip itself consists of three layers, with a middle polydimethylsiloxane (PDMS) layer patterned with these microfluidic channels and the sol-gel chamber. The microfabrication process involves soft lithography techniques, with the PDMS layer being 50 μm in depth. Park et al. presented the design of the microchip from for RNA sample preparation in Fig. 1. The microchip was composed of three microfluidic channels each of which was connected to the sample (R_S), washing solution (R_W), and elution buffer reservoir

(R_E), and the three channels were merged to the silica sol-gel chamber for RNA capture. All the flow-through solutions were collected at the outlet reservoir (R_O). A 5 mm dia. shaft hole was punched and fixed in the rotary axis. Three microfluidic channels were fabricated with 50 μm depth, but the width of each microfluidic channel was different: 120 μm for the sample reservoir (C_S), 40 μm for a washing solution (C_W), and 20 μm for an elution buffer (C_E). Such different channel dimensions allow us to dispense the solution sequentially by controlling the RPM. One microliter volume of the sol-gel chamber was packed with tetramethyl orthosilicate (TMOS) based sol-gel to capture and purify RNA from the virus sample.

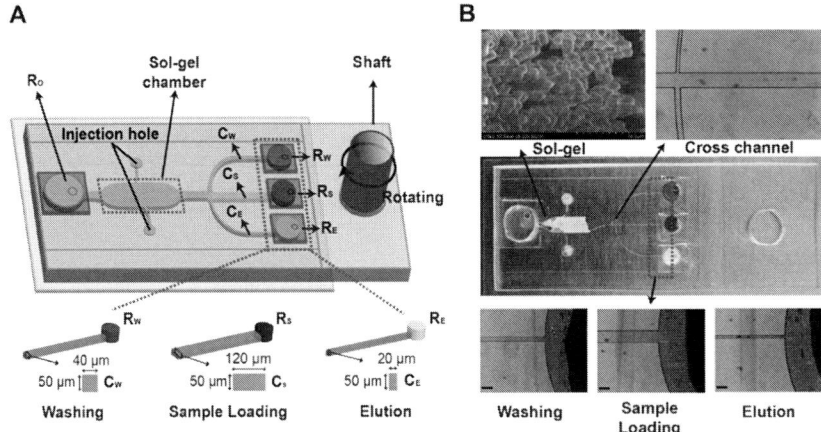

Fig. 1. Automatic RNA purification microdevice based on a centrifugal force. (A) Schematic image of the microdevice that is composed of four reservoirs (R_S, R_W, R_E, R_O), three different microfluidic channels (C_S, C_W, C_E), a sol-gel chamber, and a shaft hole. (R_S: a sample reservoir, R_W: a washing solution reservoir, R_E: an elution buffer reservoir, R_O: an outlet reservoir, C_S: 120 μm width, C_W: 40 μm width, C_E: 20 μm width) (B) A digital image of the microdevice showing the formed sol-gel, and the fabricated microfluidic channels. Scale bar: 100 μm. (Reprinted with permission from Ref. [18]. Copyright (2012) Royal Society of Chemistry.)

Jung et al. introduced an intricately designed centrifugal sample pretreatment microdevice, illustrated in their publication (Fig. 2A) [21]. The design is notable for its sophisticated arrangement of reservoirs and microchannels to facilitate efficient RNA purification. The microdevice comprises three critical reservoirs: one each for the sample, the washing solution, and the elution solution. The volumes of these reservoirs are precisely determined, with 4 μL for the sample, 6 μL for the washing solution, and 7 μL for the elution solution. These reservoirs are strategically positioned in front of the bead-bed microchannel, the central component for RNA capture and purification. The design incorporates a comb-shaped capillary valve, connecting the washing solution reservoir to the bead-bed microchannel. The dimensions of this valve are 580 μm in width, 500 μm in depth, and 2000 μm in length. Conversely, the elution solution reservoir is linked to the bead-bed microchannel via a siphon channel measuring 250 μm in width and 50 μm in depth. Both the capillary valve and the siphon channel converge into the bead-bed microchannel, where the RNA purification process takes place.

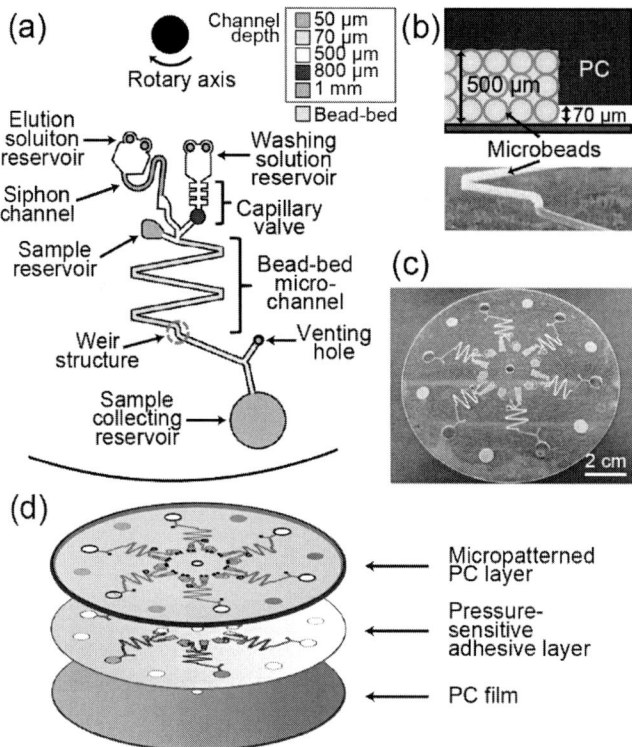

Fig. 2. (a) Schematic design of the centrifugal sample pretreatment microdevice consisting of four reservoirs (an RNA sample, a washing solution, an elution solution, and a sample collecting reservoir) and a bead-bed microchannel. The washing solution reservoir and the elution solution reservoir were connected to the bead-bed microchannel through a capillary valve and a siphon channel, respectively. (b) A cross-sectional scheme of the weir structure packed with the microbeads whose diameter was 150~212 µm (top) and a digital image at the interface of the weir structure (bottom). (c) A digital image of the centrifugal sample pretreatment microdevice. (d) Schematic illustration of a sample pretreatment microdevice which is composed of three layers: a 1-mm thick micropatterned PC layer (top), a 30-µm thick adhesive layer (middle), and a 125-µm PC film (bottom). (Reprinted with permission from Ref. [21]. Copyright (2013) Royal Society of Chemistry.)

The bead-bed microchannel itself is meticulously dimensioned, with a width of 580 µm, a depth of 500 µm, and a length of 4.2 cm. It is packed with 10 mg of glass beads, specifically placed in a designated area marked in blue in Fig. 2. To optimize the RNA capture efficiency, the microchannel is designed in a zigzag shape, which increases the loading capacity of the beads and augments the contact area with the RNA capture matrix. A crucial feature at the end of the bead-bed microchannel is a weir structure (illustrated in the interface between the blue and yellow color), which is designed to prevent the beads from flowing through during centrifugation. This structure is depicted in Fig. 2B. To achieve this, the channel depth is reduced from 500 μm in the bead-bed microchannel to 70 μm at the weir, a dimension smaller than the diameter of the microbeads, which range from 150 to 212 μm. This sophisticated design by Jung et al. reflects a deep understanding of the requirements for efficient RNA purification in a centrifugal microfluidic setting. The precise dimensions, along with the thoughtful arrangement of reservoirs, channels, and bead-bed, contribute to the high efficacy of the RNA purification process.

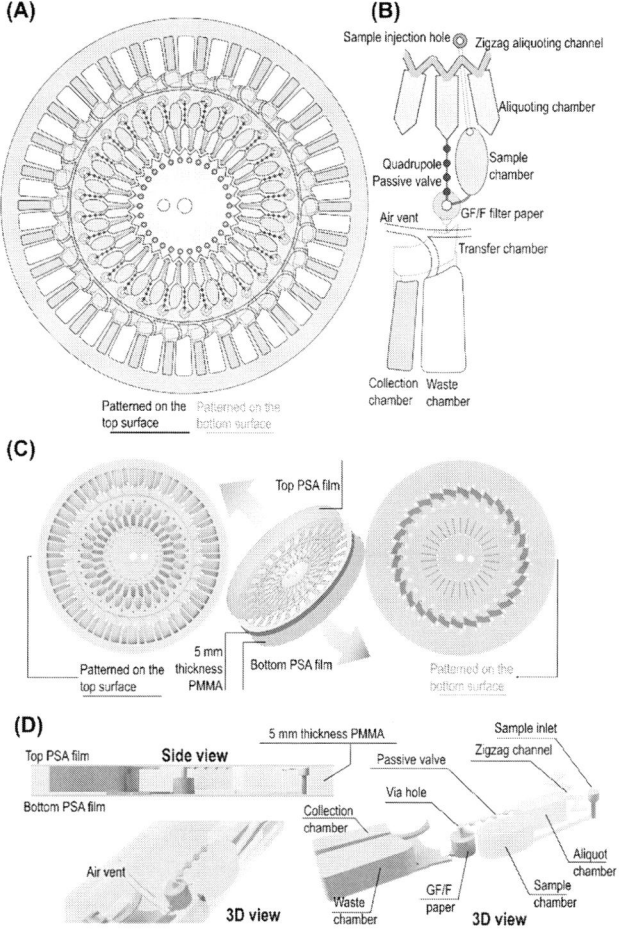

Fig. 3. (A) Schematic illustration of the high-throughput NA extraction disc. (B) Detailed structure of one extraction unit. (C) Illustration of the top and bottom pattern of the PMMA disc. The top side of the disc has an aliquoting structure, sample chambers, passive valves, collection chambers, waste chambers, and air vent lines. The bottom side of the disc has the sample injection channel, the GF/F chambers, and transfer chambers. (D) An illustration of one extraction unit. (Reprinted with permission from Ref. [22]. Copyright (2021) Elsevier.)

2.2. High-throughput centrifugal microfluidic chip for sample pretreatment

The high-throughput DNA extraction microfluidic disc was designed with a double-side micropattern on a poly(methyl methacrylate) (PMMA) disc (diameter of 130 mm and a thickness of 5 mm), featuring interconnected microchannels on the top and bottom sides via holes (Fig. 3) [22]. The aliquoting unit, sample, waste, and collection chambers, as well as air vent channels, were etched on the top side, while the sample injection channel, transfer chamber, and glass (GF/F) filter paper chamber were etched on the bottom side. This design allows for the sample, washing, and elution solutions to be transferred from the top to the bottom, passing through the GF/F paper, and then collected at the waste or collection chamber. A sophisticated aliquoting structure was developed for automatic division into 30 aliquots from one injection, consisting of a zigzag aliquoting channel, a quadrupole passive valve, and a super-hydrophobic coating. The zigzag channel connects to 30 pentagon-arrow-shaped aliquot chambers, with the passive valve positioned at the end of each chamber to prevent solution overflow during automatic injection. The quadrupole passive valves are connected by rectangular channels and circular patterns. The capillary pressure of the zigzag channel is lower than that of the passive valves, ensuring the solution prefers to fill the zigzag channel and aliquoting chambers rather than overflowing to the passive valves. To prevent bleeding of the washing solution, which contains 70% ethanol, into the GF/F filter paper during injection, a super-hydrophobic coating was utilized. This coating increases the contact angle of the ethanol on the PMMA, successfully allowing the aliquoting process of 70% ethanol into 30 aliquots from one injection. This chip represents a significant advancement in the field of microfluidic devices for high-throughput DNA extraction, catering to the needs of various applications in biomedical diagnostics, etc.

3. Chip fabrication

In the realm of microfluidic device fabrication, certain fundamental techniques are commonly employed, as seen in the works of Park et al., and Jung et al [18, 21]. These techniques typically start with the design phase, where precise layouts of microchannels and chambers are created using computer-aided design (CAD) software. This step is crucial for ensuring the accuracy of the microfluidic pathways and functionalities in the final device. The next common step involves the creation of molds for soft lithography, a technique widely used in microfabrication. This process usually employs a silicon wafer as a substrate, onto which a layer of photoresist is applied. The photoresist is then exposed to ultraviolet (UV) light through a mask that corresponds to the designed microchannel patterns. The exposed photoresist is developed, creating a mold with raised features that replicate the desired microfluidic patterns. Following the creation of the mold, the actual fabrication of the microfluidic layers begins. This typically involves pouring a polymer material onto the mold. PDMS is preferred for its flexibility, optical clarity, and biocompatibility. Once the polymer is cured and solidified, it is carefully peeled off the mold, resulting in a PDMS layer with the inverse pattern of the microchannels and chambers. The final step common to these processes is the bonding of the microfluidic PDMS layer to a substrate, which could be glass or another polymer. This bonding is often achieved using plasma treatment, which activates the surfaces of both the PDMS layer and the substrate to facilitate a strong, irreversible bond. This bonding ensures the integrity of the microchannels and prevents leaks during device operation.

The microchip developed by Park et al. consists of a three-layer structure [18]. The central layer is made of PDMS and is integral for the microfluidic channels and the sol-gel chamber. The PDMS layer is created using soft lithography, starting with the preparation of a 4-inch silicon wafer. This wafer is treated with oxygen plasma for 1 minute to enhance the adhesion of the photoresist. A negative photoresist, SU8-50, is then applied to form a layer 50 μm thick. The wafer, coated with this photoresist, is exposed to UV light through a film mask, which is intricately designed using AutoCAD software. Following UV exposure, the wafer is developed in a SU-8 developer and cleaned with isopropyl alcohol (IPA). It is then dried using nitrogen gas, ensuring the clarity and integrity of the microfluidic channels. The molding and assembling process of the microchip are critical steps. A mixture of PDMS pre-polymer and curing agent (Sylgard 184), in a 10:1 ratio, is poured over the prepared SU-8 mold. This mixture is degassed for 30 minutes and cured at 65°C for 1 hour. After curing, the PDMS layer is carefully peeled off the mold. Holes of specific diameters are punched into the PDMS layer for the rotation shaft (5 mm diameter) and the reservoirs (3 mm diameter), with injection holes measuring 1 mm in diameter. Following the punching process, the PDMS layer and the glass wafer are treated again with O_2 plasma for 1 minute and then bonded together at 85°C for an hour, ensuring a strong and leak-proof seal.

The sol-gel matrix formation within the microfluidic channels is a crucial part of the chip's functionality. The sol-gel precursor is prepared by mixing 2 mL tetramethyl orthosilicate (TMOS) (98%) with 0.44 g polyethylene glycol (PEG) in 10 mL acetic acid (0.01M). This mixture is stirred at 300 rpm in an ice bath for 45 minutes. Before introducing the sol solution into the microdevice, the chip is exposed to UV ozone to produce hydroxyl functional groups on the PDMS surface, enhancing the gelation reaction. A precise volume of 0.8 μL of the sol solution is then injected into the sol-gel chamber. The microdevice is incubated at 40°C for 12 hours, with the process repeated twice to optimize the density of the sol-gel matrix. The incubation process is tightly sealed with Parafilm® to prevent the evaporation of the sol solution.

Jung et al. adopted a unique approach for their microdevice, which incorporated TEOS-treated glass microbeads [21]. Their fabrication process utilized CNC machining to create the micropatterned polycarbonate (PC) layer, forming the core of the device. This layer was designed with specific depths to accommodate the microfluidic channels and the bead-bed for RNA purification. The device was assembled using three distinct layers: the top micropatterned PC layer, a middle layer of pressure-sensitive adhesive, and a bottom PC film. These layers were carefully aligned and bonded together to form the final microfluidic structure. The bead-bed microchannel was then packed with glass beads, essential for the RNA purification process. This method differed from traditional soft lithography, emphasizing the precision offered by CNC machining in fabricating complex microfluidic structures.

Nguyen et al.'s fabrication process for their high-throughput centrifugal microfluidic disc, designed for rapid DNA purification from forensic samples, involved a series of precise and carefully executed steps [22]. Utilizing advanced CNC machining, the team meticulously etched the disc on both the top and bottom surfaces to create 30 distinct extraction units. Each unit was designed with channels and chambers, varying in depth to serve specific functions within the DNA extraction process. The depths were strategically varied, with regions such as the zigzag aliquoting channel (sky-blue) having a depth of 0.5 mm, the aliquoting chamber (cyan) 4.2 mm, the sample chamber (green) 4.3 mm, and others according to the functional requirements of each section (Fig. 3D). A critical step in the fabrication was the application of a super-hydrophobic reagent to both surfaces of the disc. This treatment was essential for establishing controlled fluid dynamics within the disc, crucial for the accurate and efficient distribution of forensic samples and reagents across the 30 extraction units. Following the surface treatment, two circular-shaped GF/F filter papers, each 4 mm in diameter, were integrated into the disc. These filter papers served as the nucleic acid extraction matrix, fundamental to the DNA purification process. The final assembly of the disc involved sealing the etched PMMA layer with pressure-sensitive adhesive films. These films were crucial for maintaining the structural integrity of the extraction units and ensuring the effectiveness of the DNA extraction process. They prevented potential leaks and cross-contamination between the units, a vital consideration in forensic DNA analysis. Once sealed, the disc was ready for operational use, capable of processing multiple DNA samples simultaneously in a highly efficient and automated manner, when mounted on the custom-designed point-of-care testing (POCT) platform. This intricate fabrication process by Nguyen et al. reflects the complexity and innovation required to produce a high-throughput microfluidic device. The combination of precise CNC machining, thoughtful surface modification, and meticulous assembly underscores the technological advancement in creating a centrifugal microfluidic system capable of meeting the demanding requirements of forensic DNA testing.

4. Chip operation

4.1. Principle of centrifugal microfluidics

All three research groups - Park et al., Jung et al., and Nguyen et al. - innovatively utilized centrifugal force as the primary driving mechanism in their microfluidic devices [18, 21, 22]. This force played a pivotal role in overcoming the capillary pressure created by the microchannel dimensions and the hydrophobic nature of the chip materials. By finely controlling the centrifugal force, they ensured precise guidance of liquid samples to specific locations within the chip. A key feature in all three chips was the incorporation of passive capillary microvalves. These valves, formed through the strategic interplay of hydrophobic materials and microchannel geometries, acted as barriers to uncontrolled fluid flow. The applied centrifugal force was calibrated to overcome these microvalves at desired moments, allowing for the controlled movement of samples within the devices [25]. This method exemplifies a sophisticated approach to regulating fluid flow in centrifugal microfluidics.

In the high-throughput chip, particularly in Nguyen et al.'s design, syringe pumps and valves were employed to further refine the control over fluid movement [22]. The syringe pump accurately dispensed washing and elution solutions in precise volumes, while the valves directed these solutions to designated chambers within the chip. This level of control was essential for ensuring the efficiency and precision of the nucleic acid extraction process. A notable aspect of these devices was the use of super-hydrophobic surface coatings. This modification was particularly crucial in preventing the washing solution from inadvertently mixing with other components or areas, such as the filter paper, during the aliquoting process. The surface treatment ensured targeted delivery of solutions, enhancing the overall efficiency of the extraction process.

The combination of centrifugal force, passive capillary microvalves, syringe pumps, valves, and surface modifications in these chips represented a synergistic application of microfluidic principles. These elements worked together to achieve a high level of efficiency, precision, and reliability in the nucleic acid extraction process. They were foundational to the functionality of the chips and integral to attaining the desired performance in nucleic acid extraction applications. The centrifugal microfluidic platforms developed by Park et al., Jung et al., and Nguyen et al. also demonstrate a strong emphasis on automation [18, 21, 22]. The rotation of the discs and the consequent fluid movement were precisely controlled, often by automated systems, to ensure consistent and accurate processing of samples and reagents. These devices exemplify the integration of various functional units into a single format, thereby streamlining complex biochemical processes such as mixing, separation, and detection. This integration not only enhances the efficiency of the devices but also simplifies their operation, making them suitable for a range of applications, including POCT.

4.2. Operation of the low-throughput centrifugal microfluidic chip for sample pretreatment

Park et al. developed an innovative microdevice for RNA purification that operates on the principle of centrifugal force [18]. The device's functionality is enhanced by a carefully designed rotary system, which is powered by a servo motor capable of precise control over rotational speed and time. A key step in preparing the microdevice involved treating it with UV ozone for 10 minutes. This treatment likely improves the hydrophilicity of the internal surfaces of the microdevice, facilitating smoother fluid movement within the microchannels.

To validate the efficiency of controlling the sequential flow of liquids through rotational speed adjustments, Park et al. conducted an experiment using aqueous dye solutions. They loaded 5 µL of these solutions into different reservoirs: red dye in the R_W (i.e., the washing solution), blue dye in the R_S (i.e., the sample solution), and yellow dye in the R_E (i.e., the elution solution) reservoirs. A crucial aspect of this experiment was the different contact angles of the dye solutions on two surfaces: the PDMS layer and the glass layer. The contact angle of water on PDMS was measured at 79°, while on glass, it was significantly lower at 2°. Given the higher wettability of liquids on glass compared to PDMS, the flow control was predominantly influenced by the contact angle on the glass surface. In their theoretical calculations, Park et al. considered the contact angle of water on the glass to predict the theoretical burst RPM for each liquid. Experimentally, they incrementally increased the RPM to find the critical rotational speeds necessary for the controlled release of the sample, washing solution, and elution buffer. They then compared these experimental critical burst RPMs with the theoretical predictions and observed a standard deviation of 12.7% for sample loading, 6.8% for washing solution loading, and 7% for elution buffer loading. The experimental setup's results were graphically represented in Fig. 4A. At a rotational speed of 1600 RPM for 15 seconds, only the blue dye solution in the R_S was released, followed by the red dye solution in RW at 2000 RPM for 45 seconds. Finally, the yellow dye in R_E was dispensed at 2500 RPM for 120 seconds. These results, as depicted in Fig. 4B, demonstrate a strong correlation between the theoretical and experimental RPM values. This correlation

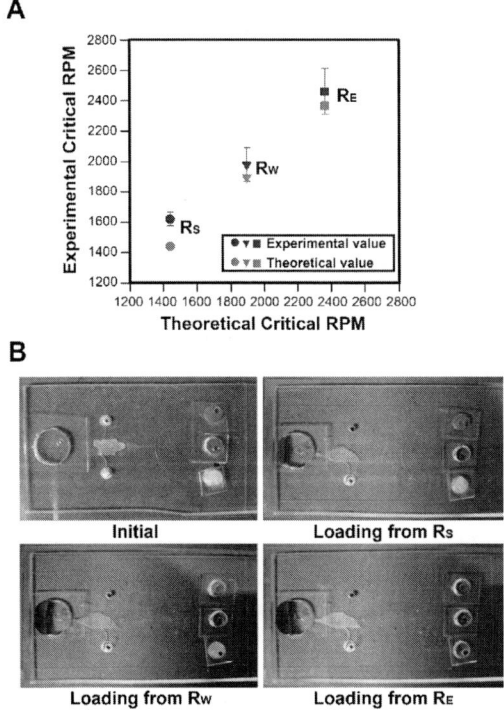

Fig. 4. (A) Theoretical versus experimental critical rotational speed. (B) RPM control subsequently pumped out the blue solution from the R_S at 1600 rpm for 15 seconds, the red solution from the R_W at 2000 rpm for 45 seconds, and finally the yellow solution from the R_E at 2500 rpm for 120 seconds. (Reprinted with permission from Ref. [18]. Copyright (2012) Royal Society of Chemistry.)

highlights the effectiveness of simple RPM control in automatically dispensing specific solutions in a sequential manner within their microdevice. Such precise control of the rotational speed enabled the successful performance of RNA purification with high efficiency and speed in Park et al.'s proposed microdevice.

Jung et al. designed their centrifugal microfluidic device for RNA sample pretreatment, emphasizing precise control over fluid movement through centrifugal forces [21]. The operation sequence of their device is intricate and begins with the RNA sample (Fig. 5). When the sample is loaded to the sample reservoir, the solution is automatically sucked to the bead-bed due to the capillary force. Once the device starts spinning at 5000 RPM for 10 seconds, the pre-loaded RNA sample, along with impurities, is captured on the packed microbeads. Immediately after, the washing solution is released to remove remaining salts and proteins adsorbed on the bead-bed, while the captured RNAs are purified. To ensure complete drying of residual ethanol, a known PCR inhibitor, the device is further centrifuged at 5000 RPM for 5 minutes. When the RPM decreases to 0, RNase-free water is primed to the siphon channel, stopping in front of the widened chamber within 30 seconds. This step ensures that both the sample waste and the washing solution accumulate in the sample collecting reservoir. Following this, the polyethylene film covering the reservoir is manually removed, and the waste and washing solution are extracted using a vacuum pump. After discarding these solutions, the reservoir is cleaned with 70% ethanol and dried. A new polyethylene film is then placed over the reservoir, and the RPM is increased to 2000 RPM for 10 seconds, allowing the elution solution to flow into the sample collecting reservoir through the bead-bed microchannel. To recover any residual water solution on the bead-bed, the RPM is increased to 5000 for 90 seconds. The final product is 4 µL of RNase-free water elution solution containing the purified RNA templates. The entire centrifugal operation of the microdevice is performed on a custom-made rotational system using a servo motor and is automatically controlled by a custom-made rotational program. The total time consumed for the sample pretreatment on the chip is 440 seconds.

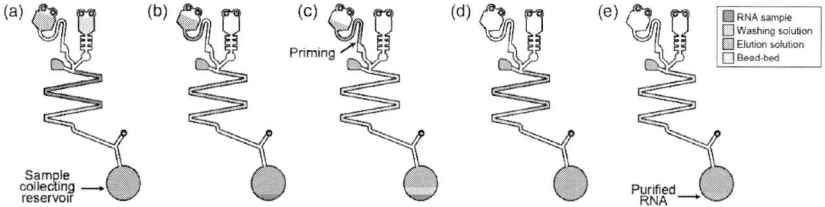

Fig. 5. The entire process for the RNA purification on the centrifugal microdevice. (a) The incubated RNA sample flows through the bead-bed microchannel at 5000 RPM for 10 seconds. (b) The 70% ethanol washing solution is loaded and then the residual ethanol in the bead-bed is dried at 5000 RPM for 5 minutes. (c) The elution solution was primed at 0 RPM for 30 seconds and the wastes in the sample collecting reservoir are removed. (d) The elution solution begins to flow through at 2000 RPM for 10 seconds to extract the captured RNAs. (e) Complete elution was performed at 5000 RPM for 90 seconds, and the purified RNAs were obtained in the sample collecting reservoir. (Reprinted with permission from Ref. [21]. Copyright (2013) Royal Society of Chemistry.)

4.3. Operation of the high-throughput centrifugal microfluidic chip for sample pretreatment

The operation of microfluidic chips for nucleic acid extraction seamlessly integrated various microfluidic principles to facilitate the automated loading of the sample, washing, and elution solution. In the single-unit chip, the sample, washing solution, and elution buffer were sequentially loaded onto the DNA capture matrix [22]. To execute an automatic division into 30 aliquots from one injection, Nguyen et al. developed a sophisticated zigzag aliquoting structure. with three components: (1) a zigzag aliquoting channel, (2) a quadrupole passive valve, and (3) super-hydrophobic coating (Fig. 3B). The zigzag channel was connected with 30 pentagon-arrow-shaped aliquot chambers. The passive valve was positioned at the end of the pentagon-arrow-shaped aliquot chamber to prevent the solution from overflowing into the GF/F filter paper chamber during the automatic injection through the aliquoting chambers. The quadrupole passive valves were linked by a 0.2 mm width × 0.2 mm depth rectangular channel and four 1.0 mm diameter × 0.5 mm depth circular patterns. The capillary pressure (0.9 kPa) of the zigzag aliquot channel was lower than the capillary force of the passive valves (3.4 kPa). The capillary pressure of the zigzag aliquot channel was calculated by the following equation [26].

$$P = -\gamma \left(\frac{\cos\theta_t + \cos\theta_b}{h} + \frac{\cos\theta_l + \cos\theta_r}{w} \right)$$

where C is the surface tension of the buffer, θ_t, θ_b, θ_l, θ_r is the contact angle of the buffer at the top, bottom, left, and right, h is the height of the channel, and w is the width of the channel, respectively. The capillary pressure of the quadrupole passive valves (C_p) was calculated as follows [27].

$$C_P = \frac{C\gamma \sin\theta}{h \times w}$$

where C is the length of the channel from the pentagon arrow chamber to the passive valve, γ is the surface tension of the buffer, θ is the contact angle, h is the height of the channel, and w is the width of the channel.

An innovative aspect of the device's operation is the use of linear movement to precisely control the position of an injection needle. Positioned 5 cm above the disc, the needle's movement is controlled by the same stepper motor system, providing an additional layer of precision to the device's operation. This careful positioning and movement of the needle are crucial for the accurate dispensing of reagents and the controlled movement of solutions within the microfluidic channels, further emphasizing the device's high-throughput efficiency and precision. The operation of Nguyen et al.'s centrifugal microfluidic device begins with the step of mounting the disc, pre-loaded with 30 forensic samples, onto the spinning base of a motor. This step is integral to the entire process, as the motor's spinning base is responsible for generating the centrifugal force necessary for fluid movement within the device. The design of the disc, with individual chambers for each sample, facilitates the simultaneous processing of multiple samples, showcasing the device's high-throughput capability. A standout feature of Nguyen et al.'s device is the automated control of the spinning speed and duration, achieved through a sophisticated stepper motor system. This automation is vital for ensuring the consistent and accurate processing of the samples throughout the operation. The custom-built software that controls the motor allows for fine-tuned adjustments to the spinning speed, ensuring that each step of the nucleic acid extraction process is executed with high precision.

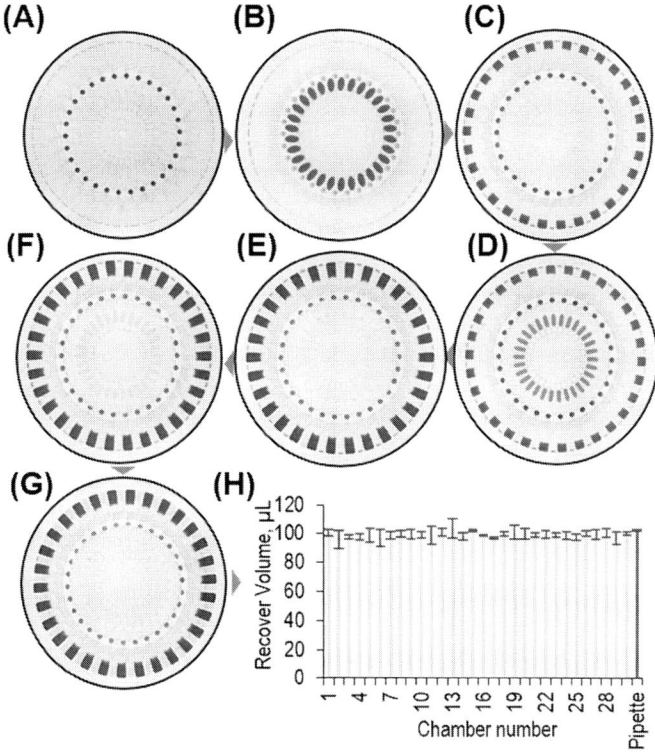

Fig. 6. Digital images of the disc during the disc operation. (A) The centrifugal disc was installed with the GF/F filter paper. (B) The sample was injected into the disc and filled in the sample chamber. (C) The disc was spun at -7000 RPM for 1 minute for releasing the sample, and the DNA was captured on the GF/F filter paper. (D) The 70% ethanol washing buffer was aliquoted into 30 aliquots from one injection by the aliquoting structure. The passive valves keep the buffer inside the chamber without leakage to the GF/F column during the injection. (E) The disc was spun at -7000 RPM for 1 minute for washing the GF/F membrane and carrying the contaminates to the waste chamber. (F) The elution buffer is introduced and filled in the aliquoting structure from one injection. (G) The disc was spun at +7000 RPM for 1 minute for eluting the captured DNA and transported the purified DNA solution to the collection chamber. (H) The recovery volume of the elution solution from three replicate discs. (Reprinted with permission from Ref. [22]. Copyright (2021) Elsevier.)

Once the disc is correctly positioned on the spinning base, it is set to rotate at -7000 RPM for one minute. This high-speed rotation is important for releasing the lysis mixture from the sample chambers and ensuring its thorough passage through the GF/F extraction paper. The centrifugal force generated by this rotation effectively moves the lysis mixture from the top to the bottom of the disc, a process that is essential for the efficient capture of human genomic DNA on the GF/F paper. This captured the target DNA and impurities on the GF/F paper, with the washing solution subsequently removing remaining salts and proteins. The washing buffer containing 70% ethanol was injected from the syringe pump, and the disc was then spun to introduce the washing buffer to the GF/F paper column. This transferred any debris to the waste chamber, leaving the genomic DNAs on the filter paper. The elution buffer was subsequently introduced from the syringe pump, and after injection, the disc was spun in a reverse direction to elute the captured DNA from the GF/F paper to the collection chamber. The purified DNA was then collected at the collection chamber for further analysis or gene amplification (Fig. 6). The operations of these three chips exemplified the versatility and adaptability of microfluidic devices in automating the nucleic acid extraction process. Each chip employed unique mechanisms and materials to optimize the extraction process, ensuring that the desired outcome was achieved in an efficient and reliable manner.

5. Performance of nucleic acid extraction

5.1. RNA extraction on the low-throughput centrifugal microfluidic chip

Park et al.'s study focused on quantifying the efficiency of RNA capture and release using their microfluidic system [18]. The effectiveness of the RNA purification process was evaluated using a fluorescence-based method. This method involved measuring the fluorescence intensity of the sol-gel chamber within their device, both before and after the RNA elution step. Sol-gel is a well-known matrix for nucleic acid capture through the electrostatic interaction on the high surface area [28]. These measurements were used in assessing the amount of RNA captured and subsequently released during the purification process. The results revealed a significant reduction in fluorescence intensity post-elution, suggesting effective RNA recovery from the chamber. Quantitatively, the fluorescence intensity in the sol-gel chamber decreased from a value of 1303 to 197 following the elution process. This substantial decrease in fluorescence, approximately 84%, indicated that a large portion of the RNA initially present in the chamber was successfully eluted. By comparing the fluorescence measurements before and after elution, they calculated an RNA capture yield of around 80% (Fig. 7). This yield was determined by dividing the amount of RNA recovered after elution by the initially added RNA amount, showcasing the device's high efficiency in RNA capture and purification. Furthermore, Park et al. performed the capture of purified H1N1 RNA and subsequent gene amplification using RT-PCR. This step was integral in demonstrating the device's practical application in processing and analyzing specific RNA samples, such as those from viral pathogens. The high capture yield and successful amplification of H1N1 RNA validated the effectiveness of their microfluidic system in RNA-based diagnostics and research applications.

Jung et al.'s study focused on evaluating the efficiency of RNA extraction using their centrifugal microfluidic device, particularly aimed at sample pretreatment for RNA purification [21]. The key aspect of their experiment was the use of real-time RT-PCR to assess the RNA extraction efficiency. They performed RNA extraction with a RT-PCR cocktail, necessitating multiple repetitions of the elution step to recover a sufficient volume of the cocktail containing the purified RNAs. This repetition ensured adequate sample volume for subsequent analysis. The real-time RT-PCR results provided valuable insights into the device's extraction efficiency. The Ct values obtained across various RNA copy numbers allowed Jung et al. to calculate RNA capture yields, which varied with the number of RNA copies.

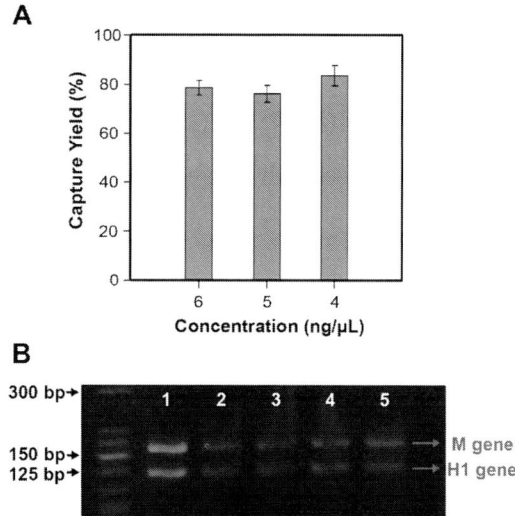

Fig. 7. (A) ~80% capture yield of RNA was obtained with the RNA samples whose concentration was 6 ng/μL, 5 ng/μL, and 4 ng/μL. (B) Agarose gel electrophoresis data showed that all the target genes (M and H1 gene) were successfully amplified from the recovered H1N1 RNA template. The initial concentration of RNA sample was 1) 1 ng/μL, 2) 0.5 ng/μL, 3) 0.25 ng/μL, 4) 0.125 ng/μL, and 5) 0.0625 ng/μL. (Reprinted with permission from Ref. [18]. Copyright (2012) Royal Society of Chemistry.)

This variability indicated a range of efficiency but generally demonstrated high effectiveness in RNA extraction. The RNA capture yields were reported as 77.2% ± 12.2, 74.8% ± 15.6, 76.6% ± 20.9, 76.4% ± 17.4, and 74.9% ± 11 for different RNA copy number ranges (Fig. 8). These yields underscore the device's capability to effectively capture and purify RNA, a critical step in diagnostic and research applications involving RNA analysis. On the contrary to the previous reports which employed a UV/Vis spectrometer for quantitative analysis with nanogram scale of nucleic acids [18, 29, 30], the use of the real-time RT-PCR in Jung et al.'s case enables us to quantify the captured RNA at femtogram scale. Additionally, Jung et al. compared the elution efficiency using RNase-free water and the RT-PCR cocktail. Interestingly, slightly higher yields were observed with water (81%) compared to the cocktail (76%), suggesting that the choice of elution solution could slightly influence the extraction efficiency. The process of extracting RNA from TEOS-treated microbeads relies on the use of positively charged, chaotropic salts. Water, which forms strong hydrogen bonds with RNA, can effectively dislodge the RNA for collection [29, 30]. Similarly, the RT-PCR mixture, though it contains some salts, predominantly acts like water in this elution role. The binding of enzymes to the beads is influenced by specific pH levels and requires extended incubation [31, 32]. A quick centrifugation step at 2000 RPM for 10 seconds minimizes unwanted enzyme binding to the beads, thereby enhancing the efficiency of RNA recovery. This comparison highlights the device's versatility and adaptability in RNA purification, allowing for optimization based on the specific requirements of the extraction process.

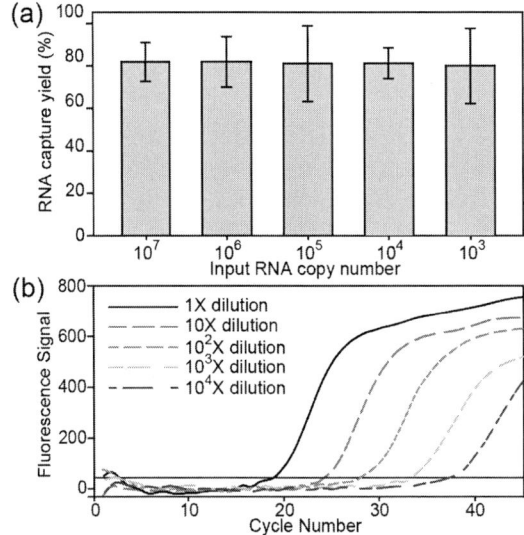

Fig. 8. (a) RNA capture yield using the purified RNAs obtained from the centrifugal sample pretreatment microdevice. The input RNA copy number ranged from 10^7 to 10^3, and the capture yield was calculated based on the calibration curve. (b) Representative real-time RT-PCR data using the purified RNAs which were obtained on a chip from the lysate of the real influenza A H3N2 virus sample. (Reprinted with permission from Ref. [21]. Copyright (2013) Royal Society of Chemistry.)

5.2. DNA extraction on the high-throughput centrifugal microfluidic chip

In their pursuit of optimizing DNA capture efficiency, Nguyen et al. focused on determining the ideal number of GF/F filter papers in the extraction matrix [22]. Their comparative study with the commercial Qiagen Investigator column revealed that two stacked GF/F filter papers resulted in the highest genomic DNA capture yield. The choice of two papers was found to be optimal, as using just one paper led to sample loss due to limited surface area, and more than four papers proved too thick for effective recovery. The GF/F filter paper, with its specific pore size and thickness, was shown to be particularly effective for capturing human chromosomes, offering an advantage in terms of porosity and thickness compared to the Qiagen silica membrane [33, 34].

Nguyen et al. extended their research to evaluate the device's performance using various human samples typical in forensic contexts, such as saliva, hair, nail, ear wax, and semen. They also tested the device's capability with dried semen samples applied to different substrates. The subsequent LAMP reaction for sex-typing, targeting the alphoid repeat sequence of the Y-chromosome and the human 18S rRNA gene, demonstrated successful amplification for both male and female samples (Fig. 9). The LAMP reaction commenced within 10 minutes, and sex-typing could be completed in about 15 minutes. Additionally, qPCR performed on purified male DNAs confirmed the successful amplification of the Y-chromosome alphoid repeat sequence, indicating that the eluted DNA was of sufficient purity for PCR-based downstream applications.

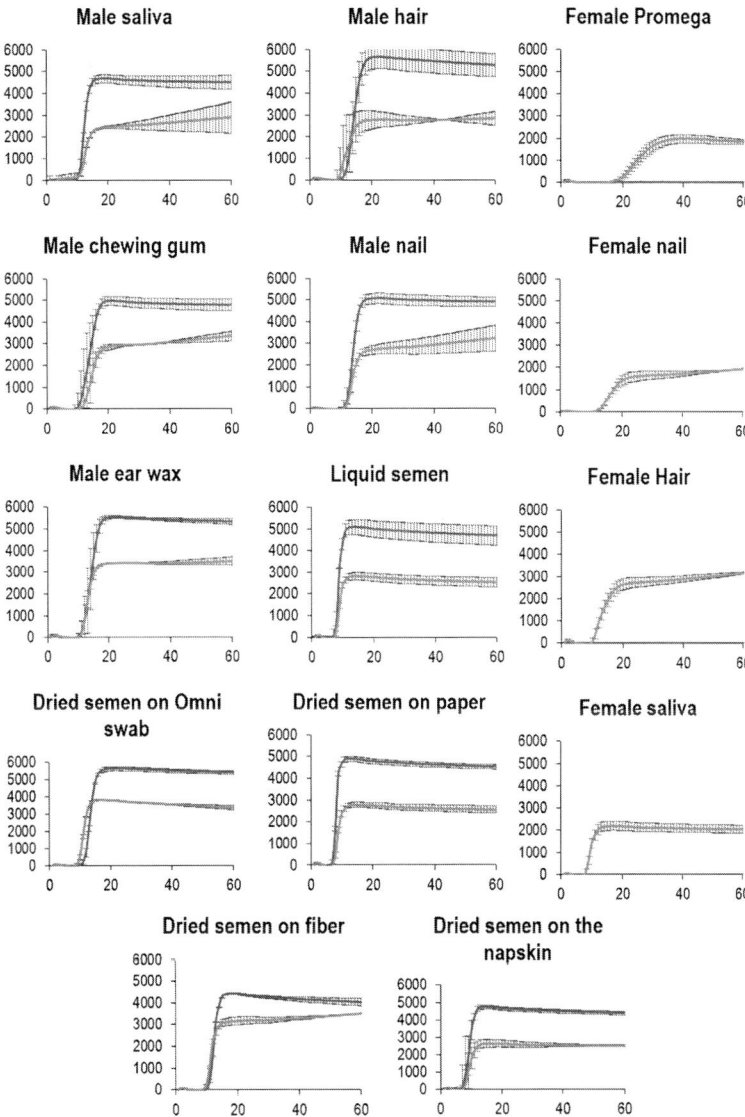

Fig. 9. The LAMP amplification plots using various male and female samples. The blue curve indicates the amplification profile of homo sapiens alphoid repeat region gene. The pink curve indicates the amplification profile of the 18S rRNA gene. The x-axis is the reaction time, and the y-axis is the fluorescent intensity. (Reprinted with permission from Ref. [22]. Copyright (2021) Elsevier.)

A key aspect of Nguyen et al.'s study was testing the reproducibility of the device. By simultaneously processing 30 samples across 30 extraction units with the same lysate solution, they were able to demonstrate the high consistency of the extraction process. The qPCR results from these 30 extractions showed similar threshold cycle numbers, with an average Cq value of 23.48 ± 0.72 and an RSD of 3.07% (Fig. 10). These results highlight the device's capability to deliver consistent and precise DNA extraction across multiple units, an essential feature for high-throughput applications.

To evaluate the LOD of the device, Nguyen et al. performed tests using serially diluted semen samples. The LAMP amplification times for the Y-chromosome at various dilutions provided key insights into the device's sensitivity. For dilutions ranging from $1:10^0$ to $1:10^3$, the amplification times were 7.2, 7.9, 10.4, and 14.1 minutes, respectively, indicating effective purification of the Y-chromosome down to a 10^3-fold dilution. At higher dilutions ($1:10^4$ and $1:10^5$), they observed more variability in the LAMP amplification curve, likely due to stochastic effects. The 18S rRNA gene amplification also correlated well with the dilution factors, further substantiating the device's efficiency in sex-typing.

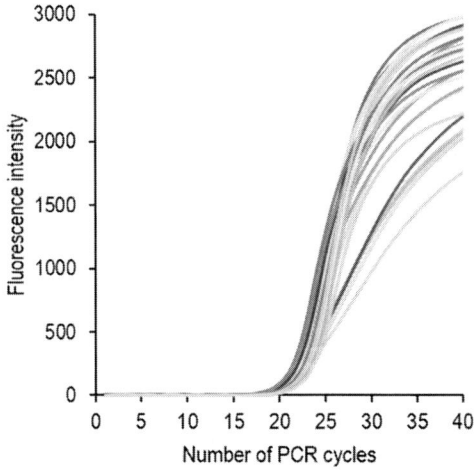

Fig. 10. The amplification curves of 30 qPCR reactions using the purified human DNA obtained from 30 extraction units on a chip. (Reprinted with permission from Ref. [22]. Copyright (2021) Elsevier.)

6. Conclusion

In conclusion, the research and development efforts of Park et al., Jung et al., and Nguyen et al. represent significant strides in the evolution of NA extraction technologies, each offering innovative solutions tailored to specific needs in molecular diagnostics.

The NA extraction on the centrifugal microfluidic device was first demonstrated by Park et al. Park et al.'s approach to RNA purification using a sol-gel incorporated microdevice on a rotary platform exemplifies a blend of simplicity, rapidity, and automation. Achieving an impressive RNA capture yield of approximately 80% within a mere 5 minutes, their system stands out for its efficiency and practicality. This level of performance, achieved in a short time span, demonstrates the potential of their system in various rapid-response scenarios, such as pathogen detection and emergency diagnostics.

Jung et al.'s centrifugal sample pretreatment microdevice, utilizing TEOS-treated microbeads as an RNA capture matrix, showcases another facet of innovation in NA extraction. With approximately 81% capture yield achieved in 440 seconds, their device not only excels in efficiency but also offers versatility, as evidenced by its effective use of an RT-PCR cocktail as an elution solution. This adaptability, yielding approximately 76% capture efficiency with the cocktail, underlines the potential of their system in diverse molecular diagnostic applications.

Nguyen et al. introduced a groundbreaking high-throughput centrifugal disc, integrated with a POCT workstation, capable of purifying 30 samples simultaneously. Completing the NA extraction process in just 10 minutes with minimal buffer usage and disposable pipette tips, their system is a paradigm shift towards high-throughput molecular diagnostics. This device's rapid processing, combined with its capacity to handle multiple samples concurrently, marks a significant advancement in the field, particularly relevant for large-scale screening and forensic genotyping.

Each of these chips represents a unique contribution to simplifying and enhancing the NA extraction process, a critical component in molecular diagnostics and personalized medicine. The integration of these microsystems with advanced liquid handling stations and further automation is a key step towards harnessing their full potential. Looking forward, the application of these technologies in various point-of-care genetic analyses, including pathogen detection, DNA sequencing, and forensic human typing, is poised to significantly impact the landscape of low-cost, automated, and high-throughput capabilities in molecular diagnostics. These innovations not only advance the field technically but also open new avenues for rapid, accurate, and accessible genetic testing in diverse settings.

References

1   A. Manz, N. Graber and H. M. Widmer, *Sens Actuators B Chem*, 1990, 1, 244–248.

2   Peng Liu, Xiujun Li, Susan A. Greenspoon, James R. Schererb and Richard A. Mathies, *Lab Chip*, 2011, 11, 1041–1048.

3   Kirsty J. Shaw, Domino A. Joyce, Peter T. Docker, Charlotte E. Dyer, Gillian M. Greenway, John Greenmanc and Stephen J. Haswell, *Lab Chip*, 2010, 11, 443–448.

4   Christopher J Easley, James M Karlinsey, Joan M Bienvenue, Lindsay

A Legendre, Michael G Roper, Sanford H Feldman, Molly A Hughes, Erik L Hewlett, Tod J Merkel, Jerome P Ferrance and James P Landers, *Proc Natl Acad Sci U S A*, 2006, 103, 19272–19277.

5   Peng Liu, Tae Seok Seo, Nathaniel Beyor, Kyoung-Jin Shin, James R. Scherer and Richard A. Mathies, *Anal Chem*, 2007, 79, 1881–1889.

6   E. Lagally, P. C. Simpson and R. Mathies, *Sens Actuators B Chem*, 2000, 63, 138–146.

7   Lindsay A. Legendre, Joan M. Bienvenue, Michael G. Roper, Jerome P. Ferrance and James P. Landers, *Anal Chem*, 2006, 78, 1444–1451.

8   P Wilding, L J Kricka, J Cheng, G Hvichia, M A Shoffner and P Fortina, *Anal Biochem*, 1998, 257, 95–100.

9   Laura Pasquardini, Cristina Potrich, Marzia Quaglio, Andrea Lamberti, Salvatore Guastella, Lorenzo Lunelli, Matteo Cocuzza, Lia Vanzetti, Candido Fabrizio Pirrice and Cecilia Pederzollia, *Lab Chip*, 2011, 11, 4029–4035.

10 Adam T. Woolley, Dean Hadley, Phoebe Landre, Andrew J. deMello, Richard A. Mathies and M. Allen Northrup, *Anal Chem*, 1996, 68, 4081–4086.

11 Yong Tae Kim, Yuchao Chen, Jong Young Choi, Won-Jung Kim, Hyun-Mi Dae, Jaean Jung and Tae Seok Seo, *Biosens Bioelectron*, 2012, 33, 88–94.

12 Nathaniel Beyor, Tae Seok Seo, Peng Liu and Richard A. Mathies, *Biomed Microdevices*, 2008, 10, 909–917.

13 Chan Joo Lee, Jae Hwan Jung and Tae Seok Seo, *Anal Chem*, 2012, 84, 4928–4934.

14 Yanju Chen, Yang Liu, Ya Shi, Jianfeng Ping, Jian Wu and Huan Chen, *Trends Analyt Chem*, 2020, 127, 115912-NA.

15 Lei Yan, Jie Zhou, Yue Zheng, Adam S. Gamson, Benjamin T. Roembke, Shizuka Nakayama and Herman O. Sintim, *Mol Biosyst*, 2014, 10, 970–1003.

16 Lu Zhang, Baozhi Ding, Qinghua Chen, Qiang Feng, Ling Lin and Jiashu Sun, *TrAC Trends in Analytical Chemistry*, 2017, 94, 106–116.

17 Jianjian Zhuang, Juxin Yin, Shaowu Lv, Ben Wang and Ying Mu, *Biosens Bioelectron*, 2020, 163, 112291-NA.

18 B. H. Park, J. H. Jung, H. Zhang, N. Y. Lee and T. S. Seo, *Lab Chip*, 2012, 12, 3875–3881.

19 Juan Astorga-Wells and Harold Swerdlow, *Anal Chem*, 2003, 75, 5207–5212.

20 Grace D. Chen, Catharina J. Alberts, William Rodriguez and Mehmet Toner, *Anal Chem*, 2009, 82, 723–728.

21 J. H. Jung, B. H. Park, Y. K. Choi and T. S. Seo, *Lab Chip*, 2013, 13, 3383–3388.

22 H. Van Nguyen and T. S. Seo, *Biosens Bioelectron*, 2021, 181, 113161.

23 Byung Hyun Park, Dahin Kim, JaeHwan Jung, Seung Jun Oh, Goro Choi, Doh C. Lee and T. S. Seo, *Sens Actuators B Chem*, 2015, 209, 927–933.

24 Frank Schwemmer, Clement E. Blanchet, Alessandro Spilotros, Dominique Kosse, Steffen Zehnle, Haydyn D. T. Mertens, Melissa A. Graewert, Manfred Rössle, Nils Paust, Dmitri I. Svergun, Felix von Stetten, Roland Zengerle and Daniel Mark, *Lab Chip*, 2016, 16, 1161–1170.

25 J Steigert, T Brenner, M Grumann, L Riegger, S Lutz, R Zengerle and J Ducrée, *Biomed Microdevices*, 2007, 9, 675–679.

26 A. Olanrewaju, M. Beaugrand, M. Yafia and D. Juncker, *Lab Chip*, 2018, 18, 2323–2347.

27 Siyi Lai, Shengnian Wang, Jun Luo, L James Lee, Shang-Tian Yang and Marc J Madou, *Anal Chem*, 2004, 76, 1832–1837.

28 P. B. Wagh, R. Begag, G. Pajonk, A. Rao and D. Haranath, *Mater Chem Phys*, 1999, 57, 214–218.

29 Kelley A. Wolfe, Michael C. Breadmore, Jerome P. Ferrance, Mary E. Power, John F. Conroy, Pamela M. Norris and James P. Landers, *Electrophoresis*, 2002, 23, 727–733.

30 Qirong Wu, Joan M. Bienvenue, Benjamin J. Hassan, Yien C. Kwok, Braden C. Giordano, Pamela M. Norris, James P. Landers and Jerome P. Ferrance, *Anal Chem*, 2006, 78, 5704–5710.

31 J.Felipe Díaz and Kenneth J. Balkus Jr, *J Mol Catal B Enzym*, 1996, 2, 115–126.

32 H.P. Yiu, Paul A. Wright and Nigel P. Botting, *Microporous and Mesoporous Materials*, 2001, 44, 763–768.

33 Daniel Brassard, Matthias Geissler, Marianne Descarreaux, Dominic Tremblay, Jamal Daoud, Liviu Clime, Maxence Mounier, Denis Charleboisc and Teodor Veres, *Lab Chip*, 2019, 19, 1941–1952.

34 R. Shi, R. S. Lewis and D. R. Panthee, *PLoS One*, 2018, 13, e0203011.

Chapter 2
Centrifugal Microfluidic Device for Polymerase Chain Reaction

1. Background

The Polymerase Chain Reaction (PCR) is a biotechnological method used to amplify a single or a few copies of a segment of DNA across several orders of magnitude, generating thousands to millions of copies of a particular DNA sequence. Developed in 1983 by Kary Mullis, PCR is now a common and often indispensable technique in medical and biological research labs for a variety of applications, including DNA cloning for sequencing, DNA-based phylogeny, or functional analysis of genes, the diagnosis of hereditary diseases, the identification of genetic fingerprints used in forensic sciences and paternity testing, and the detection and diagnosis of infectious diseases [1–3].

The fundamental requirement of PCR is the template DNA that contains the DNA region (target) to be amplified. This process is highly sensitive and specific. PCR components are mixed in a sterile, non-reactive vessel, such as a small plastic tube to avoid contaminants. Reagents include the DNA template to be amplified, two primers, which are complementary to the DNA regions at the 5' and 3'-ends of the sequence to be amplified, a thermostable DNA polymerase, such as Taq polymerase, deoxyribonucleotides (dNTPs), which are the building blocks from which the DNA polymerase synthesizes a new DNA strand, and a buffer solution providing a suitable chemical environment for optimum activity and stability of the DNA polymerase.

The key steps in PCR are denaturation, annealing, and extension [4–6]. During denaturation, the double-stranded DNA melts open to single-stranded DNA, and all enzymatic reactions stop (i.e., the DNA polymerase is inactivated). During annealing, the temperature is lowered to enable the DNA primers to attach to the template DNA. Extension is where the temperature is raised so the new DNA strand will be built by the Taq polymerase enzyme. These steps are repeated for 30 or 40 cycles. Because the amplified DNA segments also serve as templates for replication, the amount of DNA in the reaction doubles after each cycle, leading to exponential amplification of the specific DNA segment.

While PCR has been transformative, the traditional approach to PCR is not without limitations. Conventional PCR systems rely on thermal cyclers that consume significant power to heat and cool the entire volume of reaction mixtures. This results in considerable time taken to reach the required temperatures for each stage of the cycle, and thus, the total time to complete the PCR process can be lengthy. Furthermore, the handling of multiple samples simultaneously while avoiding cross-contamination requires meticulous laboratory technique and often, a sizable quantity of reagents and samples, which can be costly [7–10].

The need for rapid, cost-effective, and high-throughput DNA analysis has driven the development of PCR methodologies, leading to the integration of PCR with microfluidic technology. The microfluidic devices utilize the principles of microfluidics, manipulating small amounts of fluids in channels with dimensions of tens to hundreds of micrometers. Microfluidics offers several advantages over traditional methods, such as lower reagent consumption, reduced risk of cross-contamination, faster reaction times due to smaller thermal mass, and the ability to integrate several processes into a single device, known as lab-on-a-chip [11–15].

Microfluidic PCR (µPCR) systems, an advanced iteration of traditional PCR technology, are fundamentally categorized into two distinct types, each with unique mechanisms and applications. These systems represent a leap in PCR technology, offering more efficient, scalable, and accessible DNA amplification methods [14].

The first category is stationary PCR systems. These systems perform the PCR process within a fixed microchamber on the chip. The hallmark of stationary µPCR systems is their precise temperature control, achieved through micropatterned heaters embedded in the microchamber. This precision is crucial for the accuracy and efficiency of the PCR process, ensuring consistent thermal cycling and optimal DNA amplification conditions. However, the complexity of these systems lies in their fabrication. The integration of heaters and resistive temperature detectors (RTDs) onto a microchip requires intricate engineering and precise manufacturing techniques. Stationary µPCR systems are well-suited for applications where the thermal consistency and cycle precision are paramount, such as in diagnostic testing where the accuracy of gene amplification is critical [16].

The second type is the flow-through PCR system. Unlike stationary systems, flow-through µPCR systems facilitate the PCR process as the sample fluid dynamically flows through distinct temperature zones within the device. This approach allows for continuous thermal cycling as the sample moves through different temperature regions, significantly speeding up the PCR process. The key advantage of flow-through systems is their rapid processing capability, making them ideal for applications where time efficiency is crucial. However, these systems typically require external mechanisms, like syringe pumps, to control the flow of the sample. This dependency can introduce complexity in system setup and operation. Despite this, flow-through µPCR systems have shown great promise in high-throughput testing and situations where rapid DNA amplification is necessary, such as in urgent medical diagnostics or on-field biological testing [17, 18].

The centrifugal microfluidic device for PCR, also known as a "lab-on-a-disc", uses centrifugal force for fluid propulsion within the microchannels of the disc. The liquid samples and reagents are moved through the microchannels by spinning the disc, which controls the mixing and separation of samples, and enables the sequential delivery of reactants. This process negates the need for external pumps, simplifying device architecture and potentially reducing cost and complexity [19–21].

The Rotary PCR Genetic Analyzer is a prime example of this technology in action. It combines the benefits of stationery and flow-through PCR systems as well as centrifugal microfluidics into one integrated device. This rotary system facilitates ultrafast PCR cycles and allows for simultaneous processing of multiple samples. By rotating a microchip over a series of thermal blocks programmed at specific temperatures, the Rotary PCR Genetic Analyzer achieves precise temperature control without the need for complex fabrication processes or external pumping systems. This innovation represents a leap forward in PCR technology, enabling rapid, sensitive, and high-throughput genetic analysis, which is essential for modern diagnostics and research.

The integration of centrifugal microfluidics with PCR technology marks a critical evolution from conventional benchtop procedures to more compact, faster, and efficient methodologies. It opens up possibilities for field-based testing and point-of-care diagnostics, where quick turn-around and high-throughput are often required. As such, the centrifugal microfluidic device for PCR is not just an alternative to traditional PCR—it's an enhancement that could redefine diagnostic and analytical capabilities in molecular biology.

2. Centrifugal PCR chip

2.1. Material for microfluidic chip

In the field of microfluidic technology, especially in the context of PCR chip fabrication, the selection of materials is critical, as it directly impacts the efficiency, accuracy, and scalability of the PCR process. Among the materials commonly used are Polydimethylsiloxane (PDMS), glass, Polymethylmethacrylate (PMMA), and Cyclic Olefin Copolymer (COC), each offering unique benefits and presenting specific challenges [22, 23].

Polydimethylsiloxane (PDMS) is extensively utilized in the fabrication of microfluidic PCR chips due to its several advantageous properties. It is biocompatible, making it suitable for biological applications, and its optical transparency allows for easy monitoring of the PCR process. The gas permeability of PDMS is especially beneficial for applications involving gas exchange, such as cell culture. Its elasticity and flexibility facilitate the creation of microfluidic structures and channels. However, PDMS is not without its limitations. Its permeability, while advantageous in some applications, can lead to problems such as evaporation of solvents and absorption of hydrophobic molecules, which might interfere with the PCR process. Additionally, PDMS can suffer from deformation under prolonged exposure to high temperatures, a common requirement in PCR cycles.

Glass is another material frequently used in microfluidic devices for PCR. Its high thermal stability is a significant advantage, allowing it to withstand the repeated heating and cooling cycles of PCR without degrading. Glass's chemical inertness ensures that it does not interact with PCR reagents, thus maintaining the purity of the reactions. The optical clarity of glass is ideal for detection methods that rely on fluorescence. On the downside, glass fabrication can be complex and costly. Its rigidity, while providing structural stability, also makes it prone to cracking and breaking, posing challenges in handling and durability.

Polymethylmethacrylate (PMMA), a clear thermoplastic, is favored in many commercial applications of PCR chips due to its cost-effectiveness and ease of fabrication. It offers a reasonable degree of optical clarity for monitoring PCR reactions and is relatively strong mechanically. However, PMMA's thermal resistance is lower than that of glass, which can limit its use in high-temperature PCR applications. It can also be prone to swelling and deformation when exposed to certain chemicals, which can be a concern in PCR processes involving aggressive reagents.

Cyclic Olefin Copolymer (COC) is known for its excellent chemical resistance and low water absorption, making it ideal for PCR applications that involve harsh chemicals. Its good optical properties, including low auto-fluorescence, make it suitable for fluorescence-based detection. COC's mechanical properties ensure durability and resistance to deformation under PCR conditions. Nevertheless, COC is relatively expensive compared to other polymers like PDMS and PMMA. The complexity of its fabrication process can also be a barrier, limiting its widespread use in cost-sensitive applications.

2.2. Chip design

This book chapter draws on the work of Jung et al. as a case study to illustrate the development of a micro PCR system [24]. The focus is on the centrifugal PCR chip, a micro-engineered platform specifically designed for the rapid thermal cycling essential in PCR. Central to this chip's innovation is its PCR chamber, which holds a minimal volume of just 1 µL (Fig. 1). This design is deliberate, targeting enhanced thermal responsiveness and quicker processing of samples. The reduced volume is crucial as it facilitates rapid heating and cooling, thereby decreasing the total time required for PCR. This acceleration is achieved without compromising the efficiency or accuracy of the reaction, demonstrating the potential of miniaturization in improving PCR methodologies.

The choice of polydimethylsiloxane (PDMS) as the construction material is based on its several beneficial properties. Biocompatibility ensures that it does not interfere with the PCR reaction and gas permeability allows for the exchange of gases which is necessary for certain PCR conditions. Optical transparency facilitates monitoring of the reaction process. The integration of a 200 µm-thick glass cover slip ensures efficient heat conduction, which is paramount for maintaining the precise temperature cycles required for PCR (Fig. 1).

The chip's innovative triple microchip design allows for separate and simultaneous PCR reactions, a feature that significantly enhances its throughput. The design of the chip, accommodating multiple nanoliter PCR chambers, proposes a future where high-throughput PCR reactions can be performed, providing a scalable solution that could revolutionize genetic analysis and diagnostics. Moreover, this scalable architecture is capable of accommodating increasing sample throughput, demonstrating the chip's potential for extensive genetic analysis and screening applications.

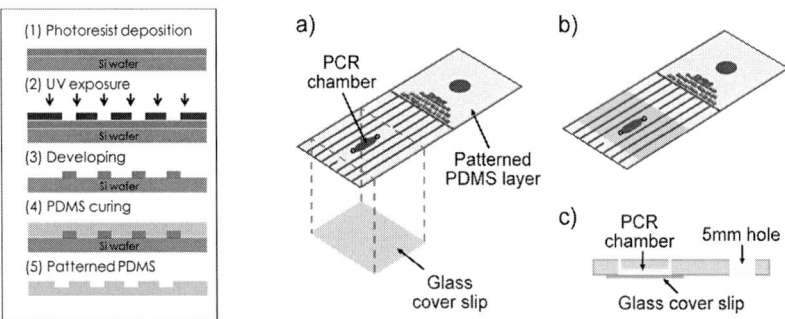

Fig. 1. (Left) Soft lithography processes. (Right) A disposable Rotary microchip: a) A patterned PDMS layer and a glass cover slip, b) schematics for an assembled Rotary microchip, and c) side view of the Rotary microchip. (Reprinted with permission from Ref. [24]. Copyright (2012) Royal Society Chemistry)

2.3. Chip fabrication

The chip is fabricated using soft lithography, a precise and cost-effective technique that involves patterning the PDMS on a master mold created from a silicon wafer. Employing soft lithography, the chip is meticulously crafted to accommodate the complex fluid dynamics and thermal requirements of PCR.

Soft lithography begins with the preparation of the master mold. Here, SU-8 50 photoresist is applied to a silicon wafer. This photoresist is chosen for its suitability in creating precise, high-resolution patterns necessary for the microfluidic channels. The wafer, coated with the photoresist, undergoes a critical step of UV exposure. This exposure is guided by a mask, intricately designed using AutoCAD software, which delineates the complex layout of the chip's microchannels and chambers. The precision in this step is paramount, as any deviation could lead to inefficiencies in fluid movement or thermal regulation within the chip. Following the UV exposure, the photoresist is developed to reveal the master mold, which now contains the negative of the chip's microfluidic network. The PDMS pre-polymer and curing agent are then mixed at a 10:1 ratio, a proportion meticulously calibrated to ensure the PDMS solidifies with the desired properties. The mixture is poured over the master mold, and upon curing at 65 °C, the PDMS forms a solid yet flexible layer replicating the microscale features of the master mold.

The bonding of the PDMS to a glass cover slip is achieved through plasma treatment, a method ensuring a robust and hermetic seal. This seal is essential for maintaining the integrity of the PCR reaction chamber during thermal cycling. The glass is specifically chosen for its thermal conductivity, facilitating efficient heat transfer, a crucial aspect for the rapid thermal cycling in PCR. The chip's final structure, comprising a 1 mm-thick PDMS layer coupled with a 200 μm-thick glass cover slip, reflects a balanced approach to material selection. The PDMS provides the necessary flexibility for the microfluidic network, while the glass ensures effective heat transfer. The fabrication process concludes with the integration of the triple microchip design, capable of conducting multiple PCR reactions simultaneously. This feature not only enhances the chip's analytical capacity but also illustrates the potential scalability of the technology for high-throughput applications (Fig. 2).

Fig. 2. A disposable Rotary microchip for the monoplex PCR (left) and the triplex PCR (right). (Reprinted with permission from Ref. [24]. Copyright (2012) Royal Society Chemistry)

3. Centrifugal PCR system

3.1. Rotational motor

The rotational motor in the centrifugal PCR chip system plays a pivotal role in the dynamic and precise control of the PCR process. This stepper motor, identified as NEMA 23 from National Instruments, is positioned beneath the rotary stage and is a crucial component for manipulating the PCR microchip's movement. It's responsible for adjusting the rotating speed of the microchip, with capabilities reaching up to 2000 revolutions per minute (rpm), a feature that highlights its efficiency and the advanced engineering behind its design.

The integration of this motor with the PCR system is a significant engineering feat. The motor's shaft is directly connected to the PCR chip, which is designed with a 5 mm diameter hole at one end to fit onto the motor's central shaft. This connection ensures that the chip is in direct contact with the heat block, an essential aspect for efficient heat transfer during thermal cycling. This direct contact is crucial for maintaining the precise temperature control necessary for PCR, as even minor deviations in temperature can significantly impact the amplification process.

A standout feature of the motor's functionality is its ability to control the designated rotational angle and the residence time of the microchip on each thermal block. This control is managed by an in-house developed LabVIEW software, interfaced through a DAQ board and a driver. The software allows for intricate programming of the motor's movements, enabling the microchip to precisely align with the different thermal blocks for the denaturation, annealing, and extension phases of PCR. The precision in this movement is paramount, as it ensures the uniformity and repeatability of the PCR cycles across different runs and samples (Fig. 3).

The transition time of the microchip from one block to another is a mere 50 milliseconds (ms), dramatically reducing the ramping rate compared to stationary PCR systems. This rapid transition is a critical component of the system's ability to perform ultrafast PCR. In traditional PCR systems, the ramping rate - the time taken to change from one temperature to another - can significantly add to the overall duration of the PCR process. However, with the rotary system's quick transition times, the overall PCR process is expedited without compromising the quality or efficiency of the DNA amplification.

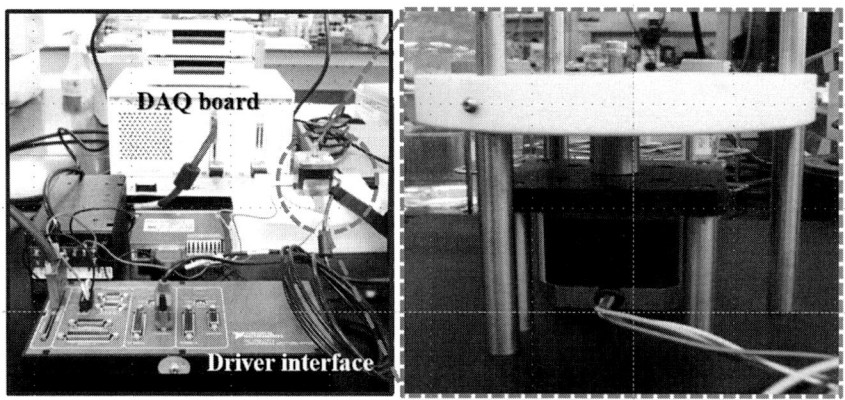

Fig. 3. A rotational motor for the rotary PCR (right) operated by a DAQ board and a LabViewTM program (left). (Reprinted with permission from Ref. [24]. (Copyright (2012) Royal Society Chemistry)

In summary, the rotational motor in the centrifugal PCR chip system is more than just a mechanical component. It is the heart of the system's dynamic operation. Its design and integration with the chip and software exemplify the innovative approach to overcoming the limitations of traditional PCR methods. The precise control of the chip's movement, the rapid transition times, and the direct interface with the thermal blocks all contribute to the system's capability to conduct ultrafast PCR, marking a significant advancement in the field of molecular diagnostics and genetic analysis.

3.2. Heater

The heater in the centrifugal PCR system is a critical component, fundamental to achieving the rapid and precise thermal cycling required for PCR. The design and integration of the heater into the system reflect a sophisticated understanding of thermal dynamics essential for molecular diagnostics.

The heater is a film type, specifically chosen for its ability to provide rapid and uniform heating. This type of heater is advantageous in PCR applications due to its rapid response time and even heat distribution, ensuring that the entire PCR chamber reaches the target temperature simultaneously. Attached to the bottom of the thermal block, the heater is instrumental in achieving the precise temperature control necessary for the denaturation, annealing, and extension phases of PCR. An integral part of the heating system is the Resistive Temperature Detector (RTD), a device critical for monitoring and controlling the temperature within the PCR system. In this design, a platinum film RTD is sandwiched between 1.5 mm-thick thermal block plates. The choice of platinum for the RTD is due to its stable and predictable resistance-temperature relationship, allowing for accurate temperature measurement over the PCR's operational range. This accurate temperature sensing is crucial in maintaining the stringent thermal conditions required for effective DNA amplification (Fig. 4).

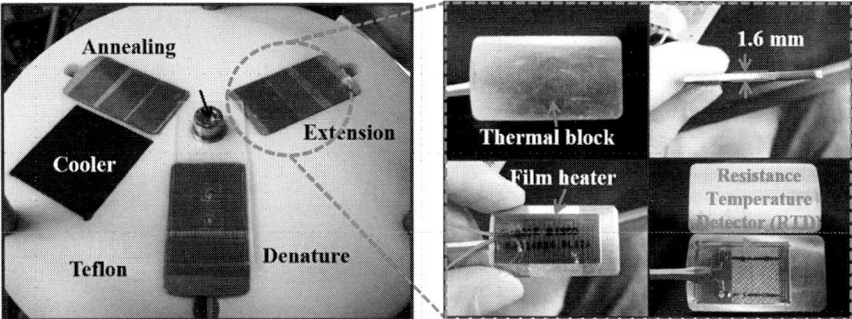

Fig. 4. Heat blocks embedded on the stage for the Rotary PCR (left) and the structure of the heater composed of the film heater, an RTD, and a heat block (right). (Reprinted with permission from Ref. [24]. Copyright (2012) Royal Society Chemistry)

The thermal blocks themselves are constructed from duralumin, a material chosen for its excellent thermal conductivity and stability. This ensures that heat from the film heater is efficiently transferred to the PCR chamber. The properties of duralumin also allow for a swift change in temperature, which is essential for

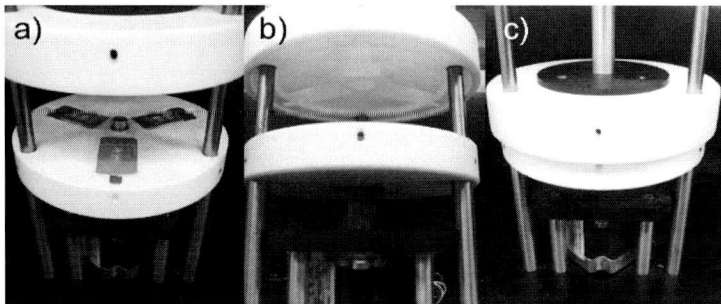

Fig. 5. Digital image for a Rotary Genetic Analyzer platform: a) Top view showing the Rotary stage which incorporates three thermal blocks, b) bottom view showing a upper teflon cover with block-shaped hollow, and c) a tightly sealed Rotary PCR system for thermal cycling. (Reprinted with permission from Ref. [24]. Copyright (2012) Royal Society Chemistry)

the rapid thermal cycling that characterizes PCR.

To complement the thermal system, Teflon is used as the material for the rotary stage due to its insulation and heat resistance properties. Teflon minimizes heat loss and contributes to maintaining a uniform temperature distribution across the thermal blocks. This feature is crucial in preventing the formation of thermal gradients that could lead to inconsistent PCR results.

The electrical circuits controlling the RTD and heater have been designed with precision, based on modifications of previous designs, to ensure accurate temperature regulation during thermal cycling. After the PCR microchip is installed onto the rotary stage, a Teflon cover with a block-shaped hollow is lowered to tightly seal the stage. This design minimizes heat loss and maintains uniform temperature distribution, a crucial factor for the consistency and reliability of the PCR process (Fig. 5).

3.3. Fluorescence detection

In the centrifugal PCR system, the fluorescence optics component is pivotal for the detection and analysis of PCR products, particularly in the Rotary PCR Genetic Analyzer system. This aspect of the system transforms PCR from a process of mere amplification to a comprehensive diagnostic tool, capable of detailed analysis and quantification of genetic material.

Following the PCR process, the amplified DNA, labeled with fluorescent markers, undergoes a precise detection process. This is where the advanced fluorescence optics of the system come into play. Specifically, the system utilizes a laser-induced confocal fluorescence microscope for the detection of the fluorescence emission signal of the separated DNA. This sophisticated microscope is a critical component of the system, employing an excitation wavelength of 488 nm from an argon laser. The chosen wavelength is pivotal for efficiently exciting the fluorescent markers without causing damage to the DNA, and the power intensity, measured at 3.6 mW from a 10× Plan Apo objective (NA 0.45), is carefully controlled to provide optimal illumination.

The scanning area is precisely defined at 0.016 mm^2 on the separation channel near the anode reservoir, which is strategically chosen to maximize the detection of DNA. The data acquisition is carried out at a scanning rate of 5 frames per second, balancing the need for rapid data collection with the requirement for accurate resolution. The emission signal of the fluorescein amidite (FAM) marker is specifically detected through a band-pass filter of 505–530 nm, ensuring that only the relevant fluorescence emission is measured. This specificity is crucial for accurately quantifying the amount of target DNA in the sample.

The peaks in the electropherogram, representing the separated DNA, are then quantified using PeakFit software. This software enables precise analysis of the peaks, which is essential for determining the concentration of the target DNA. The ability to accurately quantify these peaks is vital in various applications, such as in clinical diagnostics where the amount of pathogen DNA can indicate the stage or severity of an infection.

Incorporating such advanced fluorescence optics into the PCR system significantly enhances its capabilities. Not only does it enable the sensitive and specific detection of nucleic acids, but it also allows for real-time monitoring and quantification of the PCR process. The integration of laser-induced confocal fluorescence microscopy, with precise filtering and sophisticated software analysis, represents a significant advancement in the field of genetic analysis. This technology extends beyond basic amplification, offering comprehensive insights into the genetic material being examined, thereby opening new avenues in medical diagnostics and molecular research.

4. Chip operation

The operation of the Rotary PCR Genetic Analyzer exemplifies a blend of sophisticated engineering and precision molecular biology, utilizing the principles of both stationary and flow-through PCR systems to achieve ultrafast and efficient PCR amplification.

Sequential Rotation on Thermal Blocks: The PCR microchip is designed to be sequentially rotated over three distinct thermal blocks, each set to a specific temperature necessary for the PCR process. These temperatures are 94 °C for denaturation, 58 °C for annealing, and 72 °C for extension. The sequential rotation of the microchip aligns it with the appropriate thermal block, facilitating the rapid transition between different temperature zones. This rapid transition mirrors the flow-through concept and enables fast temperature control for the PCR reactor, similar to the low thermal mass characteristic of stationary PCR systems.

Automated Control and Precise Movement: The movement and positioning of the microchip on each thermal block are controlled by a stepper motor, regulated by custom-developed LabVIEW software. This automation allows for precise control over the residence time of the microchip on each block, ensuring that the PCR samples are exposed to the correct temperatures for the right amount of time. The precision in controlling these variables is crucial for the repeatability and reliability of the PCR results.

Initial Steps and Thermal Cycling: The PCR process begins with reverse transcription at 50 °C for 15 minutes on the annealing block, followed by initial activation at 95 °C for 5 minutes on the denature block. These initial steps are essential for preparing the RNA templates for DNA amplification. After these steps, the microchip undergoes rotational movements to align with the different thermal blocks for denaturation, annealing, and extension phases of PCR. The total number of cycles typically performed is 34, with each cycle comprising one minute-long exposure to each temperature.

Final Extension and Rapid Heating/Cooling: The final extension phase is conducted at 72 °C for 7 minutes on the extension block. The low thermal mass of the PCR chamber allows for rapid heating and natural cooling within the Rotary system, crucial for the quick transitions between different thermal cycles. The precise temperature control and the rapid heating/cooling capability are vital for the accuracy and speed of the PCR. The Rotary PCR system follows a specific thermal cycling scheme (Table 1):

The operation of the Rotary PCR Genetic Analyzer represents an innovative approach to PCR, combining the rapid sample transition of flow-through PCR with the precise temperature control of stationary PCR systems. This methodology significantly reduces the time required for PCR amplification while maintaining high accuracy and efficiency, making it a valuable asset for rapid molecular diagnostics and research applications.

Table 1. Representative thermal cycling scheme of the RT-PCR in the Rotary system

Process		Temperature	Process time	Chip position
Reverse transcription		50 °C	15 min	Annealing block
Initial activation		95 °C	5 min	Denature block
Thermo-cycling (34 cycles)	Denature	94 °C	1 min	Denature block
	Annealing	58 °C	1 min	Annealing block
	Extension	72 °C	1 min	Extension block
Final extension		72 °C	7 min	Extension block

5. Performance of PCR on a centrifugal microdevice

The performance of the Rotary PCR Genetic Analyzer in the context of a centrifugal microdevice is exemplified by its application in detecting and genotyping various subtypes of the Influenza A virus, including H1N1, H3N2, and H5N1. This application demonstrates the system's capabilities for rapid, accurate, and high-throughput nucleic acid amplification and detection, particularly in the field of infectious disease diagnostics.

Application for Influenza A Detection: The choice of Influenza A virus as a target for detection is significant, given the virus's high mutation rate and the public health implications of its various strains. The system was employed for genotyping different Influenza A subtypes, a critical task in epidemiology and vaccine development. Specific primer sequences were designed for the H1, H3, and H5 genes of the respective subtypes, and the use of fluorescently labeled primers (FAM) facilitated the detection of amplified genes. This selection allowed for the precise identification of each subtype, an essential capability during influenza outbreaks and for monitoring viral evolution.

Electrophoresis and Fluorescence Detection: After completing the PCR amplification, the products were subjected to micro-capillary electrophoresis (μCE) for separation and identification. The μCE process involves the migration of charged particles (in this case, the amplified DNA fragments) through a capillary filled with an electrolyte, under the influence of an electric field. The separation is based on the size-to-charge ratio of the DNA fragments. Following the electrophoresis, a laser-induced confocal fluorescence microscope was used to detect the fluorescently labeled DNA fragments. This method ensures high specificity and sensitivity in identifying the targeted gene sequences.

Analysis of Electropherograms: The resultant electropherograms from the μCE displayed distinct peaks corresponding to the specific genes of the influenza virus subtypes. For example, the major peak for the H3 gene (150 bp) of the H3N2 subtype virus appeared at an elution time of 275 seconds. The clarity and distinction of these peaks are indicative of the successful amplification and separation of the target genes. The electropherograms serve as a visual confirmation of the presence and quantity of specific genetic material in the samples.

Optimization of PCR Conditions: Jung et al. also demonstrated a critical optimization of PCR conditions to enhance the efficiency and accuracy of RNA detection. The study showcased an electropherogram with the major peak for the H3 gene (150 bp) at an elution time of 275 seconds. Initially, the PCR step took 102 minutes with denature (D), annealing (A), and extension (E) times set at 60 seconds each, following the protocol recommended by the Onestep RT-PCR kit used in conventional thermal cyclers. Building on this success, the team further reduced the thermal cycling time to 68 minutes (D/A/E = 30/60/30 sec), 51 minutes (D/A/E = 30/30/30 sec), 34 minutes (D/A/E = 15/30/15 sec), and finally to 25.5 minutes (D/A/E = 15/15/15 sec), to streamline the process. To accommodate the rapid thermocycling requirement, they tuned the temperature of the thermal blocks for quick heat transfer. This entailed increasing the temperature of the denature and extension blocks by 0.267 °C and decreasing the annealing block temperature by 0.2 °C for each second of reduction in cycle time. For example, in the 25.5-minute PCR reaction, the denature block temperature was set to 96 °C, and the annealing block to 47 °C. The fluorescent peaks of the H3 gene amplicon were produced under these varying conditions. Impressively, the 25.5-minute PCR cycle yielded peak heights equivalent to those obtained with longer PCR conditions, indicating the system's ability to perform RT-PCR for gene expression identification from viral RNA samples with high speed and specificity. This was achieved by finely adjusting the temperature of the heat blocks, demonstrating the Rotary Genetic Analyzer's capability for rapid and accurate RNA detection (Fig. 6, top).

Limit of Detection (LOD) Test: In a pivotal aspect of their research, Jung et al. conducted a meticulous evaluation of the Limit of Detection (LOD) for their Rotary PCR system, which is designed for ultrafast detection of multiple influenza viral RNAs (Fig. 6, middle). The LOD assessment was carried out under a 5-minute thermocycling condition, utilizing serially diluted influenza A H3N2 viral RNA templates at various concentrations, including 12 picograms, 1.2 femtograms, 120 femtograms, and 12 attograms. This range was critical for determining the system's sensitivity to varying amounts of viral RNA. The results indicated a gradual increase in peak intensity on the electropherograms proportional to the RNA template concentration. Remarkably, the target peak from the 12 attogram sample, which is approximately equivalent to 2 copies of the virus, was clearly and reproducibly detected, with a signal-to-noise ratio exceeding 3. This significant finding demonstrates the system's capability to detect extremely low levels of viral RNA, reinforcing its potential as a highly sensitive diagnostic tool for early detection of influenza. The ability to identify the influenza A H3N2 virus using the Rotary PCR platform, even at very low LODs, along with the wide dynamic range of RNA templates, highlights the system's suitability for rapid, accurate virus detection, essential for timely diagnosis and effective management of influenza outbreaks.

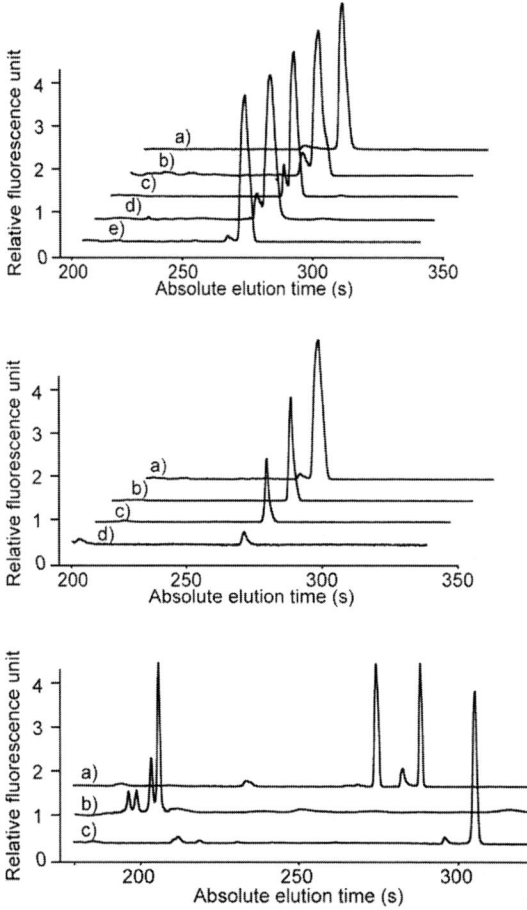

Fig. 6. (Top) Electropherogram for H3 gene amplicon separation on a μCE chip with variation of thermocycling time in the Rotary Genetic Analyzer system. Time for denature/annealing/extension steps was a) 60/60/60, b) 30/60/30, c) 30/30/30, d) 15/30/15, and e) 15/15/15 seconds. (Middle) LOD test for H3 gene expression from influenza A H3N2 virus by using the serially diluted RNA templates on the Rotary PCR system. The amount of RNA templates was a) 12 pg, b) 120 fg c) 1.2 fg, and d) 12 ag. (Bottom) Simultaneous multiple viral gene expression on the triplex Rotary RT-PCR system. a) H3 gene (150 bp) and M gene (160 bp) from influenza A H3N2 virus; b) H1 gene (102 bp) from H1N1 virus, and c) H5 gene (172 bp) from H5N1 subtype. (Reprinted with permission from Ref. [24]. Copyright (2012) Royal Society Chemistry)

Quality of Results: The high quality of the results is evidenced by the clear and reproducible detection of target genes, even at very low concentrations. The system's ability to differentiate between various influenza subtypes and to perform multiplex PCR further underscores its diagnostic accuracy and versatility.

Multiplexing and Rapid Subtyping: The capability to perform multiplex PCR was demonstrated by simultaneously targeting multiple influenza A viral RNAs and their corresponding genes in separate chambers. This approach allows for rapid subtyping of multiple virus strains, crucial in outbreak situations and for informing public health strategies (Fig. 6, bottom).

6. Conclusion

The exploration of the centrifugal microfluidic device for PCR throughout this chapter has illuminated a transformative advancement in the realm of molecular diagnostics. From the intricate design of the PCR chip to the nuanced operation of the system and its exceptional performance, this chapter has provided a comprehensive view of how this technology reshapes PCR methodologies. In summarizing the key elements, the chip's innovative design, combining PDMS and glass, has been pivotal in achieving efficient thermal cycling and multiplexing capabilities. The fabrication process, employing soft lithography, exemplifies precision engineering. Operationally, the system's automation, led by a stepper motor and advanced temperature control, facilitates rapid and accurate PCR, a significant improvement over traditional method.

The performance of the centrifugal microfluidic device, especially in detecting influenza A virus subtypes, showcases its potential in high-throughput and specific molecular analysis. This capability is crucial in rapidly evolving clinical and epidemiological landscapes where timely and accurate diagnostics are essential. The potential applications of the centrifugal microfluidic PCR chip extend beyond infectious disease diagnostics into areas such as genetic research, personalized medicine, and environmental monitoring. Its capacity for rapid processing and multiplexing opens new avenues for real-time, on-site testing, which is invaluable in resource-limited settings and in situations requiring immediate results. Moreover, ongoing research and development in this field promise further enhancements in chip design and functionality. The integration of more sophisticated detection systems, increased automation, and even greater miniaturization are on the horizon. These advancements could lead to more portable devices, expanding the utility of PCR in field applications and further democratizing molecular diagnostics.

References

1 S. A. Bustin, *J Mol Endocrinol*, 2000, 25, 169–193.

2 W. Feng, A. M. Newbigging, C. Le, B. Pang, H. Peng, Y. Cao, J. Wu, G. Abbas, J. Song, D.-B. Wang, H. Zhang and X. C. Le, *Anal Chem*, 2020, 92, 10196–10209.

3 H. A. Erlich, *J Clin Immunol*, 1989, 9, 437–447.

4   P. J. Asiello and A. J. Baeumner, *Lab Chip*, 2011, 11, 1420–1430.

5   P. Gill and A. Ghaemi, *Nucleosides Nucleotides Nucleic Acids*, 2008, 27, 224–243.

6   P. Craw and W. Balachandran, *Lab Chip*, 2012, 12, 2469–2486.

7   E. Nunez-Bajo, A. Silva Pinto Collins, M. Kasimatis, Y. Cotur, T. Asfour, U. Tanriverdi, M. Grell, M. Kaisti, G. Senesi, K. Stevenson, K. Stevenson and F. Güder, *Nat Commun*, 2020, 11, 6176.

8   G. Liu and J. F. Rusling, *ACS Sens*, 2021, 6, 593–612.

9   N. Boonham, J. Kreuze, S. Winter, R. van der Vlugt, J. Bergervoet, J. Tomlinson and R. Mumford, *Virus Res*, 2014, 186, 20–31.

10 J. Yang, V. M. Phan, C.-K. Heo, H. V. Nguyen, W.-H. Lim, E.-W. Cho, H. Poo and T. S. Seo, *Sens Actuators B Chem*, 2023, 380, 133331.

11 E. T. Lagelly, J. R. Scherer, R. G. Blazej, N. M. Toriello, B. A. Diep, M. Ramchandani, G. F. Sensabaugh, L. W. Riley and R. A. Mathies, *Anal Chem*, 2004, 76, 3162–3170.

12 P. Liu, T. S. Seo, N. Beyor, K. J. Shin, J. R. Scherer and R. A. Mathies, *Anal Chem*, 2007, 79, 1881–1889.

13 T. S. Seo, N. Beyor, L. Yi and R. A. Mathies, *Proceedings of Conference, MicroTAS 2009 - The 13th International Conference on Miniaturized Systems for Chemistry and Life Sciences*, 2009, 81, 1049–1051.

14 P. Neuzil, J. Pipper and T. M. Hsieh, *Mol Biosyst*, 2006, 2, 292–298.

15 M. Focke, F. Stumpf, G. Roth, R. Zengerle and F. Von Stetten, *Lab Chip*, 2010, 10, 3210–3212.

16 M. Allen Northrup, B. Benett, D. Hadley, P. Landre, S. Lehew, J. Richards and P. Stratton, *Anal Chem*, 1998, 70, 918–922.

17 K. H. Chung, S. H. Park and Y. H. Choi, *Lab Chip*, 2010, 10, 202–210.

18 M. U. Kopp, A. J. De Mello and A. Manz, *Science (1979)*, 1998, 280, 1046–1048.

19 H. Zhu, H. Zhang, S. Ni, M. Korabečná, L. Yobas and P. Neuzil, *TrAC - Trends in Analytical Chemistry*, 2020, 130, 115984.

20 S. F. Berlanda, M. Breitfeld, C. L. Dietsche and P. S. Dittrich, *Anal Chem*, 2021, 93, 311–331.

21 C. Dincer, R. Bruch, A. Kling, P. S. Dittrich and G. A. Urban, *Trends Biotechnol*, 2017, 35, 728–742.

22 J. B. Nielsen, R. L. Hanson, H. M. Almughamsi, C. Pang, T. R. Fish and A. T. Woolley, *Anal Chem*, 2020, 92, 150–168.

23 K. Ren, J. Zhou and H. Wu, *Acc Chem Res*, 2013, 46, 2396–2406.

24 J. H. Jung, S. J. Choi, B. H. Park, Y. K. Choi and T. S. Seo, *Lab Chip*, 2012, 12, 1598–1600.

Chapter 3
Centrifugal Microfluidic Device for Isothermal Amplification

1. Background

The field of molecular diagnostics has witnessed a transformative shift with the advent of centrifugal microfluidic devices, particularly in the realm of isothermal amplification. This technological leap has not only streamlined the process of pathogen detection but also opened new avenues for rapid, efficient, and accurate medical diagnostics. The three studies under discussion each contribute to this field, presenting innovative approaches and applications of centrifugal microfluidic technology for isothermal amplification, which is crucial in detecting a wide range of pathogens and genetic markers.

Isothermal amplification stands as a pivotal advancement in molecular biology, offering a simpler and more cost-effective alternative to traditional polymerase chain reaction (PCR) methods [1, 2]. Unlike PCR, which requires thermal cycling to amplify DNA, isothermal amplification techniques, such as Loop-Mediated Isothermal Amplification (LAMP), nucleic acid sequence-based amplification (NASBA), and Recombinase polymerase amplification (RPA), rely on a constant temperature, simplifying the equipment and process required for DNA amplification [3–8]. This technology leverages a constant temperature to amplify nucleic acids, simplifying the equipment and process requirements. This feature is especially beneficial in resource-limited settings, such as point-of-care diagnostics, where traditional PCR's complexity and cost can be prohibitive [1]. The rapid amplification capability of isothermal techniques has substantially reduced the time required for pathogen detection, enabling faster responses in clinical and field settings. The ability of isothermal amplification to rapidly amplify nucleic acids at a single temperature has significantly accelerated the diagnostic process, enabling quicker responses to infectious disease outbreaks and other medical diagnostics.

Centrifugal microfluidic devices have redefined the landscape of molecular diagnostics by integrating the ease of isothermal amplification with the precision of microfluidics. These devices use centrifugal force for fluid manipulation within microchannels, eliminating the need for external pumps and complex tubing systems. Centrifugal microdevices excel in simplicity, automation, user-friendliness, and low power consumption, enhancing their portability, reproducibility, and cost-effectiveness [9–12]. Especially, this chapter focused on the integration of the isothermal amplification assay on the centrifugal microfluidic device for molecular diagnostics.

1.1. A LAMP reaction

A LAMP reaction, developed by Notomi et al. in the early 2000s, represents a significant advancement in molecular diagnostics [13]. This novel technique, emerging as an efficient alternative to conventional PCR-based methods, operates under isothermal conditions, enabling DNA amplification at a constant temperature. The essence of LAMP lies in its use of a unique set of primers in conjunction with a polymerase capable of strand displacement. This combination allows for rapid and efficient DNA amplification, circumventing the need for the thermal cycling required in PCR. The simplicity of the LAMP process, requiring minimal equipment, renders it particularly useful in resource-limited settings and contributes to its rapid turnaround time, often yielding results within an hour [2].

The specificity and sensitivity of LAMP are noteworthy, primarily due to its use of multiple primers that target different regions of the DNA sequence. This multiplicity ensures the reliable detection of low levels of DNA, a crucial aspect for accurate diagnostics. Despite these advantages, LAMP faces certain challenges, such as the risk of carryover contamination and limitations in quantifying the amplified product. However, these limitations have not deterred its application in a wide range of settings, particularly in disease diagnosis.

In the realm of diagnostic applications, LAMP has shown exceptional utility, exemplified by the work of Jung et al. in the detection of Influenza A virus [14]. Their study demonstrates the integration of LAMP, particularly its reverse transcriptase form (RT-LAMP), into a centrifugal microdevice, combining RNA purification and amplification in a single platform. This innovation not only highlights the adaptability of LAMP but also its enhanced sensitivity and specificity compared to traditional methods.

1.2. An RCA reaction

The advent of Rolling circle amplification (RCA) has been a significant stride in the field of molecular diagnostics, particularly in the context of genetic variations and their implications in disease prognosis and diagnosis. Traditional imaging techniques like mammography, MRI, CT, PET, and ultrasound, though prevalent, face limitations in early cancer detection due to requirements for detectable tumor size, high costs, and prolonged processes [15]. Contrarily, genetic-based methods, owing to their accuracy and reliability, have gained widespread acceptance. Single Nucleotide Polymorphisms (SNPs), the most common form of human genetic variation, and mutations, which are permanent abnormal changes in nucleotide sequences, play a pivotal role in disease onset and progression [16, 17]. The ability to identify these mutations is crucial for predicting and preventing diseases and is fundamental to personalized medicine [18]. SNP typing, essential for early cancer diagnosis [19, 20], has been advanced through various methods. Techniques include primer extension reaction [21], DNA hybridization [22], SNP site ligation reaction [23], single-stranded conformational polymorphism analysis [24], and gradient gel electrophoresis [25]. Each method offers a unique approach to SNP detection.

RCA, as a technique, stands out for its sensitivity in detection, primarily due to its mechanism of amplifying circular DNA templates isothermally using DNA polymerase phi 29. This process results in the production of long, single-stranded DNA products containing thousands of sequence repeats [26, 27]. These repeats are then utilized for identifying the resultant amplicons through DNA hybridization [28–32]. This feature of RCA marks a significant departure from the conventional PCR. RCA's isothermal amplification method overcomes PCR limitations, offering a simpler and more efficient alternative for genetic analysis, especially in settings where complex equipment and precise thermal cycling are impractical. Its capacity for high sensitivity detection and the simplicity of its operational protocol renders it an invaluable tool, particularly in the realms of early cancer diagnosis and SNP typing. The future of RCA in genetic analysis and diagnostics holds significant promise, with its potential to revolutionize methodologies in the field and contribute significantly to the advancement of personalized medicine.

Heo et al. delved into the complexities and potential of RCA in the context of cancer diagnostics. The study focuses on the application of RCA in identifying point mutations, a critical aspect in understanding the genetic basis of cancer and its metastasis [33]. By integrating RCA with a novel microfluidic device, the research presents an innovative approach to SNP typing, offering a more sensitive and efficient alternative to conventional methods. This study not only highlights the technical advancements in RCA but also emphasizes its practical implications in early cancer diagnosis, underscoring the importance of genetic-based methods in medical diagnostics.

1.3. An RPA reaction

Foodborne illnesses, manifesting as fever, vomiting, and stomachache, pose a significant threat to public health. In the United States alone, millions of incidents of foodborne illness occur annually, leading to numerous hospitalizations and deaths [34]. A major cause of these illnesses is food poisoning bacteria, such as *Salmonella enterica*, *Escherichia coli* O157:H7, and *Vibrio parahaemolyticus*, which can proliferate rapidly under improper food handling or cooking conditions. The need for rapid and accurate identification of these bacteria is critical in minimizing economic loss and preventing casualties [35].

Among the isothermal methods, RPA stands out due to its rapidity and sensitivity in amplifying target DNA sequences at relatively low temperatures (around 39 °C) compared to other methods. The RPA reaction involves the formation of complexes of recombinase and primers, which induce strand exchange at homologous sequences in double-stranded DNA, and a single-stranded binding protein that stabilizes the displaced strand, preventing primer ejection. Polymerase then initiates the synthesis of the complementary DNA strand [3]. This process allows for the real-time monitoring of target gene amplification using specific detection probes. Various RPA assays have been developed to detect viruses and pathogen bacteria, highlighting the versatility and efficiency of this method in rapid genetic diagnostics [36–38].

Choi et al. (2016) explore the application of RPA in the context of food safety [39]. Their study focuses on the development of a centrifugal microdevice that employs RPA for the rapid and real-time identification of bacteria responsible for food poisoning. This innovative approach not only showcases the potential of RPA in molecular diagnostics but also addresses the critical need for quick and reliable methods to detect foodborne pathogens. The research by Choi et al. is particularly significant as it contributes to the advancement of diagnostic techniques that are vital for public health and safety, especially in the prevention and control of foodborne illnesses.

2. Centrifugal chip for isothermal amplification

2.1. Chip design for LAMP

In the research led by Jung et al., the design of an integrated RT-LAMP microdevice for detecting Influenza A virus showcases a meticulous combination of engineering precision and biological functionality [14]. This microdevice, specifically tailored for LAMP, demonstrates an innovative approach to managing fluid dynamics and biochemical reactions within a compact and efficient framework. The microdevice is strategically segmented into distinct zones, each dedicated to a specific function such as sample loading, reagent mixing, and amplification. At the forefront of the device are three reservoirs for an RNA sample, a washing solution, and an elution solution, positioned in front of a microbead bed channel that functions as an RNA capture matrix. On the right side of the device, another reservoir is designed for the RT-LAMP cocktail, containing an enzyme mix, target-specific primers, and a reaction buffer (Fig. 1A).

The sample loading zone is engineered to handle varying types and volumes of biological samples, ensuring effective processing even with minimal quantities. This attention to detail in sample handling is important for the accuracy and reliability of the diagnostic process. The RNA sample inlet includes a capillary valve with a depth of 100 μm to prevent backflow into the washing and elution solution reservoirs. Additionally, a weir structure of the same depth (100 μm) is fabricated to pack tetraethyl orthosilicate (TEOS)-treated microbeads, with diameters ranging from 150 to 212 μm [40]. The washing solution reservoir is connected to the bead-bed microchannel by a capillary valve measuring 580 μm in width and 150 μm in depth (Fig. 1B). The siphon channel, bridging between the elution solution reservoir and the bead-bed microchannel, measures 580 μm in width and 100 μm in depth. To ensure stable storage of solutions, the washing and elution solution reservoirs are designed with a ring structure of 200 μm depth and 250 μm deep patterns to prevent wetting of the top sealing tape and eliminate contamination and evaporation issues during the RT-LAMP process [41]. The waste chamber entrance contains three capillary valves, each measuring 580 μm in width, 150 μm in depth, and 1000 μm in length, to prevent backflow. The RT-LAMP chamber, directly linked with the RT-LAMP cocktail reservoir through a siphon channel, is designed with a volume of 6 μL.

This innovative microdevice design by Jung et al. represents a significant advancement in the field of molecular diagnostics. Its ability to streamline the LAMP process within a user-friendly and efficient platform opens new avenues in rapid pathogen detection and beyond, potentially transforming approaches to disease detection and monitoring. This innovative microdevice design has far-reaching implications, paving the way for future developments in biomedical diagnostics and potentially transforming the approach to disease detection and monitoring.

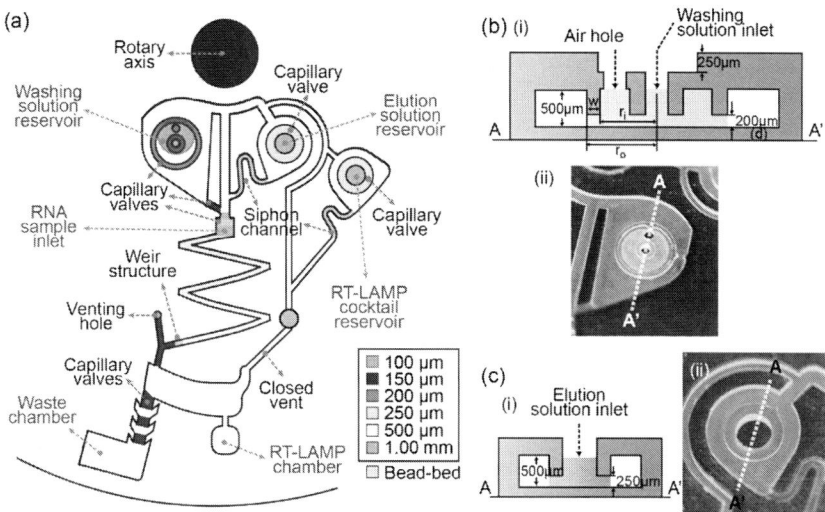

Fig. 1. (a) Schematic illustration of the integrated RT-LAMP microdevice. (b) (i) Schematic design of the washing solution reservoir, (ii) a digital image of the washing solution chamber from a bottom view. (c) (i) Schematic design of the elution solution reservoir, (ii) a digital image of the elution solution chamber from a bottom view. (Reprinted with permission from Ref. [14]. Copyright (2015) Elsevier.)

2.2. Chip design for RCA

The study by Heo et al. presents a sophisticated design for a microdevice integrating RCA [33]. This design is crucial for enhancing the molecular diagnostics process, particularly in applications such as SNP typing and point mutation identification.

The microdevice features a disk shape with symmetric fluidic channels, as depicted in Fig. 2A. The central part of this disk houses a common inlet hole, which serves as the primary entry point for samples and reagents. Around this common inlet, the microdevice includes twelve ligation solution inlets and four waste outlets. These inlets and outlets are strategically positioned within the outer channel of the device, enhancing the management of fluids and the effective disposal of waste.

The microdevice is composed of three distinct layers, each contributing to its functionality. These layers are a rotary top plate, a microfluidic channel wafer, and an RTD (Resistance Temperature Detector) wafer (Fig. 2B). The rotary top plate houses twelve reaction chambers, each with a 1 µL volume, vital for the RCA reaction. The microfluidic channel wafer, colored red in the design schematics, is equipped with twelve radial microfluidic channels that connect to the reaction chambers. Additionally, an outer circle microchannel includes four waste outlets and twelve ligation solution inlet channels, ensuring that fluids are accurately directed throughout the device. The RTD wafer plays a vital role in temperature monitoring during the RCA process. It is equipped with a Ti/Pt electrode patterned to accurately measure and control the temperature, which is crucial for the success of the RCA reaction.

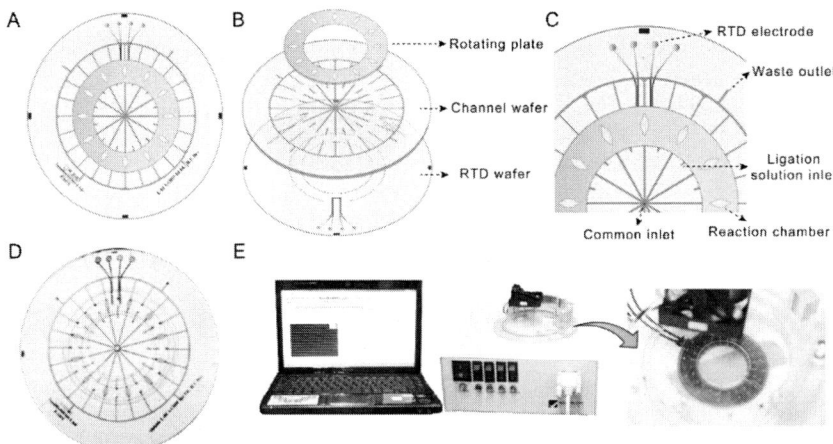

Fig. 2. (A) An assembled view of the rotary ligation-RCA microdevice. (B) Schematics of each layer of the rotary ligation-RCA microdevice. (C) Enlarged view of the rotary ligation-RCA microdevice. (D) A digital image of the assembled microdevice. (E) A digital image of the portable genetic analyser, which consists of a ligation-RCA rotary microdevice, a miniaturized operational hardware including a cooling fan and a PID controller, and a laptop computer. (Reprinted with permission from Ref. [33]. Copyright (2016) Elsevier.)

In summary, Heo et al.'s rotary RCA microdevice stands out as a significant innovation in molecular diagnostics. Its intricate design, combining fluid dynamics, precise temperature control, and efficiently configured reaction chambers, provides a comprehensive platform for genetic analysis. The device's specific measurements, including the 1 µL reaction chambers and the detailed fluidic channel design, enable accurate SNP typing and mutation detection, marking a breakthrough in the field.

2.3. Chip design for RPA

Choi et al. present a microdevice ingeniously designed for RPA, specifically aimed at detecting food poisoning bacteria in milk samples [39]. This microdevice stands as a testament to the practical application of RPA technology in a compact and efficient format. The microdevice is distinguished by its centrifugal nature, enabling the simultaneous execution of multiple RPA reactions. It features a design that allows twelve direct-RPA reactions to be conducted in parallel. Each unit of the device comprises four reaction chambers, enhancing the device's capacity for multiplex bacterial detection.

Comprising two layers, each serves a distinct purpose in the RPA process (Fig. 3). The top layer is allocated for the injection of RPA reagents. The bottom layer, on the other hand, contains sample reservoirs, specifically designed for loading milk samples spiked with food poisoning bacteria. The reaction chambers in this device are designed with a depth of 730 μm, suitable for a 10 μL direct-RPA reaction. This specific volume is critical for achieving optimal reaction conditions and amplification efficiency. The top polycarbonate (PC) layer of the microdevice is equipped with zigzag-shaped microchannels, which are instrumental in dispensing RPA reagents into the four reaction chambers [42, 43] These microchannels have dimensions of 800 μm in width, 500 μm in depth, and 7445 μm in length. They are capable of holding 25.6 μL of reagents, which are moved by capillary force. The reagents are loaded into each reactor in the bottom layer through the connecting microchannels at a controlled rotational speed of 3000 rpm, ensuring precise and efficient delivery of the reagents for the RPA reactions.

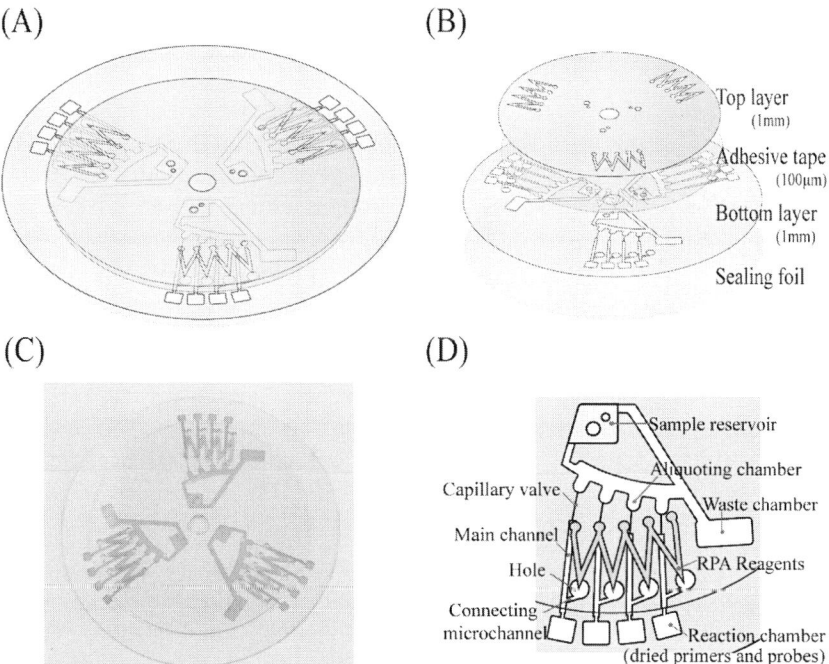

Fig. 3. Schematic illustration of the centrifugal direct-RPA microdevice. (A) An assembled image, (B) a disassembled image, (C) a real digital image of the microdevice, and (D) an enlarged schematic of each unit and its components. (Reprinted with permission from Ref. [39]. Copyright (2016) Royal Society of Chemistry.)

Choi et al.'s microdevice exemplifies the seamless integration of RPA technology into a user-friendly and efficient platform. This integration facilitates rapid on-site diagnostics, crucial in applications like food safety, where quick and accurate detection of pathogens is essential. The design of Choi et al.'s microdevice, with its specific measurements and operational details, represents a significant advancement in molecular diagnostics. Its ability to conduct RPA in a streamlined and efficient manner opens new avenues in rapid pathogen detection, demonstrating the potential of microdevices in transforming diagnostic approaches in various fields.

3. Chip fabrication process

The development of microfluidic devices in the studies by Jung et al., Heo et al., and Choi et al., illustrates a meticulous blend of preparation and fabrication techniques, each tailored to optimize molecular diagnostic procedures like LAMP, RCA, and RPA [14, 33, 39]. While each study targets a different diagnostic procedure, they all share a common thread in their fabrication approach: the integration of microfluidic channels, reagent reservoirs, and multi-layered structures. These processes are defined by an overarching emphasis on precision in both material selection and functional design. The goal is to create a device that not only accommodates the fluidic and thermal demands of molecular diagnostics but also aligns with the biological requirements of sample preparation and reaction conditions.

In their groundbreaking work on an RT-LAMP microdevice for detecting Influenza A virus, Jung et al. demonstrated precision in both preparation and fabrication [14]. The preparation process involved the strategic introduction of an RNA sample, a washing solution, an elution solution, and an RT-LAMP reaction cocktail into designated reservoirs. The RNA sample preparation was carefully calculated to include specific proportions of the virus sample, ethanol, and Gu-HCl, ensuring optimal conditions for RNA capture. The fabrication of the microdevice centered on utilizing the hydrophilic and capillary action of packed microbeads in the bead-bed microchannel, efficiently absorbing the RNA solution for subsequent purification. This feature played a pivotal role in absorbing the RNA solution effectively, setting the stage for efficient purification. Another significant aspect was the integration of adhesive films over liquid reservoirs and the assembly of a polycarbonate (PC) cover, critical steps to contain the solutions and prevent evaporation during the RT-LAMP process.

The fabrication process of the integrated RT-LAMP microdevice, as described by Jung et al., involves a multi-layered structure and several precision techniques to create a functional and efficient device for RNA amplification. The microdevice is composed of four distinct layers: a top PC cover with a thickness of 2 mm, a micropatterned PC layer of 1 mm thickness, a double-sided adhesive layer of 50 μm thickness, and a PC film with a thickness of 125 μm. The layered design is crucial for the structural integrity and functional complexity of the microdevice. The PC cover and the patterned PC layer were precisely fabricated using a computer numerical control (CNC) modeling machine. The radius of the PC cover was set at 53 mm, while the patterned PC layer had a radius of 50 mm. Additionally, the double-sided adhesive layer was patterned using a cutting plotter, and the PC film was cut to match the 50 mm radius. Prior to assembly, the siphon channels on the micropatterned PC layer were treated with a VISTEX solution to render the surface hydrophilic. This treatment was essential for ensuring proper fluid flow within the microdevice. Furthermore, dynamic passivation of the RT-LAMP chamber was carried out using a bovine serum albumin (BSA) solution for 20 minutes to prevent non-specific adhesion of biomolecules to the surface. The three layers—micropatterned PC layer, adhesive layer, and PC film—were carefully aligned and then pressed together at 40 MPa using a heating press at room temperature. This step achieved a strong and precise bond between the layers, resulting in a total thickness of 1.175 mm for the microdevice. The thick PC cover was placed on the assembled microdevice right before the RT-LAMP reaction. It includes three holes, each with a 5 mm radius, located directly above the RT-LAMP chambers to facilitate efficient heat transfer during the reaction process. A digital image of the completed integrated RT-LAMP microdevice is shown in Fig. 4B. This image illustrates the final assembled state of the microdevice, ready for use in RT-LAMP reactions.

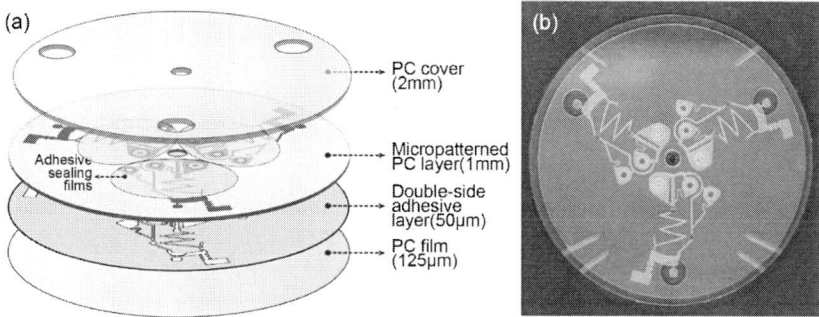

Fig. 4. (a) Schematic illustration of the integrated RT-LAMP microdevice consisting of four layers: a 2 mm thick PC cover, a 1 mm thick micropatterned PC layer, a 50 μm thick adhesive layer, and a 125 μm thick PC film. (b) A digital image of the integrated RT-LAMP microdevice. (Reprinted with permission from Ref. [14]. Copyright (2015) Elsevier.)

Heo et al. focused on fabricating a rotary ligation-RCA microdevice using advanced photolithography processes [33]. The fabrication began with coating borofloat glass with amorphous silicon, followed by priming with hexamethyldisilazane (HMDS) to enhance photoresist adhesion. The detailed application of photoresist, UV exposure, development, and etching processes, including RIE and hydrofluoric acid etching, was pivotal in creating microchannels with precise dimensions. The fabrication of the resistance temperature detector (RTD) electrode wafer further involved coating a glass wafer with titanium and platinum, followed by selective etching using aqua regia. The meticulous steps in this fabrication process highlight the complexities involved in creating a device that accommodates the fluid dynamics and temperature control necessary for RCA.

The fabrication of the rotary plate and microfluidic channel wafer in Heo et al.'s study is a demonstration of precision engineering, employing conventional photolithography. This process began with a 1.1 mm thick borofloat glass, coated with a layer of 200 nm amorphous silicon on a 4-inch glass wafer. To enhance the adhesion of the photoresist, the wafer was primed with HDMS. Following this, a positive photoresist was applied at 2500 rpm and then soft-baked at 120 °C for 90 seconds. The subsequent step involved UV exposure through a photomask, after which the channel pattern was developed using a Microposit developer solution. To form the microchannels, the exposed amorphous silicon on the wafer was removed through SF_6 reactive ion etching (RIE), and the exposed glass wafer was etched in a 49% hydrofluoric acid solution. This etching process resulted in microchannels that were 100 μm deep and 500 μm wide. Residual photoresist and silicon were then cleared away using acetone and SF_6 RIE, respectively. Additionally, the sample inlets and waste outlets were created using a CNC mill. In a further step, an RTD electrode wafer was fabricated. This involved coating a 4-inch borofloat glass wafer with 20 nm of titanium (Ti) and 200 nm of platinum (Pt). The wafer, primed again with HMDS, was covered with a positive photoresist (S1818). The Ti/Pt background, not shielded by the photoresist, was removed using an aqua regia solution (a mixture of hydrochloric acid and nitric acid in a 3:1 ratio). To complete the process, the RTD wafer was permanently bonded with the channel wafer in a vacuum furnace at a temperature of 680 °C for five hours.

This elaborate fabrication process in Heo et al.'s research highlights the intricate engineering and precision chemical processing involved in creating a functional and efficient microfluidic device. Each step, from the initial coating to the final bonding, is crucial in defining the microdevice's structural integrity and operational capabilities, especially in the context of its application in biological or chemical analysis.

The construction of Choi et al.'s centrifugal direct-RPA microdevice combined detailed preparation with multi-layered fabrication [39]. The device design, drafted in AutoCAD, included micro-reaction chambers, microchannels, and holes for multiplex detection. This integration is important for the device's functionality, allowing for simultaneous processing of multiple samples. The device is composed of four distinct layers: a top polycarbonate (PC) plate with a thickness of 1 mm, a 100 μm thick double-sided adhesive film, a bottom 1 mm thick PC plate, and a layer of poly-olefin sealing foil. This multi-layered construction is pivotal for creating the intricate pathways and chambers necessary for the direct-RPA reactions. The fabrication process involved precise mechanical and manual techniques. The PC plates were meticulously crafted using a computer-controlled milling machine, ensuring high precision in the dimensions and alignment of various components. The adhesive film was patterned using a cutting plotter, a step for defining the flow paths and reaction zones within the device. All these layers were then precisely aligned and bonded together using a press that exerted a force of 100 MPa at room temperature. This bonding process is essential to ensure the structural integrity and fluidic isolation of the different components within the microdevice. Before the final bonding, primers and probes were carefully dropped into the reaction chambers and allowed to dry at room temperature for 30 minutes. This pre-loading of primers and probes is a step in preparing the device for its intended biological reactions.

The completed device, with a diameter of 9 cm, was meticulously designed to handle the samples and reagents efficiently. The microdevice's functionality in sample distribution, reaction chamber allocation, and the use of capillary valves for fluid control showcased an innovative approach to RPA diagnostics.

Collectively, the preparation and fabrication techniques in these studies illuminate the intricate processes involved in developing microfluidic devices for molecular diagnostics. The precision in material selection, structural design, and integration of biological components is evident across all three studies, reflecting the nuanced requirements of microdevice fabrication. These advancements in microdevice technology not only enhance the efficiency and accuracy of pathogen detection but also open new avenues for rapid, on-site diagnostics, contributing significantly to the evolution of healthcare technology.

4. Chip operation

The operation of the microdevices described in the studies by Jung et al., Heo et al., and Choi et al. reflects a sophisticated integration of microfluidics, biochemical reactions, and mechanical systems, albeit with distinct functionalities tailored to their specific applications [14, 33, 39]. Common to all three devices is the use of centrifugal force as a key mechanism to facilitate fluid movement and sample processing within the microfluidic channels. This centrifugal action is crucial for directing the flow of samples, reagents, and washing solutions to designated areas of the chips without the need for external pumps. In each device, the sample introduction and subsequent movement through the microfluidic pathways are carefully controlled by the design of the channels, valves, and reservoirs. The utilization of capillary action and fluid dynamics is central to all three devices. This method of fluid control is essential for directing the flow of substances to the correct locations within the device. The capillary forces, determined by the microdevice's structural design and the materials used, are crucial for the automated and precise distribution of liquids within the microchannels.

The design ensures precise timing and sequencing of the various steps involved in the analysis for RNA amplification, SNP typing, or pathogen detection. The samples undergo a series of reactions in reaction chambers or on surfaces where reagents are pre-loaded or introduced during operation. The reactions are typically initiated or accelerated by specific temperature conditions, which are maintained by integrated heating elements or external temperature control systems. Furthermore, all three devices incorporate real-time monitoring or detection mechanisms, often using optical systems to track the progress of the reactions. This real-time analysis is essential for rapid diagnostics and allows for the immediate interpretation of results. Despite their shared use of centrifugal microfluidics and real-time analysis, each device is uniquely optimized for its specific application, whether it's detecting viral RNA, performing SNP typing, or identifying pathogenic bacteria in samples.

The integrated RT-LAMP microdevice developed by Jung et al. is a prime example of advanced microfluidic technology designed for the rapid and sensitive detection of Influenza A virus RNA [14]. The operation of this device encompasses several coordinated steps, leveraging both mechanical and biochemical principles (Fig. 5). The operation commences with the introduction of an RNA sample into the microdevice. This sample can be sourced from purified Influenza A viral RNA or lysates, accommodating different Influenza A virus strains like A/H1N1, A/H3N2, and A/H5N1. The design of the device caters to a variety of sample types, including clinical nasal swab samples, which can be processed using a viral RNA extraction kit prior to introduction. Central to the device's operation is the application of centrifugal force, generated by a custom-designed portable rotary genetic analysis microsystem. This force drives the RNA sample through the microdevice, initiating its journey through the various processing stages. The sample first enters a bead-bed microchannel, where it is absorbed onto specially designed microbeads. These microbeads play a crucial role in capturing the RNA, effectively isolating it from other sample components. Following RNA capture, the device employs a washing solution, typically composed of 70% ethanol, to purify the captured RNA. The movement of the waste solution towards the waste chamber, rather than the RT-LAMP chamber, was influenced by the combined effects of centrifugal and Coriolis forces, causing the flow direction to be oriented towards the bottom left [44–46]. This step is vital for removing impurities, such as residual salts and proteins, that could interfere with subsequent reactions. The next phase involves the sequential introduction of an elution solution (RNase-free water) and the RT-LAMP reaction cocktail into the system. The RT-LAMP cocktail is a complex mixture containing all necessary components for RNA amplification, including enzymes, primers, and buffers. This mixture is critical for the amplification process, ensuring that even minute quantities of the target RNA can be detected. The actual amplification of RNA takes place in a designated reaction chamber. Here, the conditions are carefully controlled, maintaining a temperature of 64 °C to facilitate the RT-LAMP reaction. This reaction is monitored in real-time using a miniaturized optical detector, which is sensitive enough to confirm the presence of Influenza A virus with a very low number of viral RNA copies. The real-time monitoring feature is crucial for providing immediate diagnostic results.

The operation of the rotary ligation-RCA microdevice for multiplex SNP typing, as delineated in Fig. 6 of Heo et al.'s paper, begins with the injection of a ligation mixture into twelve reaction chambers [33]. This injection is precisely controlled through twelve ligation solution inlets, each leading to a chamber on the top rotary plate of the device. The alignment of these chambers with the radial microchannels is important for ensuring accurate ligation reactions between the capture probe and the padlock probe. After loading the samples, the rotary plate is manually rotated by 7.5 degrees, a step for isolating each reaction chamber and ensuring the specificity of the ligation reaction for each SNP type. Each chamber contains a specific padlock probe, corresponding to different SNP sites. The device employs both wild and mutant targeting padlock probes, allowing for the identification of SNP sequences by comparing the fluorescence signals from each reaction chamber at the conclusion of the process. This design enables the analysis of up to five-point mutations, in addition to two negative control experiments where either a target or a padlock probe is omitted.

A key feature of the device's design is the formation of an oil layer between the

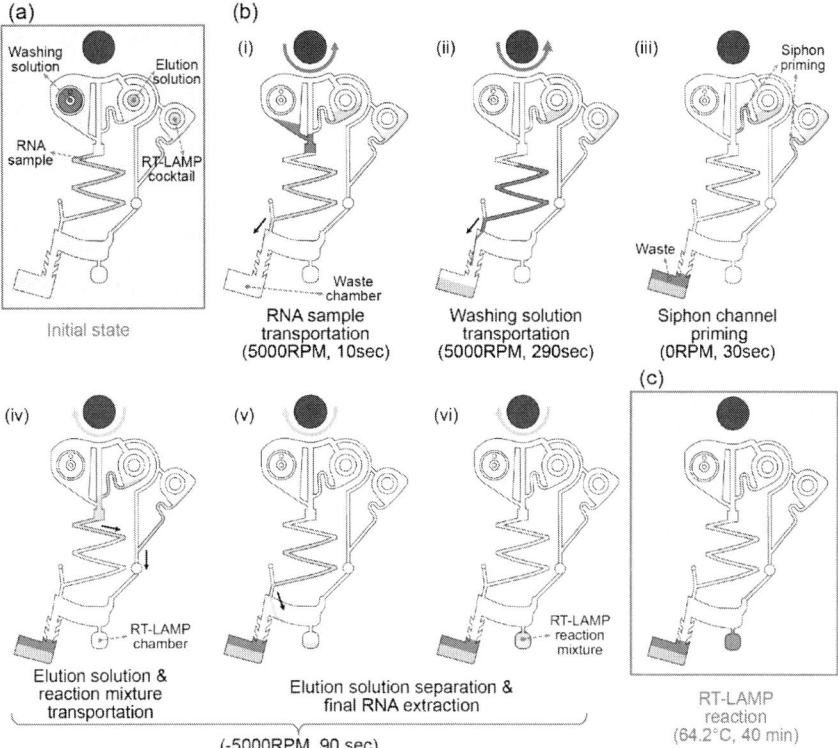

Fig. 5. The entire process for the integrated RT-LAMP microdevice. (a) An initial state of the integrated RT-LAMP microsystem with all the liquids in the designated reservoirs loaded. (b) (i) RNA sample transportation at 5000 RPM for 10 seconds, (ii) washing solution transportation at 5000 RPM for 290 seconds, (iii) siphon channel priming at 0 RPM for 30 seconds, (iv) elution solution and reaction mixture transportation at -5000 RPM for 2 seconds, (v-vi) elution solution separation and final RNA extraction at -5000 RPM for 88 seconds. (c) RT-LAMP reaction at 64.2 °C for 40 minutes. (Reprinted with permission from Ref. [14]. Copyright (2015) Elsevier.)

rotary plate and the channel wafer, which prevents the leakage of the ligation mixture and the formation of bubbles during the rotation process (Fig. 6B). This innovative design greatly simplifies the operation of the chip, eliminating the need for microvalves and thus enhancing the overall simplicity and reliability of the operation process. Following the ligation reaction, the device undergoes further clockwise rotations (Fig. 6C and 6D) for loading the RCA samples (red color). The temperature control during this phase is managed by a film heater, which is regulated by an in-house LabVIEW program (Fig. 2E), ensuring optimal conditions for the RCA process. Finally, by the counter-clockwise rotation, the detection probe loading (green color), and the clockwise rotation (Fig. 2E and 2F), the detection probe could be hybridized with the sequence of the RCA amplicons, resulting in the emission of green fluorescence signals from the FAM-labelled probes in the reaction chambers. This controlled, rotational mechanism on the hydrophobic microfluidic device allows for efficient fluidic

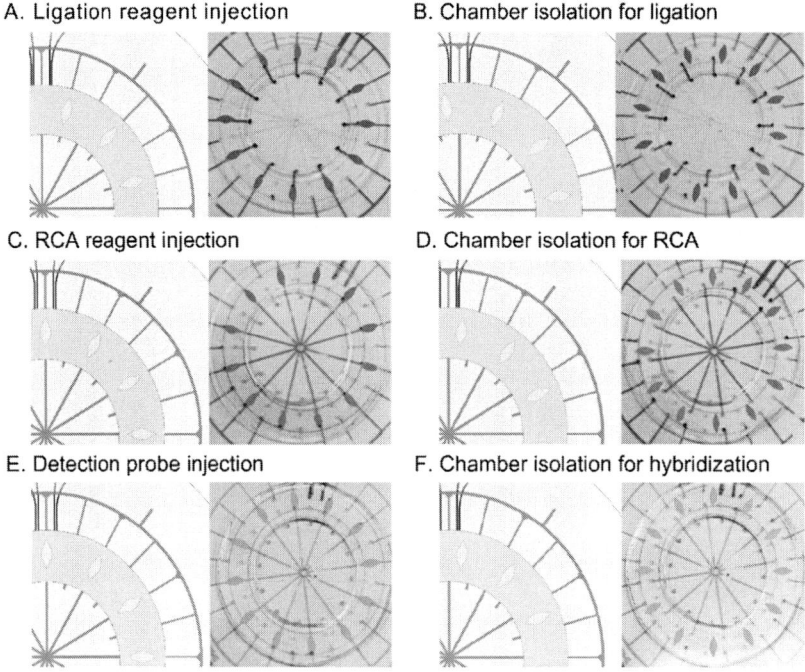

Fig. 6. Operation processes and the corresponding digital images for (A) ligation reagent injection, (B) chamber isolation for ligation, (C) RCA reagent injection, (D) chamber isolation for RCA, (E) detection probe injection, and (F) chamber isolation for hybridization reaction. (Reprinted with permission from Ref. [33]. Copyright (2016) Elsevier.)

control for multiplex reactions without the issues of sample loss, bubble generation, or evaporation, highlighting the device's advanced design and capability in genetic analysis.The operation of the centrifugal direct-RPA microdevice developed by Choi et al. for real-time detection of bacteria in milk samples is a complex process involving precise control of sample and reagent flow, as well as accurate temperature regulation [39]. The process starts off-chip, where a certain number of bacterial cells are mixed with 3.2 µL of milk and 10 µL of direct PCR buffer. This mixture leads to the lysis of the cells, releasing their genetic material. The resulting lysate, totaling 13.2 µL, is then delivered to the sample reservoir of the microdevice. Concurrently, two freeze-dried RPA reaction pellets are dissolved in 25.44 µL of a rehydration buffer and 4 µL of 280 mM magnesium acetate. This solution, along with 21 pmol of each primer and 6 pmol of probe pre-dried in the reaction chambers, constitutes the reagents for the RPA reaction.

The microdevice is mounted onto a custom-made portable genetic analyzer equipped with a miniaturized optical detector [14]. The flow control of the microdevice, as illustrated in Fig. 7, is achieved through centrifugal force. At 800 rpm, the bacterial sample is divided into four aliquoting chambers. Then, at 3000 rpm, the RPA reagents from the top layer are transferred to the reaction chambers through holes and connecting microchannels. Finally, the bacterial samples in the aliquoting chambers are injected into each reaction chamber at 5000 rpm, preparing the device for twelve simultaneous direct-RPA reactions. The reagents and sample solution are thoroughly mixed within each reaction chamber by shaking the microdevice between -600 rpm and +600 rpm for 30 seconds [47]. This step ensures uniform distribution of the reagents with the sample. The direct-RPA reaction and the real-time analysis are performed with precise control of temperature and rotation profiles, managed by an in-house LabVIEW software. The thermal blocks of the microsystem are heated up to 39 °C to facilitate the RPA reaction. The miniaturized optical detector monitors fluorescence signals from the reaction chambers every 2 minutes, and fluorescence images are captured every 10 minutes using a fluorescence microscope, providing real-time data on the reaction progress. The design of the microdevice allows for simultaneous performance of twelve reactions under identical conditions. This multiplex capability is essential for comprehensive and rapid bacterial detection in milk samples.

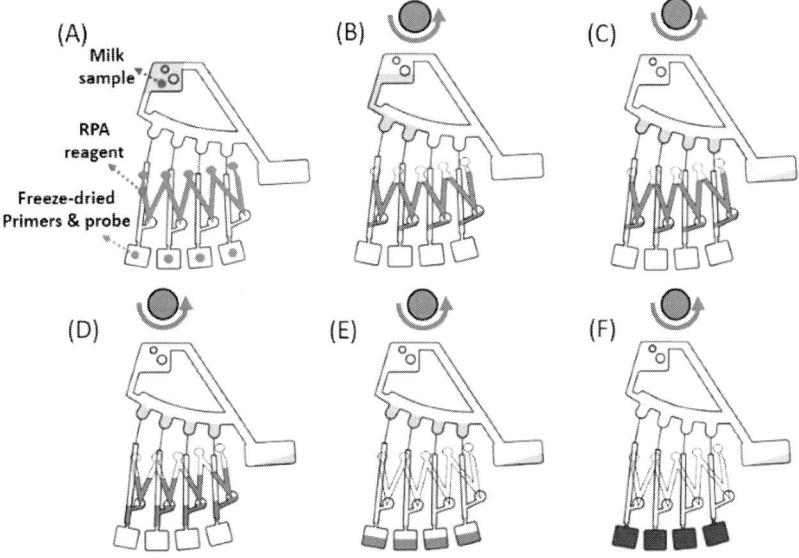

Fig. 7. Illustration of the entire flow control on the centrifugal direct-RPA microdevice. (A) An initial state of the direct-RPA microdevice with the injected milk sample, the RPA reagents, and the dried primers and probe set. (B) & (C) Aliquoting process of the sample at 800 rpm. (D) & (E) Dispensing process of the RPA reagents at 3000 rpm. (F) Injection of the sample in the aliquoting chamber into four reaction chambers at 5000 rpm and mixing of the injected samples and reagents. (Reprinted with permission from Ref. [39]. Copyright (2016) Royal Society of Chemistry.)

Across these devices, sealing mechanisms are employed to prevent contamination and evaporation. This is achieved through the application of adhesive films or covers that secure the reservoirs and chambers, maintaining the integrity of the reactions. The importance of these sealing methods cannot be overstated, as they are vital for preserving the reaction environment and ensuring accurate diagnostic results. The operation of these microfluidic devices, as described in the studies, highlights the sophisticated blend of engineering precision and biological understanding. While there are common operational strategies like fluid management and contamination prevention, each study introduces unique features tailored to their specific diagnostic goals. The integration of these operational elements enhances the efficiency and accuracy of molecular diagnostics, contributing significantly to the advancement of rapid, on-site diagnostics and healthcare technology.

5. Performance of isothermal amplification on a centrifugal microdevice

In their respective studies on microfluidic devices for molecular diagnostics, Jung et al., Heo et al., and Choi et al. achieved significant results in molecular amplification, each demonstrating the effectiveness of their devices in detecting various biological targets [14, 33, 39].

The research conducted by Jung et al. on their integrated RT-LAMP microdevice yielded significant results in the detection and analysis of Influenza A virus RNA [14]. The integrated microdevice successfully conducted the RT-LAMP reaction at a precisely controlled temperature of 64 °C for 40 minutes. This reaction was

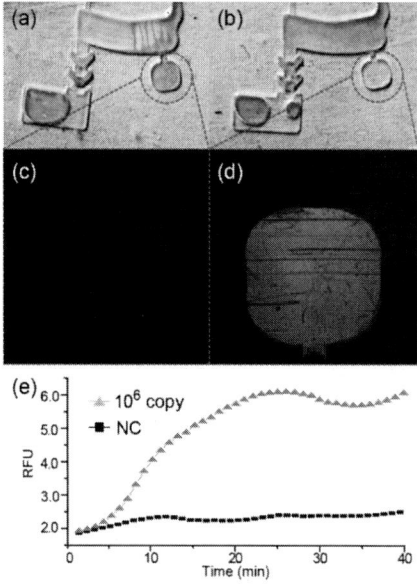

Fig. 8. Digital images of the RT-LAMP chamber (a) before and (b) after the RT-LAMP reaction. Confocal images of the RT-LAMP chamber (c) before and (d) after the RT-LAMP reaction. (e) Real-time RT-LAMP profiles obtained from the RT-LAMP chamber. (RFU: Relative fluorescence units, NC: Negative control) (Reprinted with permission from Ref. [14]. Copyright (2015) Elsevier.)

crucial for amplifying the viral RNA. The efficiency of the reaction was visually indicated by a color change in the RT-LAMP chamber, shifting from orange to yellow (Fig. 8). This color transition was due to the interaction of the quenched calcein metal indicator with manganese ions initially and magnesium ions during the RT-LAMP reaction, resulting in a strong fluorescence signal (λ_{ex} = 488 nm/ λ_{em} = 515 nm) [48]. A critical aspect of the study was establishing the limit of detection (LOD) for the device. The researchers performed tests using serially diluted Influenza A H1N1 viral RNA templates, ranging from 10^6 to as few as 10 copies. The results demonstrated that the device could detect even 10 copies of viral RNA, as evidenced by the gradual increase in fluorescence signals (Fig. 9). This sensitivity is significantly better than that of conventional real-time RT-PCR methods, marking a tenfold enhancement in detection capability. The real-time RT-LAMP profiles indicated that positive fluorescence signals were observed within 20 minutes, and signal intensities reached a plateau after 30 minutes.

For quantitative analysis, a standard curve was established by plotting the threshold time against the logarithm of RNA copy number. This curve showed a high correlation coefficient (R^2=0.9924), underscoring the device's accuracy in quantitative analysis. Interestingly, while the total processing time was 47 minutes, including RNA purification and RT-LAMP reaction, the saturation of the fluorescence signal in the RT-LAMP reaction was achieved within 30 minutes. This observation suggests that the total processing time could

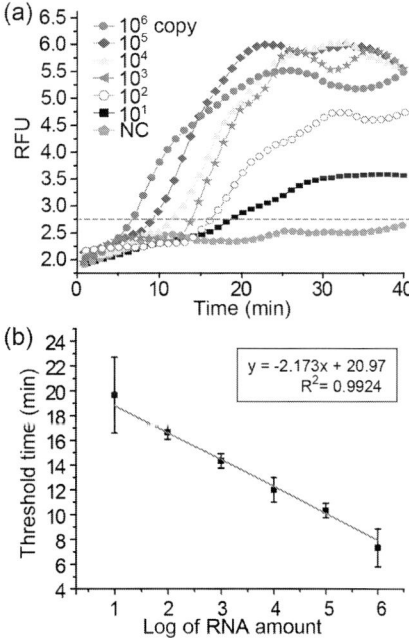

Fig. 9. (a) Real-time RT-LAMP profiles using purified RNAs ranging from 10^6 to 10 copies. (b) A standard curve by plotting the threshold time versus the logarithm of RNA copy number. (RFU: Relative fluorescence units, NC: Negative control) (Reprinted with permission from Ref. [14]. Copyright (2015) Elsevier.)

potentially be reduced to 37 minutes. An important feature of the microdevice was its ability to perform rapid subtyping of Influenza A virus, critical for early medical intervention. The device successfully amplified both the H1 and M genes of Influenza A H1N1 virus, confirming the presence of this specific subtype.

Further tests for specificity revealed that the H1 targeting primer sets did not produce amplicons for H3N2 and H5N1 viral lysates, demonstrating their specificity to the H1 gene (Fig. 10). In contrast, the M gene targeting primer sets showed amplification for H3N2 and H5N1 viruses, indicating that they were targeting a conserved region of the M gene, universal for identifying the Influenza A virus. The microdevice was also tested with clinical samples to confirm its practical application. Three clinical nasal swab samples previously confirmed to be infected with the Influenza A virus through conventional real-time RT-PCR were analyzed. The RT-LAMP microdevice successfully detected the M gene in all these samples, demonstrating its effectiveness in clinical settings.

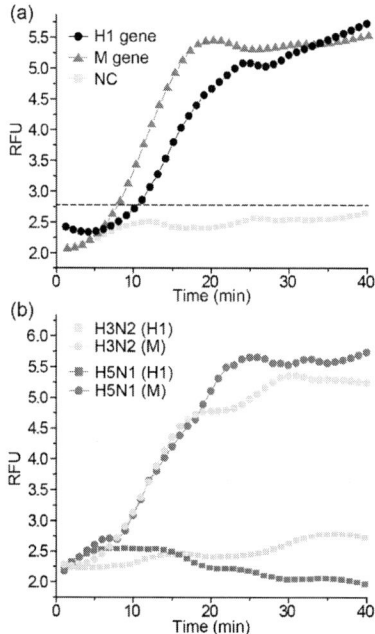

Fig. 10. The real-time RT-LAMP profiles for (a) subtyping of Influenza A H1N1 virus and (b) specificity. (RFU: Relative fluorescence units, NC: Negative control) (Reprinted with permission from Ref. [14]. Copyright (2015) Elsevier.)

In the study by Heo et al., the rotary microdevice designed for ligation-RCA exhibited remarkable sensitivity in genotyping at mutation points [33]. The study focused on genotyping five-point mutations of the *tp53* gene, along with a negative control, using the rotary microdevice. The protein expressed by the *p53* gene functions as a tumor suppressor, it plays a crucial role in regulating cell division, ensuring that cells do not grow and divide too rapidly or uncontrollably

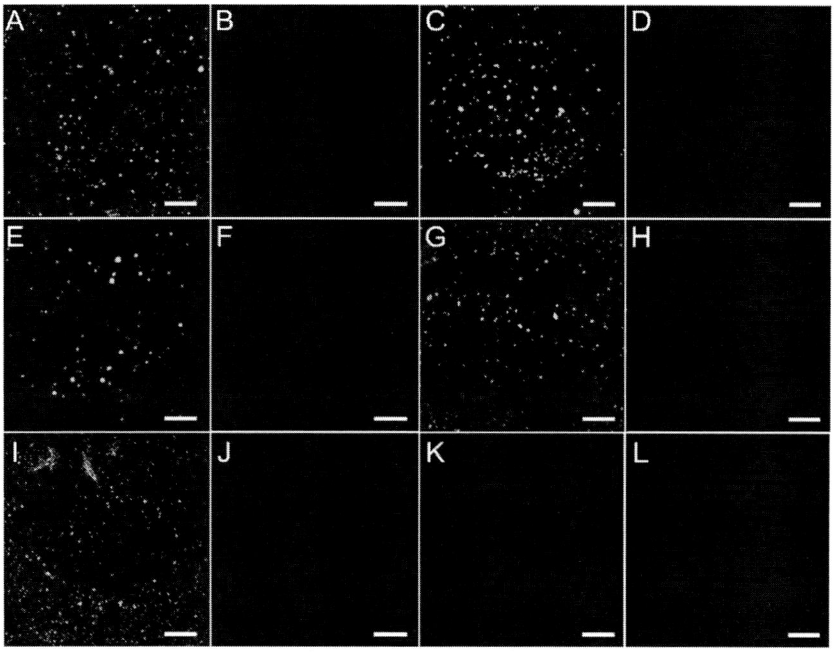

Fig. 11. Fluorescence images of the L-RCA reaction chambers. The used padlock probes were designed for genotyping (A) wild-type of target 1, (B) mutant-type of target 1, (C) wild-type for target 2, (D) mutant-type for target 2, (E) wild-type for target 3, (F) mutant-type for target 3, (G) wild-type for target 4, (H) mutant-type for target 4, (I) wild-type for target 5, and (J) mutant-type for target 5. (K) and (L) show the results of the negative controls, in which a template or a padlock probe was omitted, respectively. (Scale bar = 200 μm) (Reprinted with permission from Ref. [33]. Copyright (2016) Elsevier.)

[49]. Templates targeting the wild-type sequence were employed for all experiments. The resultant fluorescence signals in the reaction chamber were used to distinguish between wild-type and mutant-targeting padlock probes. The results showed that the wild-type targeting padlock probe produced a fluorescence signal for each target SNP site, while the mutant-targeting padlock probe did not produce any FAM signal. For instance, in case of target 1 SNP, the wild type revealed fluorescence (Fig. 11A), but the mutant did not (Fig. 11B). This pattern was consistent across all SNP positions (targets 2, 3, 4, and 5), with only the wild type leading to L-RCA reactions and higher fluorescence intensity, compared to the mutant targeting cases. Negative control experiments were also performed, where either a target or a padlock probe was omitted. These controls showed results in Fig. 11K and 4L, further confirming the specificity of the reaction. In these cases, direct comparisons of fluorescence intensity allowed for identification of the sequence of the point mutation. The study quantitatively compared the fluorescence intensities of wild type and mutant types. It was found that the fluorescence intensities of the wild type were higher than those of the mutant type by at least 22-fold. A limit-of-detection study was conducted using template concentrations of 100 fmol, 10 fmol, and 1 fmol (Fig. 12). Even at the lowest concentration of 1 fmol, a recognizable fluorescence signal was

generated in the chamber, demonstrating the device's capability to genotype at mutation points even with a low number of template copies. The entire process was completed within 120 minutes, emphasizing the efficiency of the rotary microdevice in performing multiplex SNP analysis without the need for complex fluidic control components like microvalves, micropumps, and tubing lines.

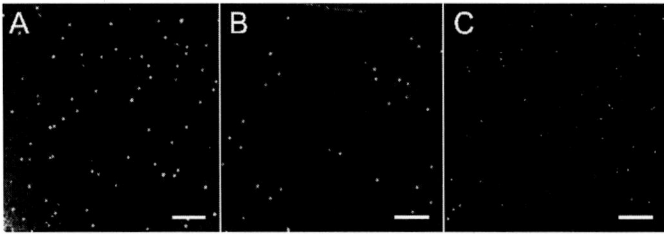

Fig. 12. A limit-of-detection test of the SNP typing on a rotary microdevice. SNP position of 19260 (target 5) was targeted with different concentration of the target. (A) 100 fmol, (B) 10 fmol, (C) 1 fmol. (Scale bar = 200 μm) (Reprinted with permission from Ref. [33]. Copyright (2016) Elsevier.)

Choi et al.'s study focused on the detection of food poisoning bacteria in milk samples using a centrifugal direct-RPA microdevice [39]. The microdevice was utilized for the monoplex real-time detection of three types of food-poisoning bacteria: *S. enterica, E. coli* O157:H7, and *V. parahaemolyticus*. Cultured bacteria cells (4×10^4 cells) were spiked into 3.2 μL of milk and mixed with direct PCR buffer, then introduced into the microdevice's sample reservoir. The RPA reagents, primers, and probes specific for each bacterial target were processed in the device. The reaction was conducted at 39 °C for 30 minutes. The fluorescence intensity of the reaction chambers was measured every 2 minutes, and images were captured every 10 minutes. The results showed a gradual enhancement of fluorescence intensity over the 30-minute reaction time, indicating successful amplification of the target genes (Fig. 13). The real-time

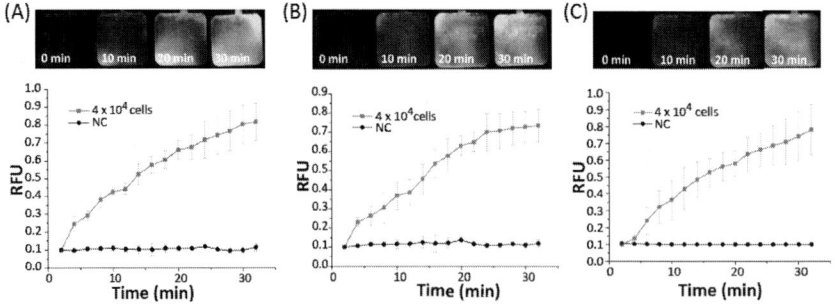

Fig. 13. Results of the on-chip direct-RPA reaction for targeting (A) *S. enterica*, (B) *E. coli* O157:H7, and (C) *V. parahaemolyticus*. Top panels: Fluorescence images of the reaction chamber at the reaction time of 0-, 10-, 20-, and 30-minutes. Bottom panels: Real-time amplification profiles obtained from triplicate experiments. (RFU: relative fluorescence unit, NC: negative control) (Reprinted with permission from Ref. [39]. Copyright (2016) Royal Society of Chemistry.)

fluorescence signals showed a significant increase compared to the negative control experiments.

A limit-of-detection study was performed, where *S. enterica* cells were serially diluted from 4×10^4 cells to 4×10^0 cells and mixed with milk. The results demonstrated that the microdevice could detect as few as 4 cells of target bacteria

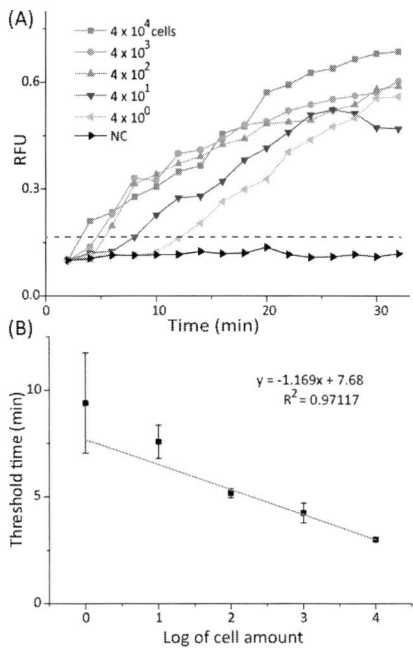

Fig. 14. A detection sensitivity test of the on-chip direct-RPA for analysing *S. enterica*. (A) Real-time direct-RPA profiles depending on the number of cells. (B) A standard curve by plotting the threshold time versus the logarithm of cell number. The data were obtained from triplicate experiments. (RFU: relative fluorescence unit, NC: negative control) (Reprinted with permission from Ref. [39]. Copyright (2016) Royal Society of Chemistry.)

within 30 minutes (Fig. 14). The detection sensitivity of the platform was established to be 4 cells/3.2 µL which corresponds to 1,250 cells/ml, significantly lower than the standard amount of the infectious bacteria for milk set by the FDA (20,000-50,000 cells/ml). A standard curve was created by plotting the threshold time against the logarithm of cell number, with a linearity of $R^2=0.97117$, indicating the potential for quantitative analysis. This standard curve was based on triplicate experiments, and the standard deviation of the threshold time was shown by error bars.

For the specificity test, the different target-specific primers and probes were separately dried in the three reaction chambers of each unit. A negative control reaction was set up by drying DEPC water instead of primer sets in the first reaction chamber. The bacterial cells, diluted to 4×10^3 cells, were spiked into milk and used as samples. The samples and reagents were dispersed into the four reaction chambers, and the fluorescence signals were measured every 2 minutes. The results showed positive outcomes only in the reaction chambers containing specific primers and probes for each bacteria type. For instance, in the test for *E. coli* O157:H7, the fluorescence images and signals indicated a positive reaction only in the chamber with specific primers and probe sequence for *E. coli* O157:H7 (Fig. 15B). Duplex and triplex bacterial detection tests were also performed. For the duplex detection, *S. enterica* and *V. parahaemolyticus* cells were spiked into the milk sample. The results showed increased fluorescence intensities and signals in the reaction chambers where specific primer and probe sets were present for targeting *S. enterica* and *V. parahaemolyticus* (Fig. 15C). For the triplex detection test, milk samples spiked with *S. enterica*, *E. coli* O157:H7, and *V. parahaemolyticus* were used. The fluorescent signals of the reaction chambers increased significantly, except in the negative control chamber, demonstrating the device's ability to precisely and specifically detect multiple types of food-poisoning bacteria in a single test (Fig. 15D). Additional negative control experiments were conducted with fresh milk samples to ensure no contamination issues. These controls further confirmed the specificity and reliability of the direct-RPA reaction on the microdevice.

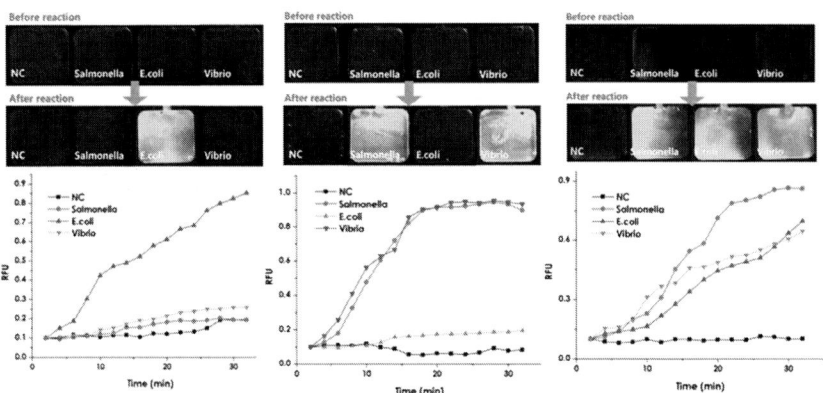

Fig. 15. Specificity and multiplicity tests of the centrifugal direct-RPA microdevice. (A) Image of the direct-RPA microdevice with the dried different target specific primers and probe in each reaction chamber. (B) Specificity test for detecting 4×10^3 cells of *E. coli* O157:H7. (C) Duplex test for detecting 4×10^3 cells of *S. enterica* and *V. parahaemolyticus*. (D) Triplex test for detecting 4×10^3 cells of three bacteria (*S. enterica*, *E. coli* O157:H7, and *V. parahaemolyticus*). Top panels: before and after fluorescence images of reaction chambers. Bottom panels: real-time amplification profiles of reaction chambers. (RFU: Relative Fluorescence Unit, NC: Negative Control) (Reprinted with permission from Ref. [39]. Copyright (2016) Royal Society of Chemistry.)

6. Conclusion

The studies by Jung et al., Heo et al., and Choi et al. have made significant contributions to the field of molecular diagnostics through the development of advanced centrifugal microfluidic devices. These devices, integrating sophisticated biochemical processes like LAMP, RCA, and RPA, demonstrate the potential of microfluidic technology in enhancing the sensitivity, specificity, and speed of diagnostic tests. Jung et al.'s study, focusing on the detection of Influenza A virus, showcased a remarkable 10-fold increase in sensitivity compared to conventional PCR methods. This advancement is particularly notable for its potential to significantly reduce the time and resources required for viral detection, making rapid, on-site testing more feasible, especially in urgent or resource-limited settings. Similarly, the work of Heo et al. in developing a rotary microdevice for ligation-RCA highlights the potential for precise genotyping and mutation detection. Their device's ability to identify specific genetic mutations quickly and accurately has implications for personalized medicine, particularly in cancer diagnosis and treatment. Choi et al.'s centrifugal direct-RPA microdevice for detecting food poisoning bacteria in milk samples demonstrates the adaptability of these technologies for applications beyond healthcare, such as food safety. The device's high sensitivity in detecting low levels of bacterial contamination aligns with regulatory standards and underscores its potential for ensuring food safety in various settings.

These studies collectively indicate a promising future for microfluidic devices in molecular diagnostics. The integration of such technologies could lead to the development of more compact, automated, and user-friendly diagnostic tools. These advancements could have a profound impact on global health, particularly in the rapid identification and management of infectious diseases, as well as in personalized healthcare approaches. Furthermore, the potential for these microfluidic devices extends beyond healthcare and into fields like environmental monitoring and food safety, demonstrating their versatility and wide range of applications. As technology continues to advance, we can expect more innovative solutions that will further enhance the efficiency, accuracy, and accessibility of molecular diagnostics, making a significant impact in both clinical and field settings. In conclusion, the contributions of Jung et al., Heo et al., and Choi et al. are pivotal in the evolution of molecular diagnostics. Their work exemplifies the technical feasibility and practical implications of centrifugal microfluidic devices in diagnostics, paving the way for future innovations that promise to revolutionize this field. As these technologies continue to develop, they hold the potential to significantly improve diagnostic processes, impacting a broad spectrum of applications in healthcare and beyond.

References

1   P. J. Asiello and A. J. Baeumner, *Lab Chip*, 2011, 11, 1420–1430.

2   M. M. Parida, S. Sannarangaiah, P. K. Dash, P. V. L. Rao and K. Morita, *Rev Med Virol*, 2008, 18, 407–421.

3   O. Piepenburg, C. H. Williams, D. L. Stemple and N. A. Armes, *PLoS Biol*, 2006, 4, 1115–1121.

4   B. A. Rohrman and R. R. Richards-Kortum, *Lab Chip*, 2012, 12, 3082–3088.

5   Y. Shin, A. P. Perera, K. W. Kim and M. K. Park, *Lab Chip*, 2013, 13, 2106–2114.

6   J. Compton, *Nature*, 1991, 350, 91–92.

7   L. M. Zanoli and G. Spoto, *Biosensors (Basel)*, 2013, 3, 18–43.

8   P. Craw and W. Balachandran, *Lab Chip*, 2012, 12, 2469–2486.

9   Y.-K. Cho, J.-G. Lee, J.-M. Park, B.-S. Lee, Y. Lee and C. Ko, *Lab Chip*, 2007, 7, 565–573.

10 J. P. Lafleur, A. A. Rackov, S. McAuley and E. D. Salin, *Talanta*, 2010, 81, 722–726.

11 M. Madou, J. Zoval, G. Jia, H. Kido, J. Kim and N. Kim, *Lab on a CD*, 2006, vol. 8.

12 A. Kazarine, M. C. R. Kong, E. J. Templeton and E. D. Salin, *Anal Chem*, 2012, 84, 6939–6943.

13 T. Notomi, H. Okayama, H. Masubuchi, T. Yonekawa, K. Watanabe, N. Amino and T. Hase, *Nucleic Acids Res*, 2000, 28, e63.

14 J. H. Jung, B. H. Park, S. J. Oh, G. Choi and T. S. Seo, *Biosens Bioelectron*, 2015, 68, 218–224.

15 S. W. Duffy, L. Tabar, B. Vitak and J. Warwick, *Breast Journal*, 2006, 12, S91–S95.

16 F. S. Collins, M. S. Guyer and A. Chakravarti, *Science (1979)*, 1997, 278, 1580–1581.

17 K. Nakatani, *ChemBioChem*, 2004, 5, 1623–1633.

18 A.-C. Syvanen, *Nat Genet*, 2005, 37, S5–S10.

19 S. Kim and A. Misra, *Annual Review of Biomedical Engineering*, 2007, 9.

20 P.-Y. Kwok, Annu Rev Genomics Hum Genet., 2001, 2.

21 B. P. Sokolov, *Nucleic Acids Res*, 1990, 18, 3671.

22 H. Matsuzaki, S. Dong, H. Loi, X. Di, G. Liu, E. Hubbell, J. Law, T. Berntsen, M. Chadha, H. Hui, S. P. A. Fodor and R. Mei, *Nat Methods*, 2004, 1, 109–111.

23 U. Landegren, R. Kaiser, J. Sanders and L. Hood, *Science (1979)*, 1988, 241, 1077–1080.

24 M. Orita, Y. Suzuki, T. Sekiya and K. Hayashi, *Genomics*, 1989, 5, 874–879.

25 S. G. Fischer and L. S. Lerman, *Proc Natl Acad Sci U S A*, 1983, 80, 1579–1583.

26 D. Liu, S. L. Daubendiek, M. A. Zillman, K. Ryan and E. T. Kool, *J Am Chem Soc*, 1996, 118, 1587–1594.

27 W. Zhao, M. M. Ali, M. A. Brook and Y. Li, *Angewandte Chemie - International Edition*, 2008, 47, 6330–6337.

28 P. M. Lizardi, X. Huang, Z. Zhu, P. Bray-Ward, D. C. Thomas and D. C. Ward, *Nat Genet*, 1998, 19, 225–232.

29 D. Y. Zhang, W. Zhang, X. Li and Y. Konomi, *Gene*, 2001, 274, 209–216.

30 X. Qi, S. Bakht, K. M. Devos, M. D. Gale and A. Osbourn, *Nucleic Acids Res*, 2001, 29, 116.

31 M. Nilsson, H. Malmgren, M. Samiotaki, M. Kwiatkowski, B. P. Chowdhary and U. Landegren, *Science (1979)*, 1994, 265, 2085–2088.

32 G. J. Hafner, I. C. Yang, L. C. Woiter, M. R. Stafford and P. M. Giffard, *Biotechniques*, 2001, 30, 852–867.

33 H. Y. Heo, S. Chung, Y. T. Kim, D. H. Kim and T. S. Seo, *Biosens Bioelectron*, 2016, 78, 140–146.

34 E. Scallan, R. M. Hoekstra, F. J. Angulo, R. V. Tauxe, M.-A. Widdowson, S. L. Roy, J. L. Jones and P. M. Griffin, *Emerg Infect Dis*, 2011, 17, 7–15.

35 X.-W. Wang, L. Zhang, L.-Q. Jin, M. Jin, Z.-Q. Shen, S. An, F.-H. Chao and J.-W. Li, *Appl Microbiol Biotechnol*, 2007, 76, 225–233.

36 M. Euler, Y. Wang, O. Nentwich, O. Piepenburg, F. T. Hufert and M. Weidmann, *Journal of Clinical Virology*, 2012, 54, 308–312.

37 Z. A. Crannell, B. Rohrman and R. Richards-Kortum, *Anal Chem*, 2014, 86, 5615–5619.

38 A. Abd El Wahed, A. El-Deeb, M. El-Tholoth, H. Abd El Kader, A. Ahmed, S. Hassan, B. Hoffmann, B. Haas, M. A. Shalaby, F. T. Hufert, F. T. Hufert and M. Weidmann, *PLoS One*, 2013, 8(8), e71642.

39 G. Choi, J. H. Jung, B. H. Park, S. J. Oh, J. H. Seo, J. S. Choi, D. H. Kim and T. S. Seo, *Lab Chip*, 2016, 16, 2309–2316.

40 J. H. Jung, B. H. Park, Y. K. Choi and T. S. Seo, *Lab Chip*, 2013, 13, 3383–3388.

41 M. Focke, F. Stumpf, B. Faltin, P. Reith, D. Bamarni, S. Wadle, C. Müller, H. Reinecke, J. Schrenzel, P. Francois, R. Zengerle and F. Von Stetten, *Lab Chip*, 2010, 10, 2519–2526.

42 P. Andersson, G. Jesson, G. Kylberg, G. Ekstrand and G. Thorsén, *Anal Chem*, 2007, 79, 4022–4030.

43 B. H. Park, D. Kim, J. H. Jung, S. J. Oh, G. Choi, D. C. Lee and T. S. Seo, *Sens Actuators B Chem*, 2015, 209, 927–933.

44 J. Kim, H. Kido, R. H. Rangel and M. J. Madou, *Sens Actuators B Chem*, 2008, 128, 613–621.

45 T. Brenner, T. Glatzel, R. Zengerle and J. Ducrée, *Lab Chip*, 2005, 5, 146–150.

46 J. Ducrée, S. Haeberle, S. Lutz, S. Pausch, F. Von Stetten and R. Zengerle, *J. Micromech. Microeng.*, 2007, 17, S103.

47 M. Grumann, A. Geipel, L. Riegger, R. Zengerle and J. Ducrée, *Lab Chip*, 2005, 5, 560–565.

48 N. Tomita, Y. Mori, H. Kanda and T. Notomi, *Nat Protoc*, 2008, 3, 877–882.

49 C. C. Harris, *J Natl Cancer Inst*, 1996, 88, 1442–1455.

Chapter 4
Colorimetric Loop-mediated Isothermal Amplification Reaction on a Centrifugal Microfluidic Device

1. Background

1.1. Colorimetric assay for genetic analysis

The integration of colorimetric assays into the realm of genetic analysis heralds a transformative approach to visualizing and interpreting complex molecular phenomena. Colorimetric assays are analytical techniques that enable the detection and quantification of substances through color changes in a solution. These assays exploit the interaction between chemical compounds and specific biomolecules that result in a measurable change in color. This change is often indicative of the presence, absence, or quantity of a particular entity, such as nucleic acids, proteins, or ions. At their core, colorimetric assays are grounded in the principles of chemical reactivity and molecular absorption spectrometry. They typically involve the addition of reagents that react with the test substance to produce a colored compound, the intensity of which correlates with the concentration of the substance in question. This color production can be the result of a simple pH change, a reduction-oxidation reaction, or the formation of a complex between the assay reagent and the target molecule [1].

The beauty of colorimetric assays lies in their simplicity and ease of use. By visually representing the presence of a target biomolecule as a color change, these assays make the detection of specific genetic sequences accessible even in the absence of sophisticated laboratory equipment. This is particularly advantageous in settings where resources are limited, or rapid responses are necessary, such as in-field diagnostics or in developing countries [1–4]

In genetic analysis, colorimetric assays have been ingeniously paired with amplification techniques to detect and quantify DNA or RNA sequences. The robustness of this method is evident in its broad application range, from medical diagnostics, where it aids in identifying pathogens and diagnosing genetic disorders, to environmental monitoring, where it helps in the detection of pollutants and the assessment of biodiversity. The advent of colorimetric assays marked a significant leap in genetic analysis, as it replaced more cumbersome, time-consuming, and technically demanding methods. Prior to their introduction, genetic analyses often required radiolabeling and autoradiography, which not only involved radioactive materials but also necessitated extensive safety precautions and lengthy development times for visualizing results. Colorimetric assays emerged as a safer, quicker, and more environmentally friendly alternative [1, 2, 5].

The colorimetric assays can also be semi-quantitative or quantitative. Semi-quantitative results give a rough estimate or indication of concentration, often through a gradient of color intensity that can be compared to a standard color chart. Quantitative assays, on the other hand, require the measurement of the absorbance or transmittance of the colored product using a spectrophotometer to determine the exact concentration of the analyte. Moreover, colorimetric assays have the advantage of versatility. They can be adapted to a wide variety of targets by changing the reagent or the substrate upon which the assay is based. This adaptability has led to the development of numerous colorimetric tests for different genetic markers, each tailored to the specific absorption spectrum of the resulting-colored product.

1.2. Colorimetric PCR and colorimetric LAMP

Colorimetric PCR and colorimetric isothermal amplification represent two distinct approaches to nucleic acid detection, each with its unique challenges and mechanisms.

Starting with colorimetric PCR, the concept of integrating a visual element into the polymerase chain reaction poses significant challenges. In its conventional form, PCR is a sophisticated technique that relies on thermal cycling to amplify specific DNA sequences. The process involves repeated cycles of denaturation, annealing, and elongation, requiring precise temperature control [6–8]. Introducing a colorimetric aspect into this finely tuned process is complex, particularly when aiming for a system that allows for direct visual interpretation without specialized equipment.

One prominent approach in colorimetric PCR involves the use of SYBR Green I. This dye is well-known for its ability to bind to double-stranded DNA. During the PCR process, as the target DNA is amplified, SYBR Green I intercalate with the newly formed double-stranded DNA, leading to enhanced fluorescence. The fluorescence intensity increases proportionally to the amount of DNA synthesized, allowing for a semi-quantitative assessment of the amplification. However, this method necessitates the use of a fluorescence reader, making it less accessible for settings without advanced laboratory equipment. The reliance on fluorescence detection, while offering sensitivity and specificity, diverges from the traditional concept of colorimetric detection, which is typically visual and more straightforward.

The development of a purely colorimetric PCR assay, one that can provide a clear visual indication of DNA amplification without the need for fluorescence detection, remains a challenging endeavor. This challenge stems from the nature of the PCR process itself. PCR does not inherently produce a significant change in the chemical environment that could be easily translated into a visual cue, such as a color change. The reaction's buffer system is designed to maintain a constant pH to ensure optimal conditions for the DNA polymerase enzyme. Any significant alteration in this pH, which could potentially serve as a basis for colorimetric detection, might adversely affect the efficiency and accuracy of the

PCR.

In contrast, colorimetric isothermal amplification techniques operate on a different principle that naturally lends itself to colorimetric detection. Loop-mediated isothermal amplification (LAMP), for example, is known for its robust and efficient DNA amplification at a constant temperature, without the need for thermal cycling [9–16]. This isothermal process results in the accumulation of large amounts of pyrophosphate ions as a by-product of DNA synthesis.

The production of these ions leads to a noticeable change in the pH of the reaction mixture. Utilizing pH-sensitive dyes such as Phenol Red or Cresol Red, colorimetric LAMP assays capitalize on this pH shift to indicate DNA amplification. Phenol Red, for instance, shifts from red to yellow as the reaction environment becomes more acidic due to the accumulation of pyrophosphate ions. This visual change is easily observable, making colorimetric LAMP particularly suited for rapid and accessible DNA detection in resource-limited settings [3, 4].

The ease of developing colorimetric assays for isothermal amplification methods like LAMP, as opposed to PCR, is primarily due to the natural by-products of the amplification process that can be visually detected. The simplicity and rapidity of these isothermal methods, coupled with their minimal equipment requirements, make them highly attractive for point-of-care testing, field applications, and educational purposes [13–17].

The coupling of colorimetric assays with isothermal amplification methods has further expanded the potential of these assays [3, 12, 18–20]. By integrating a colorimetric readout into the LAMP protocol, researchers have created a powerful tool that combines the high specificity and rapid amplification of LAMP with the direct and intuitive visualization of colorimetric assays. This combination presents a paradigm shift in the rapid, on-site detection of genetic material, enabling even non-specialists to perform sophisticated diagnostic assays with minimal training and equipment [5, 21–23].

This chapter presents a comprehensive analysis of how colorimetric assays are practically applied in genetic analysis, particularly for detecting nucleic acids linked to pathogens. It focuses on the use of Eriochrome Black T (EBT) as a colorimetric marker in LAMP reactions. This allows for visual monitoring of the reaction's progress and results. The color change mechanism in EBT, from purple to sky blue, is driven by its interaction with magnesium ions, a crucial component in the LAMP process, indicating the successful amplification of the desired nucleic acid [18].

The integration of colorimetric assays into centrifugal microfluidic devices represents a significant innovation, particularly in the field of molecular diagnostics [24, 25]. These devices combine the precision of microfluidics with the simplicity of colorimetric detection, facilitating the execution of complex laboratory protocols in a streamlined and user-friendly manner [26–28]. This chapter shows the ability of such integrated systems to perform fully automated DNA extraction, amplification, and detection on a single compact disc, with the entire process from sample introduction to final detection being conducted without manual intervention.

In the broader context of genetic assays, colorimetric methods have been harnessed for a variety of applications that go well beyond pathogen detection, playing a crucial role in diverse fields such as environmental monitoring, food safety, and agriculture [29]. In food safety, these assays are not just limited to detecting the presence of foodborne pathogens, and they are essential tools for the identification of genetically modified organisms (GMOs). As the public demand for transparency in food labeling grows, colorimetric assays provide a clear method for verifying the genetic makeup of crops and food products. Similarly, allergens present in food—which can pose serious health risks to susceptible individuals—can also be detected rapidly and effectively, allowing for better management of allergen-free production processes and consumer safety. Environmental science, besides, greatly benefits from the specificity and sensitivity of colorimetric assays. They are used extensively to assess water quality by detecting microbial contaminants that can affect both human health and ecosystems. For instance, the presence of coliform bacteria as indicators of water contamination can be swiftly identified using colorimetric methods, facilitating prompt water safety evaluations and remedial actions [29, 30].

The application of colorimetric assays in diagnostics, especially for foodborne pathogens, is particularly noteworthy. The development of colorimetric LAMP assays has revolutionized the rapid detection of pathogens such as *E. coli*, *Salmonella*, and *Listeria*, which are responsible for many foodborne illnesses. The assays' simplicity and the ability to observe results without specialized equipment make them ideal for field tests and quick checks within the food production chain, ensuring the safety and integrity of the food supply.

1.3. Principle of EBT-based colorimetric detection

The EBT-based colorimetric detection method is a prime example of the ingenuity behind colorimetric assays in genetic analysis. EBT, a complexometric indicator commonly used in titrations to detect the presence of metal ions, has found a novel application in the realm of molecular diagnostics, particularly in LAMP assays.

EBT functions in colorimetric LAMP assays by capitalizing on its ability to form colored complexes with magnesium ions, a key component in LAMP reactions. In the presence of high concentrations of free magnesium ions, EBT exhibits a distinct purple color. However, as the LAMP reaction proceeds, the concentration of magnesium ions in the solution diminishes, leading to a noticeable color change.

This decrease in magnesium ion concentration is attributed to the mechanics of the LAMP process. During LAMP, nucleotides are incorporated into the amplified DNA strands, a process that liberates pyrophosphate ions. These pyrophosphate ions subsequently react with the free magnesium ions to form magnesium pyrophosphate, an insoluble compound. As the concentration of free magnesium ions decreases due to this reaction, the EBT-magnesium ion complex is disrupted, resulting in a color shift from purple to sky blue (Fig. 1). This color change is a visual indicator of successful DNA amplification [5, 18, 31].

The EBT-based colorimetric detection offers several advantages in genetic analysis. Firstly, it allows for real-time monitoring of the LAMP reaction without the need for specialized detection equipment. The color change can be observed with the naked eye, making it an accessible method for various settings, including field diagnostics and low-resource laboratories.

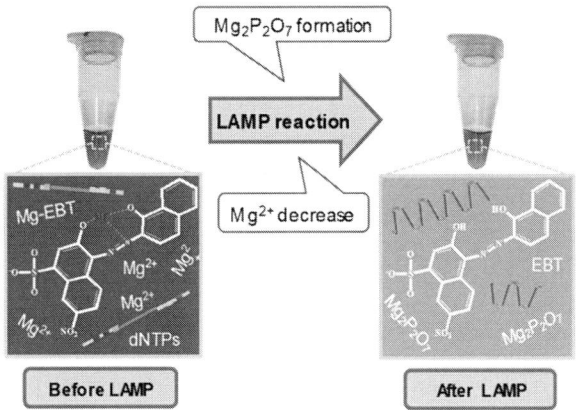

Fig. 1. Principle of the colorimetric detection method using EBT. Initially, the high concentration of Mg^{2+} causes the purple-coloring mixture as the complexes of EBT-Mg^{2+} are dominant. When pyrophosphates are released during the LAMP reaction, Mg^{2+} ions are combined with pyrophosphates to change the structure of EBT-Mg^{2+}, turning the color from violet to sky blue. (Reprinted with permission from Ref. [5]. Copyright (2016) Springer.)

Additionally, the EBT-based method enhances the specificity of LAMP assays. Since the color change directly correlates with the amplification of the target DNA, false positives are minimized. This specificity is particularly crucial in clinical diagnostics and food safety testing, where accurate detection of pathogens or contaminants can have significant health implications. Moreover, EBT-based colorimetric detection is highly versatile and can be adapted to various LAMP assays targeting different genetic sequences. This adaptability extends the utility of colorimetric LAMP assays across a wide range of applications, from detecting infectious diseases and foodborne pathogens to environmental monitoring and agricultural testing.

Despite its advantages, EBT-based colorimetric detection has limitations that must be considered. The intensity of the color change can be influenced by factors such as reaction conditions, including pH and temperature, and the concentration of reagents. Therefore, careful optimization and standardization of assay conditions are necessary to ensure consistent and reliable results.

2. Centrifugal microdevice for colorimetric LAMP reactions

2.1. Chip design

In the pursuit of advancing point-of-care diagnostics, two works by Oh et al.

present pioneering designs of centrifugal microfluidic devices tailored for colorimetric detection and high-throughput analysis. The intent behind these designs is to streamline the complex processes of pathogen detection into automated, user-friendly, and rapid assays that can be performed outside conventional laboratory settings [18, 31].

Common to the designs are the core principles that underpin the operation of these devices. Both chip designs proposed by Oh et al. utilize a centrifugal force applied through a rotating disc to move and mix reagents, which eliminates the need for external pumps [24, 25, 28, 32–34]. The discs are engineered with channels and chambers that guide the fluids through different stages of the diagnostic process, from sample introduction to the final detection step. Another critical point in both designs is the use of capillary forces to facilitate passive fluid control within the microchannels. This design consideration is crucial for the automated and sequential delivery of reagents without the need for active pumping mechanisms.

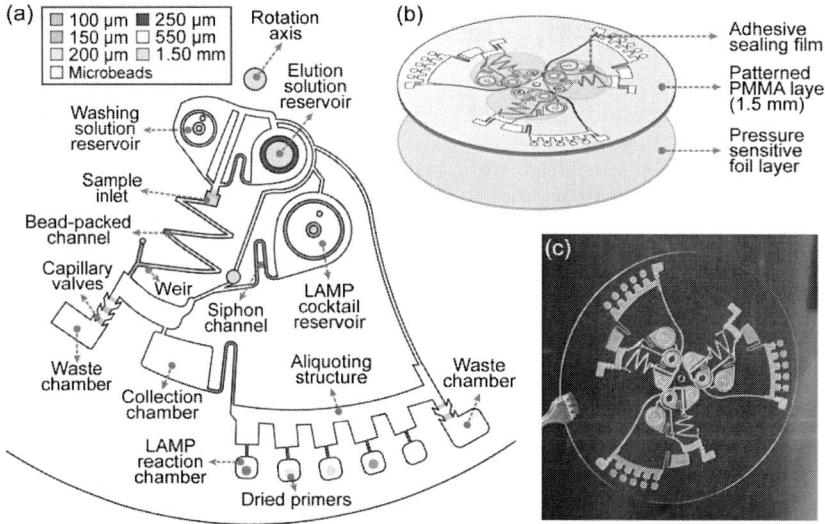

Fig. 2. (a) Chip design includes: i) a sample inlet, a washing solution reservoir, an elution solution reservoir and a bead-packed channel for DNA extraction and purification; ii) a LAMP cocktail reservoir, a collection chamber, an aliquoting structure and five LAMP reaction chambers for multiplexed LAMP; iii) waste chambers, capillary valves and siphon channels for programmed fluid manipulation. (b) Assembly of the lab-on-a-disc consisting of a 1.5 mm-thick patterned PMMA layer, a pressure sensitive foil layer and three adhesive sealing films. (c) Photograph of the assembled lab-on-a-disc. (Reprinted with permission from Ref. [31]. Copyright (2016) Royal Society of Chemistry.)

In details, Oh et al. [31] introduced a lab-on-a-disc that integrates DNA extraction, purification, amplification, and colorimetric detection within a single disc, revolutionizing the detection of foodborne pathogens (Fig. 2). This lab-on-a-disc, with a diameter of 13 cm, incorporates a sample inlet, washing and elution solution reservoirs, and a microbead-packed channel designed to harness centrifugal forces for fluid control. The microbead-packed channel, ending in a weir structure, ensures the sequential DNA binding, washing, and elution [16, 35–37]. The microbeads, too large to pass through the weir, are stacked in the channel by vacuum-suction.

The intricate design includes a LAMP cocktail reservoir, strategically positioned for containing the LAMP reaction components, including the Bst 2.0 polymerase and EBT for colorimetric detection. Ring structures with depths of 200 μm for the washing solution and 150 μm for the LAMP cocktail are engineered to utilize capillary forces, preventing the highly wetting liquids from spilling [16]. Siphon channels, each 400 μm wide and 100 μm deep, function as valves, releasing the solutions in sequence during the operation.

A waste chamber to collect debris and a collection chamber for the LAMP cocktail are also integrated into the design. The collection chamber is connected to an aliquoting structure leading to five LAMP reaction chambers, allowing the simultaneous preparation of multiple tests. The aliquoting chambers are linked to the LAMP reaction chambers by shallow channels, each 300 μm wide and 100 μm deep, facilitating the transfer of the aliquots under centrifugal force. The disc is assembled from a 1.5-mm thick PMMA layer, combined with a pressure-sensitive adhesive foil, and sealed with adhesive films to prevent contamination and evaporation [12, 38].

In a different approach, Oh et al. [18] developed a microdevice optimized for high-throughput detection (Fig. 3). This 12 cm diameter device incorporates five identical units, each capable of independent LAMP reactions. The design features zigzag-shaped dispensing microchannels [26, 29], the dimensions of which—1000 μm wide by 800 μm deep for the LAMP cocktail, and 750 μm wide by 600 μm deep for the LAMP primer mixture—facilitate the aliquoting process through capillary action. The connecting channels are precisely dimensioned for controlled sequential loading of the LAMP cocktail, primer mixture, and DNA sample via RPM control.

The microdevice consists of four layers: a 1 mm thick micropatterned PC layer, a 30 μm double-sided adhesive film, a second 1 mm thick PC layer, and a PSA foil. This multilayer assembly is carefully bonded using a heating press to ensure integrity during centrifugal operations. The operation process is finely tuned, with a rotational speed control protocol based on the Young-Laplace equation, ensuring that the LAMP cocktail, primer mixture, and DNA samples are accurately and successively injected into the 25 reaction chambers.

The design considerations in both studies are not only a showcase of microengineering but also a testament to the commitment to simplify and expedite the pathogen detection process. These devices reflect a profound understanding of fluid dynamics, materials science, and molecular biology, culminating in tools that make rapid, on-site detection of pathogens a tangible reality. Their potential to impact public health, particularly in resource-limited settings where rapid response is essential, is immense.

The work of Oh et al. stands as a significant milestone in the journey towards accessible and rapid diagnostics. It illustrates the power of microfluidic technology to bring complex laboratory procedures to the point of need, aligning with the chapter's theme of colorimetric LAMP reactions on centrifugal microfluidic devices. The implications of these designs extend beyond the immediate applications, suggesting a future where such devices are commonplace in combating infectious diseases and ensuring food safety.

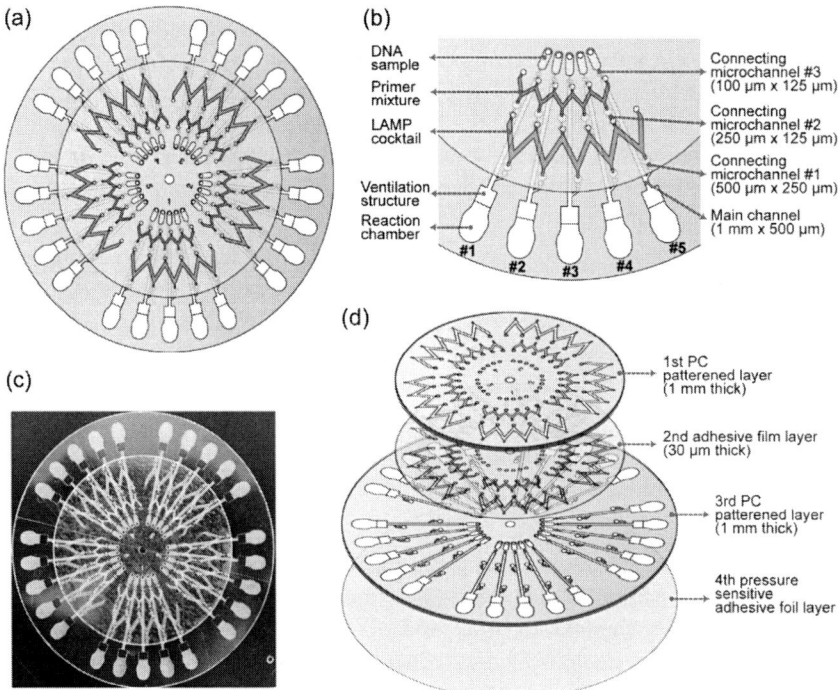

Fig. 3. (a) Schematic illustration of the centrifugal LAMP chip. (b) Detailed structure of one unit of the microdevice. (c) A digital image of the assembled microdevice. (d) Schematic illustration of the dissembled microdevice composed of four layers. From top to bottom, a 1 mm-thick PC layer patterned with zigzag-shaped dispensing microchannels, a 30 μm-thick adhesive layer, a 1 mm-thick PC layer patterned with sample reservoirs, main and connecting channels, and reaction chambers, and a pressure sensitive adhesive foil layer. (Reprinted with permission from Ref. [18]. Copyright (2016) Elsevier.)

2.2. Chip fabrication

The advancements in centrifugal microfluidic devices for pathogen detection are not just limited to their sophisticated design but also to their intricate fabrication process. The studies by Oh et al. [18, 31], provide us with insights into the meticulous fabrication processes that translate complex designs into functional devices.

The fabrication of the device begins with a 1.5-mm thick poly(methyl methacrylate) (PMMA) layer (Fig. 2). This layer forms the backbone of the lab-on-a-disc, a disc with a diameter of 13 cm containing three identical units. The PMMA is crafted using a computer numerical control (CNC) milling machine, ensuring precision and reproducibility. After the CNC milling, the siphon channels on the disc are coated with a hydrophilic coating reagent to modify surface properties and facilitate fluid movement [40]. LAMP primer sets, essential for the amplification process, are then applied to the reaction chambers and left to dry naturally. The assembly of the patterned PMMA layer with a pressure-sensitive adhesive (PSA) foil layer is achieved by applying pressure with a plastic roller. This process is critical as it secures the layers together, preventing leaks and contamination. Final assembly includes the use of a double-sided adhesive film and a thin polycarbonate (PC) film, which are cut and sealed to enclose the loaded solutions within the disc.

In Oh et al. earlier paper [18], the construction of a high-throughput device, which consists of four layers: a micropatterned PC layer, a double-sided adhesive film, a second micropatterned PC layer, and a PSA foil. The PC layers, each 1 mm thick, are fabricated on the PC sheet using a CNC milling machine. The adhesive layer, carefully patterned with a cutting plotter, plays a crucial role in the chip's integrity by bonding the layers together. The layers are assembled with precision and bonded using a heating press at room temperature, applying a force of 100 MPa to ensure a tight seal (Fig. 3).

The commonalities in these fabrication processes reflect an emphasis on precision engineering and material selection. Both devices utilize materials that are not only compatible with the biochemical assays but also conducive to the structural and mechanical demands of centrifugal microfluidics. The use of CNC milling machines for shaping and the application of hydrophilic coatings to modify surface properties are indicative of the detailed attention paid to each step of the fabrication process. Additionally, the adhesive techniques employed, whether through pressure-sensitive or double-sided adhesive films, are crucial for maintaining the structural integrity of the devices during operation. The innovation in fabrication extends beyond the materials and machinery; it also lies in the alignment of the design features with the fabrication steps to ensure that the devices are not only manufacturable but also reliable in their performance. The integration of design and fabrication exemplifies the complexity of creating such devices and the precision required to ensure their functionality. As the field of centrifugal microfluidics advances, the fabrication processes described by Oh et al. will undoubtedly serve as a benchmark for future developments. The precision and care taken in each step of the fabrication process underscore the dedication to creating devices that are both innovative and practical, pushing the boundaries of what is possible in point-of-care diagnostics.

3. Chip operation

The operation of centrifugal microfluidic devices demonstrates a meticulous interplay of engineering precision and biochemical process control, essential for colorimetric LAMP reactions and high-throughput pathogen detection.

In Oh et al. study [31], the operation of the lab-on-a-disc is initiated with the precise allocation of a real sample, washing and elution solutions, and a LAMP cocktail into designated reservoirs. The disc's design strategically positions these reservoirs to utilize centrifugal forces for sequential DNA processing. Microbeads loaded in the bead-packed microchannel enhance DNA binding, a critical step for the LAMP reaction. The custom-made centrifugal system, equipped with a spinning motor and heating blocks, regulates the rotational speed, ensuring the meticulous flow of samples through the microchannels [16]. The disc's operation, visualized in Fig. 4 and 5 and Table 1, showcases the dynamic flow control pivotal for the DNA purification process.

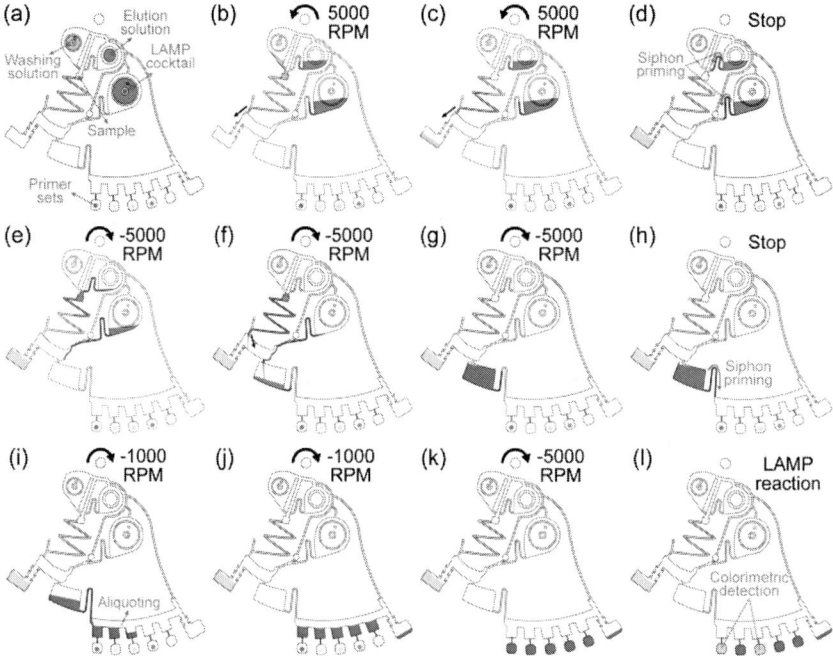

Fig. 4. Schematic depiction of the chip operation mediated by centrifugal forces. (a) Initial state of the lab-on-a-disc containing microbeads, a real sample, a washing solution, an elution solution, a LAMP cocktail and pre-stored LAMP primer sets. (b) Binding of genomic DNAs to microbeads. (c) Washing of the microbeads. (d) Siphon valve priming. (e) Elution of the purified DNAs. (f-g) Collection of the purified DNA and the LAMP cocktail. (h) Siphon priming. (i-j) Aliquoting of the LAMP mixture. (k) Transfer of the aliquots of the LAMP mixture to the LAMP reaction chambers. (l) Colorimetric detection. (Reprinted with permission from Ref. [31]. Copyright (2016) Royal Society of Chemistry.)

The operation proceeds with the real sample flowing through the bead-packed channel at a controlled speed of 5000 RPM. This stage is critical for binding DNAs to the microbeads. The subsequent release of the washing solution, following the sample, serves to purify the bound genomic DNAs, removing adsorbed salts and proteins. The elution solution is then released, flowing through the beads and eluting the purified DNAs, a process facilitated by the Coriolis force generated in the spinning disc [41]. The resultant mixture of elution solution and LAMP cocktail, containing the purified DNAs, is then collected and mixed in the collection chamber. This mixing step is crucial for ensuring a homogeneous LAMP reaction.

The operational complexity of the device is further exemplified in the aliquoting process. The device uses a rotational speed of -1000 RPM to divide the LAMP mixture into five aliquots, a vital step for multiplexed pathogen detection. The aliquots are then transferred into individual LAMP reaction chambers, where they are mixed with pre-stored LAMP primer sets. This step marks the beginning of the LAMP reactions, with the results being observable through colorimetric detection, a key feature of the device enabling rapid visual analysis of the pathogen presence.

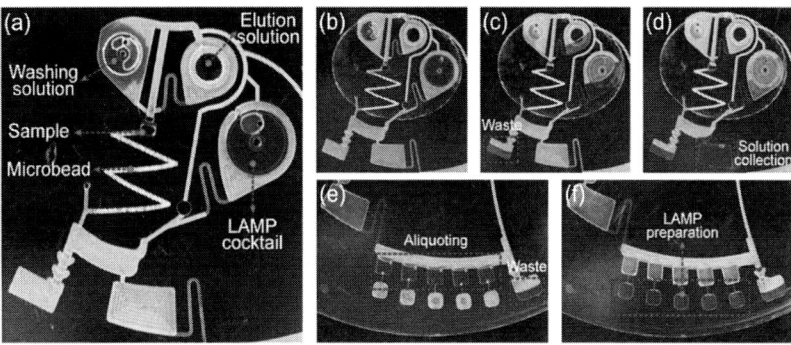

Fig. 5. Flow control of the lab-on-a-disc. (a) Loading of the microbeads, a real sample, a washing solution, an elution solution and a LAMP cocktail into the lab-on-a-disc. (b) Sealing of the disc. (c) Washing of the beads and collection of the waste. (d) Collection of the LAMP cocktail and the elution solution containing purified DNAs. (e) Aliquoting of the LAMP mixture into five parts and wastes. (f) Flushing of the aliquots to the LAMP reaction chambers ready for LAMP for pathogen detection. (Reprinted with permission from Ref. [31]. Copyright (2016) Royal Society of Chemistry.)

Table 1. Rotational speed control protocol for disc operation

Step	Speed [RPM]	Time [s]	Operation
1	5000	50	DNA binding and bead washing
2	0	10	Siphon priming
3	-5000	10	Elution and collection
4	0	10	Siphon priming
5	1000~-1000	10	Mixing of the LAMP mixture by a shake-mode
6	-1000	20	Aliquoting of the LAMP mixture
7	-5000	10	Transfer of the LAMP mixture aliquots to reaction chambers

The high-throughput microdevice operation of Oh et al. [18] is equally intricate (Fig. 6). A LAMP cocktail, prepared with EBT and DNA polymerase, is injected into dispensing microchannels. The primer mixture and bacterial genomic DNA samples are introduced into their respective channels. The device is then mounted on a custom-made centrifugal system, where the RPM is carefully controlled to divide and deliver the solutions into the reaction chambers. This process involves an advanced understanding of fluid dynamics, adhering to the

Young-Laplace equation for burst RPM in the connection microchannels. The solutions are subsequently mixed through alternating rotational speeds, ensuring an effective LAMP reaction [42].

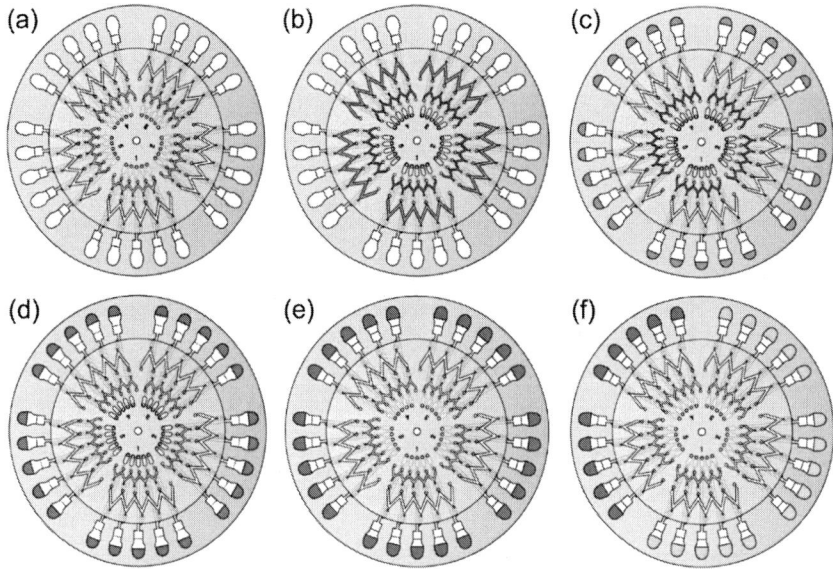

Fig. 6. Schematics for the entire operation process for the multiplex LAMP reaction on the centrifugal microdevice. (a) An initial state of the microdevice. (b) Loading of the LAMP cocktail (orange), the primer mixture (blue), and the DNA sample solution (yellow). (c) Ejection of the LAMP cocktail at 500 RPM by centrifugal force. (d) Ejection of the primer mixture at 1000 RPM. (e) Ejection of the DNA sample solution at 2000 RPM. (f) The LAMP reaction at 65 °C for 1 hour and colorimetric detection. (Reprinted with permission from Ref. [18]. Copyright (2016) Elsevier.)

The operational protocol of this high-throughput device emphasizes the sequential ejection of the LAMP cocktail, primer mixture, and genomic DNA solution. This process highlights the device's capacity for parallel processing of multiple samples, a critical feature in scenarios requiring rapid and large-scale pathogen screening. Post-reaction, the color change in the reaction chambers is monitored, providing a straightforward visual indication of the LAMP reaction results. The device's capability to correlate colorimetric changes with UV-Vis absorbance measurements of the LAMP products further underlines the precision and reliability of the detection method.

These operational processes of both devices showcase a sophisticated interplay of mechanical and biochemical components in centrifugal microfluidic devices. The precision in controlling fluid dynamics, coupled with the strategic design of the microchannels and chambers, ensures the successful execution of complex LAMP reactions. These devices exemplify the potential of microfluidics in enhancing the efficiency and accessibility of pathogen detection, offering a promising tool for point-of-care diagnostics.

4. Application for food-borne pathogen detection

The integration of colorimetric detection methods and microfluidic technology in identifying foodborne pathogens has marked a transformative approach in the field of molecular diagnostics. These centrifugal microfluidic devices blend the precision of microfluidics with the simplicity and efficacy of colorimetric detection, targeting pathogens like *Escherichia coli* O157:H7, *Salmonella typhimurium*, and *Vibrio parahaemolyticus*.

The cornerstone of these devices lies in their application of colorimetric LAMP reactions, utilizing EBT for direct visual interpretation of results. This approach not only simplifies the detection process but also enhances its reliability, as colorimetric changes can be easily observed and interpreted even in resource-limited settings. The centrifugal microfluidic design of these devices further augments this process by facilitating the rapid and precise manipulation of minute volumes of reagents and samples, critical for the accuracy and sensitivity of pathogen detection.

Moreover, the colorimetric LAMP reaction on these microfluidic platforms presents a paradigm shift in foodborne pathogen detection. Traditional methods, often cumbersome and time-consuming, are replaced by this rapid, user-friendly, and highly sensitive approach. The microfluidic channels, intricately designed within the disc, ensure the efficient flow and mixing of reagents, leading to consistent and reliable LAMP reactions. This integration of microfluidics with colorimetric detection not only streamlines the analytical process but also significantly reduces the time and cost involved in pathogen detection.

These devices underscore the importance of microfluidic technology in enhancing diagnostic capabilities. By miniaturizing and automating the complex processes involved in pathogen detection, these devices offer a portable and efficient solution, pivotal for on-site testing. The ability to conduct multiplex detection adds another layer of functionality, enabling the simultaneous screening of multiple pathogens, a feature particularly beneficial in outbreak scenarios or routine safety checks in the food industry.

Furthermore, the application of these devices in food safety transcends laboratory boundaries, bringing high-precision diagnostics directly to the point of need. This aspect is crucial in managing public health risks associated with foodborne diseases, where rapid response and accurate detection can significantly mitigate the impact on communities.

In extending these technological advances, future iterations of these devices could see enhancements in detection limits, specificity, and multiplexing capabilities. The potential for these devices in revolutionizing food safety is immense, offering a glimpse into a future where rapid, accurate, and accessible pathogen detection is the norm in safeguarding public health.

4.1. Monoplex detection of pathogen

The study conducted by Oh et al. [31], presents a methodical and innovative approach to the singleplex colorimetric detection of *E. coli* O157:H7 in real samples, specifically milk. This approach, which integrates centrifugal microfluidic technology with colorimetric detection, marks a significant milestone in the field of food safety diagnostics (Fig. 7).

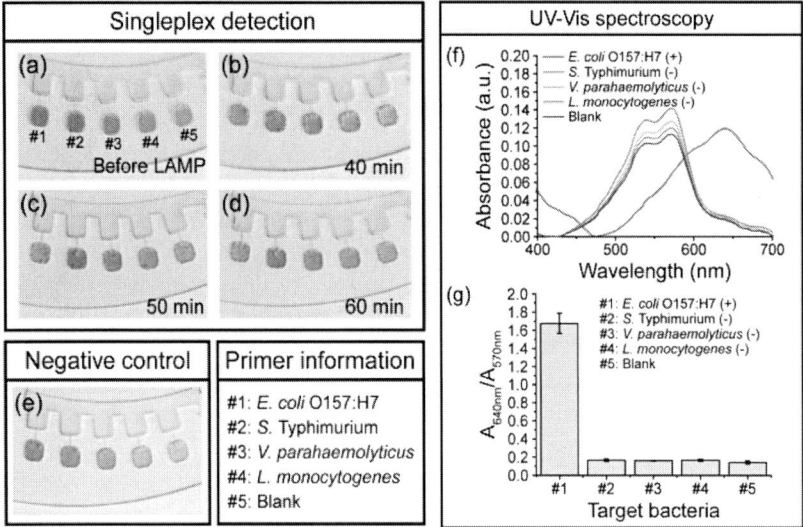

Fig. 7. Monoplex detection of *E. coli* O157:H7 in real samples by the colorimetric assay. (a) Five LAMP reaction chambers where four kinds of primer sets were pre-coated individually. #1: *E. coli* O157:H7, #2: *S. Typhimurium*, #3: *V. parahaemolyticus*, #4: *L. monocytogenes* and #5: blank. Colorimetric detection with the LAMP reaction time of (b) 40 minutes, (c) 50 minutes, and (d) 60 minutes. (e) A negative control. (f) UV-Vis absorption spectra of the LAMP products. (g) A_{640nm}/A_{570nm} values of the LAMP products. Error bars represent standard deviations. (Reprinted with permission from Ref. [31]. Copyright (2016) Royal Society of Chemistry.)

The researchers began by inoculating a milk sample with approximately 10^4 cells of *E. coli* O157:H7, resulting in a final bacterial concentration of about 2.7×10^7 cells/mL. This meticulous preparation was crucial for ensuring a realistic representation of a contaminated milk sample. Following this, a cell lysis step was performed, and the lysate was used directly for the LAMP reaction, called Direct-LAMP reaction.

Oh et al. utilized a centrifugal microfluidic device equipped with individual LAMP reaction chambers. Each chamber was pre-coated with a specific LAMP primer set, with one dedicated to *E. coli* O157:H7 and others to different pathogens, including *Salmonella Typhimurium*, *Vibrio parahaemolyticus*, and *Listeria monocytogenes*. A blank chamber served as a control. This arrangement was critical in isolating the detection of *E. coli* O157:H7, ensuring specificity and minimizing cross-reactivity.

Upon the initiation of the LAMP reaction within the device, the chamber designated for *E. coli* O157:H7 exhibited a distinct color transition from purple to navy and eventually to sky blue. This color change was a direct result of the

LAMP reaction, where the Mg^{2+} concentration decreased, altering the Mg-EBT complex and hence the solution's color. The absence of any color change in the other reaction chambers, including the blank, reaffirmed the assay's specificity and eliminated the possibility of cross-reaction among the targeted pathogens.

To validate the assay's specificity and reliability, negative control experiments were essential. These included tests with non-pathogenic *E. coli* strains and samples devoid of bacterial cells. The lack of color change in these controls provided strong evidence of the assay's ability to accurately differentiate between pathogenic and non-pathogenic strains and to confirm the absence of false positives.

A critical aspect of the study was the use of UV-Vis spectrophotometry to analyze the reaction mixtures post-LAMP reaction. The shift in UV-Vis absorbance peaks from 570 nm to 640 nm in the *E. coli* O157:H7-targeted mixture provided a quantifiable confirmation of the pathogen's presence. This spectral shift correlated with the observed color change and was quantified through the A_{640nm}/A_{570nm} absorbance ratio. This ratio, significantly higher in the *E. coli* O157:H7 mixture compared to the controls, served as a numerical confirmation of the pathogen's detection.

The monoplex colorimetric detection of *E. coli* O157:H7 in real samples is a testament to the innovative integration of colorimetric detection methods with advanced microfluidic technology. This approach not only ensures high specificity and reliability but also offers a visually interpretable and quantifiable method for pathogen detection in food safety applications. The study's methodologies and findings underscore the potential of such integrated systems in revolutionizing pathogen diagnostics in various fields, especially in ensuring food safety and public health.

4.2. Multiplex detection of pathogen

Studies by Oh et al. demonstrate the power and reliability of multiplex and colorimetric detection in identifying multiple foodborne pathogens in a single test run. The use of colorimetric assays, particularly the EBT-mediated detection, provides a simple yet effective method for pathogen detection without the need for complex analytical instrumentation. The ability to visually interpret the results, backed by quantitative UV-Vis spectrophotometry, offers a rapid and accessible solution for pathogen detection, crucial in food safety and public health contexts. The multiplexing capabilities of the centrifugal microfluidic devices, as evidenced by these studies, highlight the potential for significant advancements in diagnostic microbiology, particularly in the rapid and reliable detection of foodborne pathogens.

In their pivotal study, Oh et al. [31] tested the multiplexing capability of their proposed disc by inoculating a milk sample with multiple foodborne pathogens (Fig. 8). Specifically, they spiked a real milk sample with 10^4 cells each of *E. coli* O157:H7 and *V. parahaemolyticus*. The assay was designed such that color changes in the reaction chambers would indicate the presence of these pathogens.

As expected, color changes were observed in chambers #1 and #3, corresponding to *E. coli* O157:H7 and *V. parahaemolyticus*, respectively. This simultaneous detection was visually confirmed by the change from purple to sky blue in these chambers, which was further supported by UV-Vis analyses. The absorption spectra for the LAMP products of *E. coli* O157:H7 and *V. parahaemolyticus* showed major peaks at 640 nm, indicative of positive signals, while negative signals were marked by peaks at 570 nm. The experiment, conducted in triplicate, provided consistent A_{640nm}/A_{570nm} values, aligning with the visual interpretation of sky blue for positive signals.

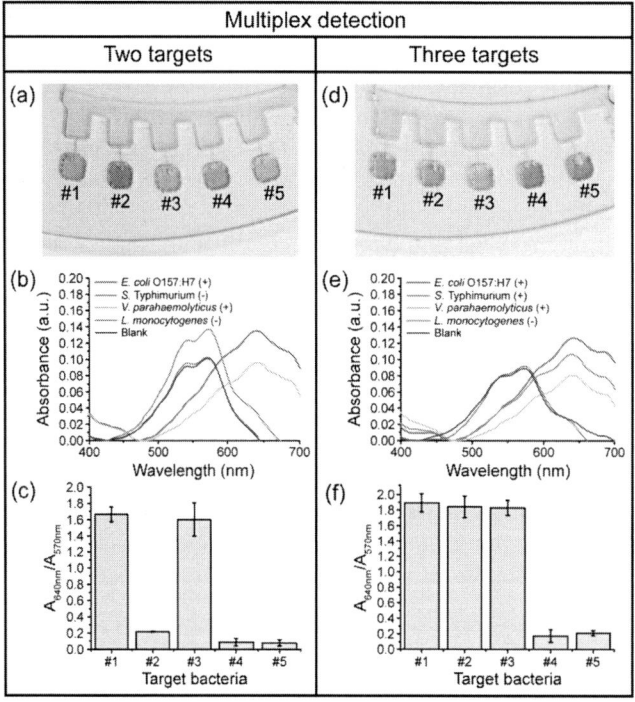

Fig. 8. Multiplex detection of foodborne pathogens in real samples by the colorimetric assay. (a-c) Detection of two kinds of pathogens and (d-f) three kinds of pathogens. (a) Colorimetric detection of *E. coli* O157:H7 and *V. parahaemolyticus*. (b) UV-Vis absorption spectra of the LAMP products. (c) A_{640nm}/A_{570nm} values of the LAMP products. (d) Colorimetric detection of *E. coli* O157:H7, *S. Typhimurium* and V. *parahaemolyticus*. (e) UV-Vis absorption spectra of the LAMP products. (f) A_{640nm}/A_{570nm} values of the LAMP products. All the error bars represent standard deviations. (Reprinted with permission from Ref. [31]. Copyright (2016) Royal Society of Chemistry.)

Expanding the scope, the study then introduced three pathogens – *E. coli* O157:H7, *S. Typhimurium*, and *V. parahaemolyticus* – each at 10^4 cells, into a milk sample. The color change observed in reaction chambers #1, #2, and #3 confirmed the specific detection of all three pathogens. The UV-Vis absorption spectra reinforced these findings, showing major peaks at 640 nm for the targeted pathogens and 570 nm for the negative controls. This consistent result in triplicate experiments confirmed the reproducibility and reliability of the assay on a single disc, validating the multiplexing capability of the disc-based assay for specific pathogen identification.

In their other study, Oh et al. [18] further explored multiplex pathogen detection using a centrifugal LAMP microdevice (Fig. 9). This study targeted three foodborne pathogens with two negative control experiments to ensure no DNA template contamination. The experimental setup involved injecting DNA samples, corresponding primer mixtures, and LAMP cocktails into designated reservoirs and microchannels. The target bacteria were *E. coli* O157:H7, *S. typhimurium*, and *V. parahaemolyticus*, with negative controls including pure water in place of the genomic DNA sample or the LAMP primer mixture. Post-LAMP reaction, color changes from purple to sky blue were observed in parts 1-3 of the microdevice, indicating successful amplification of the target genes and the effectiveness of EBT-mediated colorimetric detection. No color change in the negative controls indicated the absence of contamination problems.

The study's results were further validated using UV-Vis spectrophotometry, which showed maximum absorption peaks at 640 nm for the EBT-containing final product solutions of parts 1-3, aligning with positive results. The negative controls exhibited maximum absorption peaks at 570 nm, attributed to the Mg-EBT complexes. The ratio value of A_{640nm}/A_{570nm} was set as an indicator for the LAMP reaction, with higher values in the positive results than the negative

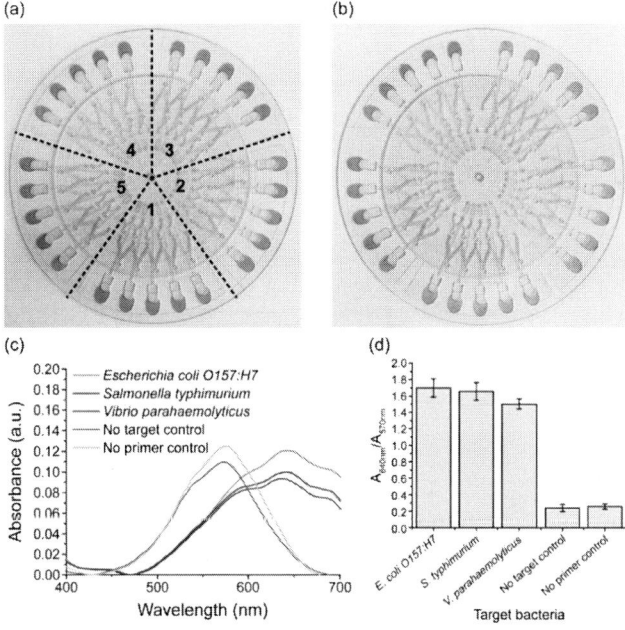

Fig. 9. Digital images of the centrifugal LAMP microdevice for multiplex food-borne pathogen detection (a) before and (b) after the LAMP reaction. Part 1: *E. coli O157:H7*, part 2: *S. typhimurium*, part 3: *V. parahaemolyticus*, part 4: a negative control without target DNA, and part 5: a negative control without primers. (c) UV-Vis absorption spectra of the LAMP product solution. (d) Intensity ratio of $A_{640\,nm}/A_{570\,nm}$ of the LAMP products obtained from part 1-5. The data were generated by quintuplex experiments. (Reprinted with permission from Ref. [18]. Copyright (2016) Elsevier.)

controls, confirming the color change to sky blue as an indicator of LAMP amplicon production.

4.3. Limit-of-detection test

Oh et al. delved into the detection sensitivity for *E. coli* O157:H7 in milk samples [31]. The researchers prepared a range of samples with bacterial counts from a single cell to 10^4 cells. This wide range allowed for an exhaustive evaluation of the assay's sensitivity across a spectrum of concentrations. The critical finding from this study was establishing a LOD of 10 bacterial cells for *E. coli* O157:H7. This level of sensitivity is noteworthy, especially considering the potential for foodborne pathogens to cause infection at relatively low doses (Fig. 10).

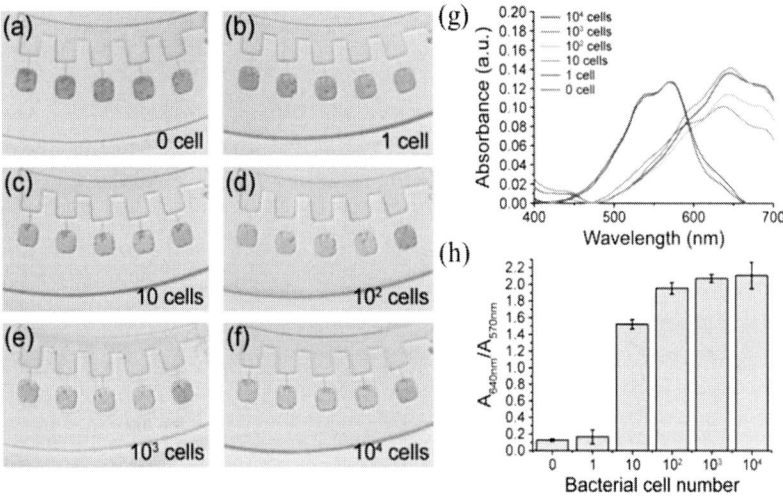

Fig. 10. A limit of detection of the assay carried out with varying concentrations of *E. coli* O157:H7 in the range of 0 to 10^4 cells: (a) Negative control, (b) 1 cell, (c) 10 cells, (d) 10^2 cells, (e) 10^3 cells, and (f) 10^4 cells. (g) UV-Vis absorption spectra of the LAMP products. (h) A_{640nm}/A_{570nm} values of the LAMP products. Error bars represent standard deviations. (Reprinted with permission from Ref. [31]. Copyright (2016) Royal Society of Chemistry.)

The methodological rigor in this experiment is evident from the careful preparation and handling of the samples to the precise calibration of the colorimetric LAMP assay. The ability of the assay to detect as low as 10 cells per test, even after the division of the sample into aliquots, highlights the high sensitivity of this approach. However, the study also brings to light the limitations at the lower end of the concentration spectrum. The absence of color change at the single-cell level raises questions about the assay's effectiveness in detecting extremely low concentrations of pathogens.

Oh et al.'s other study expanded the scope of LOD assessment by using serially diluted DNA templates ranging from 3.8×10^4 to 3.8×10 copies [18]. The LOD was determined to be 380 copies of *E. coli* O157:H7 DNA, based on the color change observed in the LAMP reaction and confirmed by UV-Vis spectrophotometry. This finding is significant in the context of early detection of pathogens [43–45], where even a small number of bacterial cells can have severe implications for food safety and public health (Fig. 11).

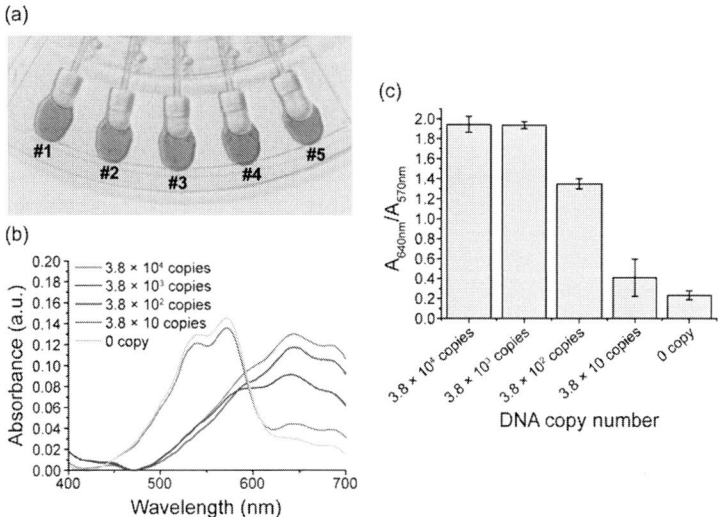

Fig. 11. (a) LOD test for *E. coli O157:H7* on the centrifugal LAMP microdevice. DNA copy number was 3.8×10^4 for #1, 3.8×10^3 for #2, 3.8×10^2 for #3, 3.8×10 for #4, and 0 for #5, respectively. (b) UV-Vis absorption spectra of the LAMP products #1~#5. (c) Intensity ratio of A_{640nm}/A_{570nm} of the LAMP products #1~#5. The data were generated by quadruplex experiments. (Reprinted with permission from Ref. [18]. Copyright (2016) Elsevier.)

The application of colorimetric detection in these studies is a noteworthy aspect. The visual interpretation of results, facilitated by the color change from purple to sky blue, provides an intuitive and straightforward method for identifying the presence of pathogens. This approach, complemented by quantitative UV-Vis spectrophotometry data, offers a dual validation of the assay's results, enhancing its reliability and practicality for real-world applications.

While these studies showcase the effectiveness of the centrifugal microfluidic device in pathogen detection, they also illuminate the challenges in interpreting results at very low bacterial concentrations. The ambiguity in color change at lower DNA copy numbers points to the need for further refinement in the assay's design and sensitivity. Enhancing the assay's ability to detect low concentrations of pathogens will be crucial, especially for pathogens known to be infectious at low doses.

Looking ahead, the findings from Oh et al.'s research lay a strong foundation for future advancements in pathogen detection technology. Enhancing the sensitivity and accuracy of such assays could lead to more robust and efficient diagnostic tools, significantly impacting food safety monitoring and public health surveillance. The potential for integrating these assays into a broader range of diagnostic platforms also opens up new avenues for rapid, on-site testing in various settings, from food processing facilities to clinical

5. Conclusion

This chapter represents a significant stride forward in the field of diagnostic microbiology, particularly in the rapid and accurate detection of foodborne pathogens. Through the innovative application of colorimetric LAMP reactions on centrifugal microfluidic devices, these studies have carved a pathway for the development of point-of-care testing platforms that are both efficient and user-friendly.

The specificity and sensitivity of these colorimetric LAMP assays, alongside their ability to provide high-throughput and multiplex detection, align well with the urgent needs of food safety and public health. The assays' capacity to detect pathogens at low concentrations and with high specificity is critical in preventing the spread of foodborne diseases and managing public health crises effectively.

Furthermore, the colorimetric aspect of these LAMP assays offers a visual confirmation of pathogen presence, which is a substantial advantage in settings where complex lab equipment and specialized technical expertise may be lacking. This visual cue, backed by quantitative data from spectrophotometric analyses, ensures that the assay's results are both accessible and reliable.

As we look to the future, the application of these centrifugal microfluidic devices for pathogen detection promises not only to enhance food safety monitoring but also to broaden its utility across various domains, including environmental monitoring and clinical diagnostics. The potential for adaptation and scale-up of this technology could see it becoming an integral part of global efforts to safeguard against infectious diseases.

References

1. B. Hu, J. Guo, Y. Xu, H. Wei, G. Zhao and Y. Guan, *Anal Bioanal Chem*, 2017, 409, 4819–4825.

2. B. Khanal, P. Pokhrel, B. Khanal and B. Giri, *ACS Omega*, 2021, 6, 33837–33845.

3. M. Goto, E. Honda, A. Ogura, A. Nomoto and K.-I. Hanaki, *Biotechniques*, 2009, 46, 167–172.

4. M. Safavieh, M. U. Ahmed, E. Sokullu, A. Ng, L. Braescu and M. Zourob, *Analyst*, 2013, 139, 482–487.

5. D. Van Nguyen, V. H. Nguyen and T. S. Seo, *Biochip J*, 2019, 13, 158–164.

6. P. J. Asiello and A. J. Baeumner, *Lab Chip*, 2011, 11, 1420–1430.

7. P. Gill and A. Ghaemi, *Nucleosides Nucleotides Nucleic Acids*, 2008,

27, 224–243.

8   P. Craw and W. Balachandran, *Lab Chip*, 2012, 12, 2469–2486.

9   T. Notomi, Y. Mori, N. Tomita and H. Kanda, *Journal of Microbiology*, 2015, 53, 1–5.

10 T. Notomi, H. Okayama, H. Masubuchi, T. Yonekawa, K. Watanabe, N. Amino and T. Hase, *Nucleic Acids Res*, 2000, 28, e63.

11 M. M. Parida, S. Sannarangaiah, P. K. Dash, P. V. L. Rao and K. Morita, *Rev Med Virol*, 2008, 18, 407–421.

12 N. Tomita, Y. Mori, H. Kanda and T. Notomi, *Nat Protoc*, 2008, 3, 877–882.

13 X. Fang, H. Chen, L. Xu, X. Jiang, W. Wu and J. Kong, *Lab Chip*, 2012, 12, 1495–1499.

14 X. Fang, H. Chen, S. Yu, X. Jiang and J. Kong, *Anal Chem*, 2011, 83, 690–695.

15 J. H. Jung, B. H. Park, S. J. Oh, G. Choi and T. S. Seo, *Lab Chip*, 2015, 15, 718–725.

16 J. H. Jung, B. H. Park, S. J. Oh, G. Choi and T. S. Seo, *Biosens Bioelectron*, 2015, 68, 218–224.

17 C. Liu, M. G. Mauk, R. Hart, M. Bonizzoni, G. Yan and H. H. Bau, *PLoS One*, 2012, 7(8), e42222.

18 S. J. Oh, B. H. Park, J. H. Jung, G. Choi, D. C. Lee, D. H. Kim and T. S. Seo, *Biosens Bioelectron*, 2016, 75, 293–300.

19 K. Hsieh, A. S. Patterson, B. S. Ferguson, K. W. Plaxco and H. T. Soh, *Angewandte Chemie - International Edition*, 2012, 51, 4896–4900.

20 Y. Mori, K. Nagamine, N. Tomita and T. Notomi, *Biochem Biophys Res Commun*, 2001, 289, 150–154.

21 J. E. Lee, H. Mun, S. R. Kim, M. G. Kim, J. Y. Chang and W. B. Shim, *Biosens Bioelectron*, 2020, 151, 111968.

22 M. Dou, D. C. Dominguez, X. Li, J. Sanchez and G. Scott, *Anal Chem*, 2014, 86, 7978–7986.

23 J. Luo, X. Fang, D. Ye, H. Li, H. Chen, S. Zhang and J. Kong, *Bioelectron.*, 2014, 60, 84–91.

24 M. Madou, J. Zoval, G. Jia, H. Kido, J. Kim and N. Kim, *Lab on a CD*, 2006, vol. 8.

25 O. Strohmeier, M. Keller, F. Schwemmer, S. Zehnle, D. Mark, F. Von Stetten, R. Zengerle and N. Paust, *Chem Soc Rev*, 2015, 44, 6187–6229.

26 J. Ducrée, S. Haeberle, S. Lutz, S. Pausch, F. Von Stetten and R. Zengerle, *Journal of Micromechanics and Microengineering*, 2007, 17, S103.

27 D. Mark, S. Haeberle, G. Roth, F. V. Stetten and R. Zengerle, *Chem Soc Rev*, 2010, 39, 1153–1182.

28 R. Gorkin, J. Park, J. Siegrist, M. Amasia, B. S. Lee, J.-M. Park, J. Kim, H. Kim, M. Madou and Y.-K. Cho, *Lab Chip*, 2010, 10, 1758–1773.

29 B. Liu, J. Zhuang and G. Wei, *Environ Sci Nano*, 2020, 7, 2195–2213.

30 J. Das and H. N. Mishra, *European Food Research and Technology*, 2022, 248, 1125–1148.

31 S. J. Oh, B. H. Park, G. Choi, J. H. Seo, J. H. Jung, J. S. Choi, D. H. Kim and T. S. Seo, *Lab Chip*, 2016, 16, 1917–1926.

32 J. P. Lafleur, A. A. Rackov, S. McAuley and E. D. Salin, *Talanta*, 2010, 81, 722–726.

33 C. E. Nwankire, M. Czugala, R. Burger, K. J. Fraser, T. M. Connell, T. Glennon, B. E. Onwuliri, I. E. Nduaguibe, D. Diamond and J. Ducrée, *Biosens Bioelectron*, 2014, 56, 352–358.

34 O. Strohmeier, N. Marquart, D. Mark, G. Roth, R. Zengerle and F. Von Stetten, *Analytical Methods*, 2014, 6, 2038–2046.

35 G. Czilwik, T. Messinger, O. Strohmeier, S. Wadle, F. Von Stetten, N. Paust, G. Roth, R. Zengerle, P. Saarinen, J. Niittymäki, J. O'Leary and D. Mark, *Lab Chip*, 2015, 15, 3749–3759.

36 F. Stumpf, F. Schwemmer, T. Hutzenlaub, D. Baumann, O. Strohmeier, G. Dingemanns, G. Simons, C. Sager, L. Plobner, F. Von Stetten, R. Zengerle and D. Mark, *Lab Chip*, 2016, 16, 199–207.

37 J. H. Jung, B. H. Park, Y. K. Choi and T. S. Seo, *Lab Chip*, 2013, 13, 3383–3388.

38 K. Hsieh, P. L. Mage, A. T. Csordas, M. Eisenstein and H. T. Soh, *Chemical Communications*, 2014, 50, 3747–3749.

39 P. Andersson, G. Jesson, G. Kylberg, G. Ekstrand and G. Thorsén, *Anal Chem*, 2007, 79, 4022–4030.

40 M. Focke, F. Stumpf, G. Roth, R. Zengerle and F. Von Stetten, *Lab Chip*, 2010, 10, 3210–3212.

41 T. Brenner, T. Glatzel, R. Zengerle and J. Ducrée, *Lab Chip*, 2005, 5, 146–150.

42 M. Grumann, A. Geipel, L. Riegger, R. Zengerle and J. Ducrée, *Lab Chip*, 2005, 5, 560–565.

43 N. Arunrut, J. Kampeera, R. Suebsing and W. Kiatpathomchai, *J Virol Methods*, 2013, 193, 542–547.

44 T.-T. Dai, C.-C. Lu, J. Lu, S. Dong, W. Ye, Y. Wang and X. Zheng, *FEMS Microbiol Lett*, 2012, 334, 27–34.

45 J. K. F. Wong, S. P. Yip and T. M. H. Lee, *Small*, 2014, 10, 1495–1499.

Chapter 5
Lateral Flow Strip Assay-Incorporated Centrifugal Microfluidic Device for Genetic Analysis

1. Background

1.1. Lateral flow strip assay

The lateral flow strip assay (LFA) or an immunochromatographic strip assay (ICS), a technology that dates back several decades, has become a mainstay in point-of-care testing due to its ease of use, rapid turnaround time, ancd the ability to be used in a variety of settings without the need for sophisticated equipment. Originating from the first generation of home pregnancy tests, LFAs have since been adapted to detect a broad range of substances, from pathogens to biomarkers of disease [1–5].

The need for such technology arises from the global demand for rapid, accurate, and accessible diagnostic methods, particularly in low-resource settings [5–7]. LFAs meet this need by providing a means to conduct on-the-spot testing with minimal training. The basic principle involves the application of a liquid sample onto a test strip that wicks the fluid through capillary action, interacting with reagents that produce a colorimetric readout. This simplicity belies the intricate design and optimization of antibodies and antigens that allow for the specific and sensitive detection of target molecules.

Historically, the evolution of LFAs has been shaped by the pursuit of enhanced diagnostic performance, leading to the integration of novel materials, such as nanoparticles, and advanced manufacturing techniques. Modern LFAs can provide qualitative, semi-quantitativc, or even quantitative results. While lateral flow strip assays have proven to be invaluable in various settings, challenges persist. Achieving higher sensitivity for certain analytes, maintaining consistent performance across different environmental conditions, and integrating multiplexed detection capabilities are areas that researchers continue to explore.

Early lateral flow assays faced limitations in terms of sensitivity and specificity which were rooted in various factors. Conjugate materials, such as colloidal gold or latex beads, used in the initial assays were limited in their ability to provide high sensitivity, especially in detecting low concentrations of target analytes. For instance, colloidal gold particles were initially employed due to their visibility, but they posed limitations in terms of signal amplification. Visual interpretation of results introduced subjectivity, and the lack of quantitative data hampered precision. Binding affinity issues, particularly the selection of antibodies with lower affinities for their targets, resulted in false positives or negatives. For example, early assays for infectious diseases might cross-react with related pathogens, compromising specificity. Cross-reactivity challenges were further exacerbated by the potential for non-specific binding. Environmental conditions, including temperature and humidity variations, affected the stability of reagents, impacting assay reliability.

Relentless research and development efforts in recent years have propelled lateral LFAs into a new era of enhanced sensitivity and reliability. Innovations in materials, including the integration of nitrocellulose membranes, as well as advancements in conjugate technologies and signal detection methods, have collectively refined these assays. This progress enables the detection of low-concentration analytes with remarkable accuracy, positioning LFAs as reliable tools for both clinical and non-clinical applications. Furthermore, the integration of LFAs with molecular techniques such as polymerase chain reaction (PCR) or loop-mediated isothermal amplification (LAMP) offers a powerful strategy to further boost both specificity and sensitivity in diagnostic applications [8, 9]. To achieve this integration, the process typically involves a two-step approach. Initially, the nucleic acid amplification technique (PCR or LAMP) is employed to selectively and exponentially amplify the target genetic material. This step significantly increases the amount of the target analyte, making it more easily detectable [2, 10–12]. Once the amplification is complete, the products from the molecular amplification are then utilized as inputs for the LFA. By leveraging the precision of nucleic acid amplification, this combined approach enhances the overall capabilities of LFAs, making them invaluable in various fields, particularly in the rapid and accurate detection of pathogens for point-of-care applications.

In the case of PCR, the amplified DNA fragments, often labeled with reporter molecules, are introduced to the lateral flow strip. The strip is designed with specific capture probes that can recognize and bind to the amplified DNA. The interaction triggers a visual signal, usually in the form of colored lines, indicating the presence of the target analyte. This combined approach, often referred to as PCR-LFA, allows for the molecular precision of PCR to be seamlessly integrated with the rapid and user-friendly features of lateral flow assays.

Similarly, with LAMP, the amplified nucleic acids can be employed in conjunction with LFAs. LAMP is an isothermal amplification technique that rapidly synthesizes a large amount of DNA under constant temperature conditions [10, 13]. The resulting amplicons, typically labeled, can be applied to the lateral flow strip. The interaction with specific probes on the strip produces visible signals, providing a qualitative and rapid readout. The integration of nucleic acid amplification techniques with LFAs offers several advantages. Firstly, it significantly enhances the sensitivity of the assay, allowing for the detection of even low concentrations of genetic material. Secondly, it improves specificity by selectively amplifying the target nucleic acid, reducing the likelihood of false positives. Additionally, the combination of molecular techniques with LFAs maintains the user-friendly and portable nature of lateral flow assays, making them suitable for point-of-care applications.

The chapter provided expand on this history by showcasing how LFAs can be incorporated with advanced molecular techniques like LAMP, a nucleic acid amplification method offering high specificity and sensitivity, ideal for genetic analysis [14–16]. Such integration represents the cutting edge of LFA evolution, where the traditional simplicity of the assay is maintained while significantly amplifying its diagnostic capabilities.

1.2. Principle of amplicon detection on a lateral flow strip

In the field of genetic diagnostics, the detection of amplicons after post-amplification is crucial. Common methods include gel electrophoresis, which sorts DNA fragments by size but can be cumbersome; real-time PCR that uses real-time fluorescence to monitor DNA amplification; and spectrophotometry for DNA quantification. These traditional techniques, while effective, can be bulky and costly, often requiring specialized laboratory settings. LFAs, on the other hand, provide a more accessible alternative, particularly notable for their compact, cost-effective nature, allowing for easy on-site deployment. LFAs become especially powerful when paired with LAMP, a method that enhances specificity and sensitivity. This combination mitigates the inherent limitations of each approach: LAMP's requirement for bulky detection system and LFAs' historically lower sensitivity. Together, they allow for the differentiation of amplicons with similar patterns that might otherwise be indistinguishable, as is often the case with multiplex RT-LAMP in gel electrophoresis. The integration of LFA and LAMP technologies thus presents a promising advancement in point-of-care diagnostics, facilitating rapid, accurate, and field-deployable testing for

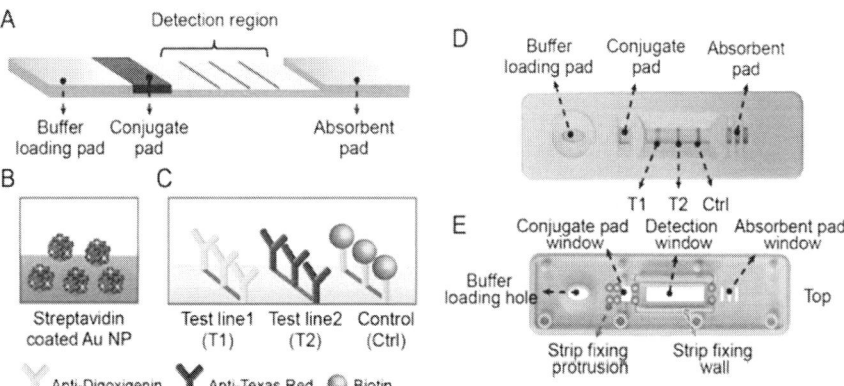

Fig. 1. (A) Schematic of a packaged paper fluidic-based analytical microdevice (PFAM). The PFAM consists of a buffer loading pad, a conjugate pad, a detection region, and an absorbent pad. (B) Streptavidin coated gold nanoparticles on the conjugation pad. (C) In the detection region, anti-Digoxigenin, anti-Texas Red, and biotin are immobilized on a T1, a T2, and a Ctrl line, respectively. (D) Photograph of the packaged PFAM. (E) Photograph of the inside structure of the PFAM. The inner structure was composed of strip fixing protrusions (red circle), strip fixing walls (green line), and strip laid protrusions (blue rectangle). (Reprinted with permission from Ref. [8]. Copyright (2014) Elsevier.)

a range of applications [8, 17–22].

The principle of amplicon detection in LFAs aligns closely with the standard LFAs approach, with the key difference being the nature of the sample [9]. In this case, the sample is an amplicon – a piece of DNA or RNA that has been amplified through techniques like PCR or LAMP. Once the target genetic material is amplified and labeled, it is applied to the LFA strip. Overall, this sample then migrates to the conjugate pad, which contains antibodies or antigens labeled with a colored marker, typically gold nanoparticles or colored latex beads [23–25]. These labeled antibodies bind to the target analyte if present in the sample. The mixture then flows into the nitrocellulose membrane, the core of the assay, where specific antibodies or antigens are immobilized at the test line [26, 27]. Here, the target-labeled antibody complexes accumulate, forming a visible line that indicates a positive result. The control line, further down the membrane, captures any excess labeled antibodies, ensuring the assay is functioning correctly. The absorbent pad at the end of the strip aids in drawing the fluid through the membrane (Fig. 1).

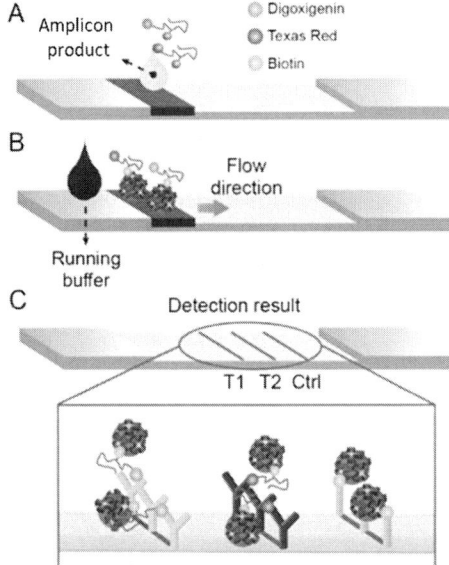

Fig. 2. Scheme of the multiplex gene expression using a paper fluidic-based analytical microdevice (PFAM). (A) The multiplex amplicon products, which were labeled with a Digoxigenin or a Texas Red at the end and biotin in the middle of the DNA strand, were loaded in the conjugate pad to be linked with streptavidin coated Au NPs. (B) The Au NP labeled amplicons are moved to the detection region upon addition of a running buffer in the buffer loading pad. (C) While the excess of the streptavidin coated Au NPs is immobilized on the biotin in the Ctrl line, the Digoxigenin or the Texas Red labeled amplified products are captured by anti-Digoxigenin and anti-Texas Red on the T1 and T2 lines, respectively, producing the violet band which is derived from Au NPs. (Reprinted with permission from Ref. [8]. Copyright (2014) Elsevier.)

For instance, it utilizes a one-step RT-LAMP process to amplify target genes, which are then labeled with distinct markers-Digoxigenin for a target gene of a pathogen and Texas Red for other specific target genes, while biotin is integrated throughout the DNA strands. These LAMP products are applied to a pad embedded with streptavidin-coated gold nanoparticles (Au NPs). Upon application to the lateral flow strip, the sample migrates via capillary action. The gold nanoparticles facilitate visual detection as the sample progresses along the strip, encountering antibodies that bind the labeled amplicons, triggering a color change that signifies the presence of the pathogens (Fig. 2). The intensity of this line can sometimes be used to estimate the amount of the target nucleic acid present in the sample, thanks to the proportionate relationship between the amplicon concentration and the signal intensity.

This method combines molecular amplification's specificity with the ease of use of lateral flow assays, leading to a powerful diagnostic tool. The simplicity of the lateral flow assay and the robustness of the amplification process together allow for rapid, on-site testing without the need for complex laboratory equipment, making it highly suitable for field diagnostics and point-of-care applications.

2. Lateral flow assay-incorporated centrifugal microfluidic chip

2.1. Design of a LAMP-lateral flow strip chip

The research conducted by Jung et al. (2015) [20] represents an innovative combination of integrating LAMP with a lateral flow strip and a centrifugal microfluidic chip for molecular diagnostics. LAMP's rapid, isothermal amplification capabilities are harnessed within a closed microfluidic system to minimize contamination risks and simplify sample processing, making it ideal for point-of-care testing and resource-limited settings. The integration of LAMP amplification steps within the microfluidic chip streamlines the diagnostic process, while the visual readout provided by lateral flow strips offers user-friendly, on-the-spot results. This synergy enables highly sensitive and specific nucleic acid detection with reduced turnaround times, ensuring timely and efficient patient care.

As an example, the so-called RT-LAMP-ICS (immunochromatographic strip) microdevice was designed for the rapid detection of influenza A virus, showcasing an array of innovative features (Fig. 3). The device features a sample inlet strategically positioned to introduce the RT-LAMP solution (including RT-LAMP cocktails and viral RNA templates) into the microfluidic system. This inlet is directly connected to a zigzag-shaped dispensing microchannel that, due to its geometric design and depth differential, precisely aliquots the solution into three separate RT-LAMP chambers without the risk of cross-contamination that is a critical factor for avoiding false positives or negatives in diagnostic testing.

Each RT-LAMP chamber is allocated a portion of the dispensing channel measuring 2.4 cm in length, allowing for an exact injection volume of 2 µL of the RT-LAMP solution. The introduction of the solution into the sample inlet automatically fills the dispensing microchannel by capillary force. The device also includes a circular air chamber, positioned around the rotary axis to apply air pressure for controlled transportation of the solution to the RT-LAMP chambers, mitigating the risk of mechanical failure associated with pump-driven systems.

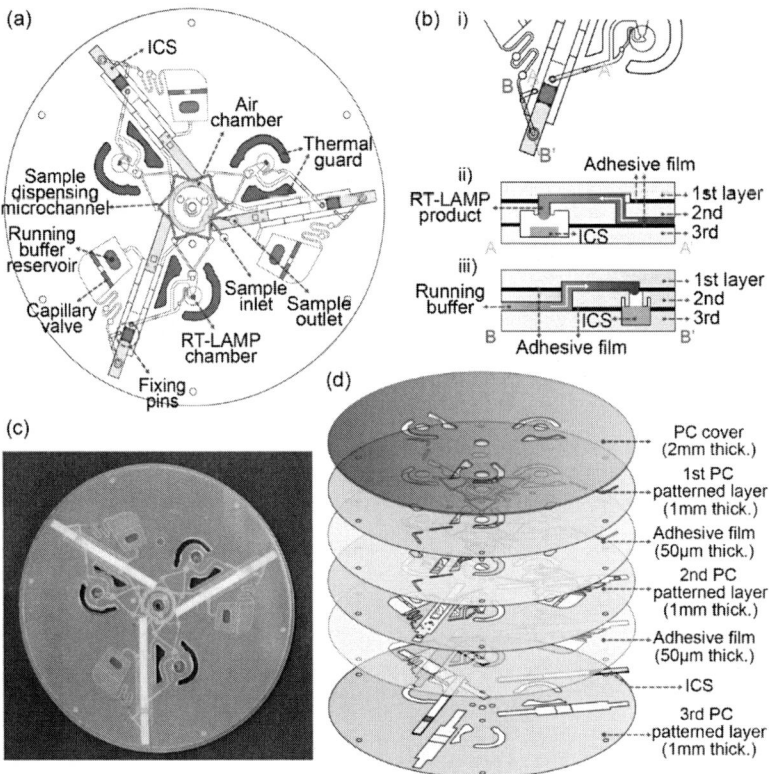

Fig. 3. (a) Schematic illustration of the centrifugal multiplex RT-LAMP-ICS microdevice. (b) (i) A floor plan of the microchannels connected to the ICS, (ii) Schematic design of the connecting microchannel between the RT-LAMP chamber and the ICS. (iii) Schematic design of the connecting microchannel between the running buffer reservoir and the ICS. (c) A digital image of the centrifugal multiplex RT-LAMP-ICS microdevice. (d) Schematic illustration of the RT-LAMP-ICS microdevice consisting of seven layers: a 2 mm thick PC cover, a 1 mm thick 1st micropatterned PC layer, a 50 μm thick adhesive layer, a 1 mm thick 2nd micropatterned PC layer, a 50 μm thick adhesive layer, ICSs, and a 1 mm thick 3rd micropatterned PC layer. (Reprinted with permission from Ref. [20]. Copyright (2015) Royal Society of Chemistry.)

Furthermore, the inclusion of a siphon channel and a round-shaped capillary valve at the channel's end demonstrates a novel approach to overflow prevention. These components work in tandem to inhibit leakage during the RT-LAMP reaction, even at high rotational speeds (4000 RPM), ensuring that the amplification process remains uncontaminated, and the reaction components are not prematurely transferred to the ICS.

The design also features an innovative arrangement of connecting microchannels across three layers, guiding the RT-LAMP products and running buffer towards the ICS. The buffer reservoir, with its capillary valves, exemplifies an advanced solution for fluid retention, utilizing capillary forces to prevent unintended flow, a critical feature for the staged processing of samples. The hydrophilic treatment of the siphon channel ensures consistent fluid movement, reflecting a deep understanding of material properties and their role in microfluidic dynamics [28].

2.2. Design of a sample pretreatment-LAMP-lateral flow strip chip

The integration of multiple analytical steps, which include important components like sample pretreatment (involving cell lysis and DNA/RNA extraction), amplification, and detection, onto a unified microfluidic platform, has emerged as a central focus within the domain of point-of-care diagnostics. However, achieving the requisite level of precision and automation in fluidic control while ensuring the efficient transfer of samples and products between various functional units within the microdevice presents a formidable challenge. In the context of the existing "LAMP-lateral flow strip chip," it's essential to note that, despite its capabilities, the initial setup still relies on a purified RNA/DNA template.

However, this established design has the potential to undergo further refinement, evolving into an innovative "sample pretreatment-LAMP-lateral flow strip chip". In this advanced iteration, a sample pretreatment step has been seamlessly integrated into the previous design, facilitating direct DNA/RNA extraction within a single chip. This groundbreaking development not only streamlines the genetic analysis process but also significantly enhances its overall efficiency. The result is a sophisticated sample-to-answer diagnostic tool that promises to catalyze a paradigm shift in genetic analysis within point-of-care settings [29].

The microdevice comprises three identical units arranged on a five-layer stacked disc. Each unit houses three functional components: a solid-phase DNA extraction module, a LAMP reaction chamber, and a lateral flow strip (Fig. 4). The innovation lies in the integration of a sample pretreatment step, enabling direct DNA/RNA extraction on the chip itself. Within the solid-phase DNA extraction module, glass microbeads serve as the solid-phase matrix. The DNA extraction process relies on the adsorption of DNA onto the silica bead surface in the presence of chaotropic salts, which enhance hydrogen bonding interactions between the silica and nucleic acids. The microbeads are effectively packed into a silica microbead-bed channel using a weir structure. This innovation simplifies the DNA extraction process, minimizing the need for complex, labor-intensive, and time-consuming off-chip pretreatment procedures.

Within the LAMP reaction chamber, eluted DNA from the microbead bed is mixed with a LAMP cocktail. Incubation at a constant temperature of 66 °C facilitates the generation of LAMP products, which are labeled with haptens (Texas red or Digoxigenin) and biotin-dUTP. To enable colorimetric detection by the naked eye, a lateral flow strip is employed [9]. This strip incorporates critical components, including a buffer loading pad, a conjugate pad housing streptavidin coated Au NPs for LAMP product conjugation, and a detection zone where anti-Digoxigenin, anti-Texas Red, and biotin are immobilized in the test lines and control line, respectively. The design also incorporates thermal guards to maintain the integrity of the detection zone. This integrated approach enhances the efficiency of genetic analysis by enabling straightforward, visual detection without the need for specialized equipment. Notably, the microdevice incorporates siphon channels and capillary valves actuated by centrifugal forces, which enable precise fluid control during DNA extraction, amplification, and detection. This eliminates the need for complex microvalve and pump systems, simplifying the chip's operation.

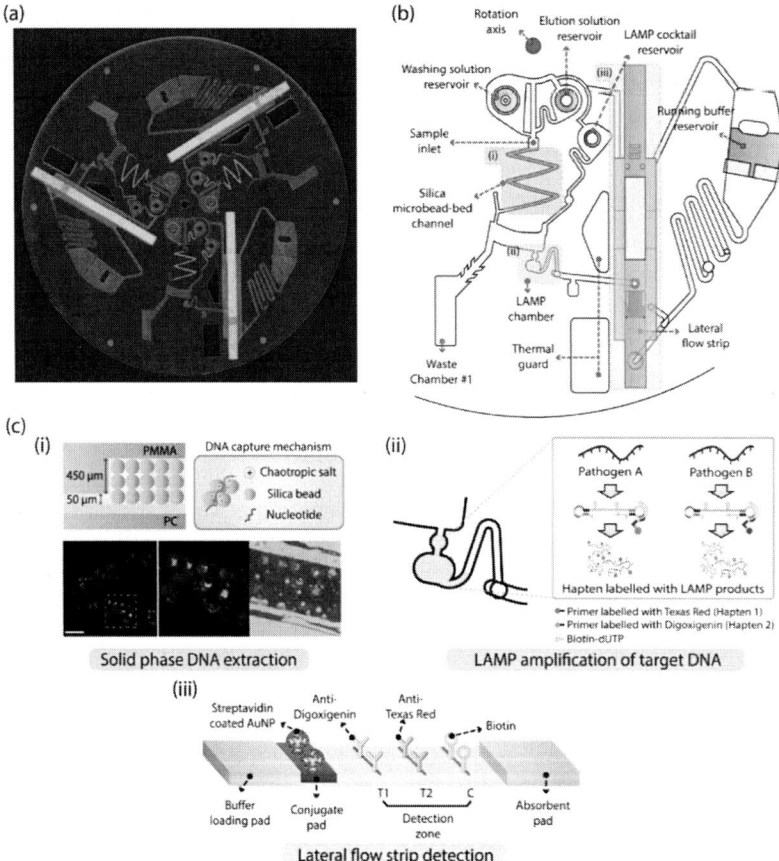

Fig. 4. (a) A digital image of the chip design (b) Schematic illustration of the integrated centrifugal microdevice for the DNA extraction, the LAMP reaction, and the lateral flow strip detection. (c-i) Schematic design of the solid phase DNA extraction unit and the fluorescence images of the FAM-labelled DNA adsorbed glass microbeads, (c-ii) Schematic image of the LAMP amplification of target DNA (c-iii) Schematics of a lateral flow strip which consists of a buffer loading pad, a conjugate pad that contains streptavidin coated AuNPs, a detection zone where an anti-Digoxigenin, an anti-Texas red, a biotin are immobilized in the test line 1, the test line 2 and the control line, respectively, and an absorbent pad. (Reprinted with permission from Ref. [29]. Copyright (2017) Elsevier.)

2.3. Chip fabrication

The fabrication of centrifugal microfluidic devices for genetic analysis and pathogen detection represents a pivotal step in the development of advanced diagnostic tools. In this chapter, we will explore the fabrication processes of two seminal microdevices: the RT-LAMP-ICS microdevice developed by Jung et al. [20] and the Sample Pretreatment-LAMP-Lateral Flow Strip Chip created by Park et al. [29].

Jung et al. developed a sophisticated RT-LAMP-ICS microdevice, which consisted of seven layers. (Fig. 3C and D). The central layer was a 1 mm thick polycarbonate (PC) sheet, and it formed the core of the disc-shaped device. The PC layer underwent precision micropatterning using a CNC milling machine, ensuring accurate and reproducible channel structures. Additionally, siphon channels on the disc were coated with a hydrophilic reagent to modify surface properties, facilitating fluid movement. An essential step in the fabrication process was the application of RT-LAMP primer mix (comprising outer, inner, and loop primers) to the reaction chambers, followed by natural drying. The microdevice assembly was achieved by layering patterned PC sheets and pressure-sensitive adhesive (PSA) foil, firmly pressed together to prevent leaks. Finally, a double-sided adhesive film and a thin PC film sealed the device, enclosing the loaded solutions within the disc.

On the other hand, Park et al. introduced a Sample Pretreatment-LAMP-Lateral Flow Strip Chip with a distinct fabrication process. This microdevice featured three primary microfluidic layers. The top layer held injection holes and microfluidic channels, the second layer housed DNA extraction and amplification micropatterns, and the third layer embedded the lateral flow strip for colorimetric detection. Design and structural engineering were executed using AutoCAD software, with micropatterns on the first and third layers fabricated on 0.5 mm and 1.0 mm thick PC sheets, respectively, utilizing a CNC milling machine. The second layer, responsible for DNA extraction and amplification, was crafted on a 1.5 mm thick poly(methyl methacrylate) (PMMA) sheet. Assembly involved the use of double-sided adhesive film, aligning the layers and bonding them under pressure with a hot press (Fig. 5).

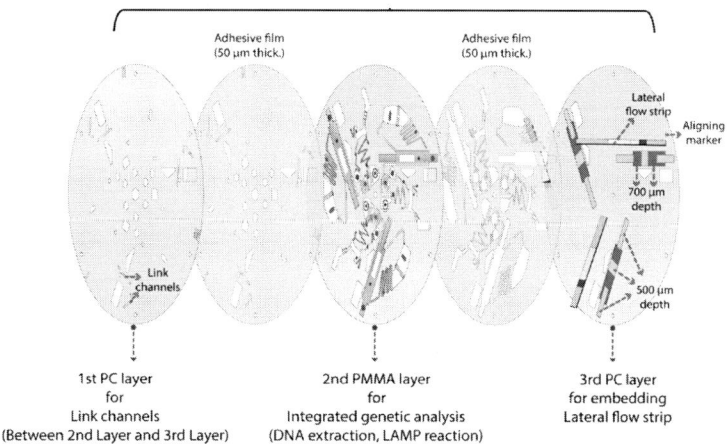

Fig. 5. Chip fabrication for the Sample Pretreatment-LAMP-Lateral Flow Strip Chip. This microdevice featured three primary microfluidic layers. (Reprinted with permission from Ref. [29]. Copyright (2017) Elsevier.)

Both fabrication processes underscore the significance of precision engineering and material selection in creating functional diagnostic microdevices. The careful integration of design features with fabrication steps ensures manufacturability and reliable performance.

3. Chip operation

The evolution from the "LAMP-lateral flow strip chip" (Jung et al. [20]) to the "sample pretreatment-LAMP-lateral flow strip chip" subsequent design (Park et al. [29]) illustrates significant progress in microfluidic technology, particularly in chip operation complexity. While the former chip requires pre-purified RNA/DNA, the latter advanced version incorporates an integrated sample pretreatment stage, allowing for direct DNA/RNA extraction on the chip. This additional functionality adds operational complexity, yet it significantly streamlines the diagnostic process. This progression not only demonstrates technical advancement but also highlights the growing sophistication in the design and functionality of microfluidic devices for medical diagnostics.

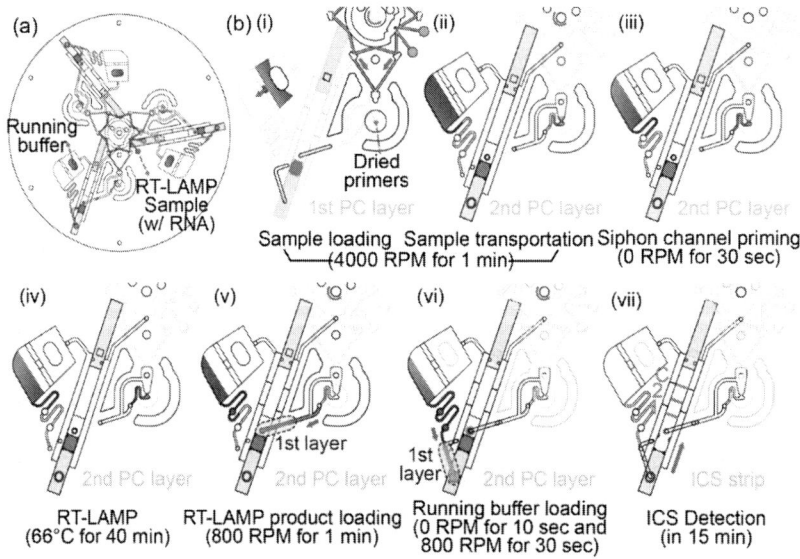

Fig. 6. The entire process for the multiplex RT-LAMP-ICS microdevice. (a) An initial state of the multiplex RT-LAMP-ICS microdevice with the RT-LAMP sample containing RNA template and the running buffer in the designated reservoirs. (b) (i) Sample loading and (ii) sample transportation at 4000 RPM for 1 minute, (iii) siphon channel priming at 0 RPM for 30 seconds, (iv) RT-LAMP reaction at 66 °C for 40 minutes, (v) RT-LAMP product loading at 800 RPM for 1 minute, (vi) running buffer loading at 0 RPM for 10 seconds and at 800 RPM for 30 seconds, and (vii) ICS colorimetric detection at 0 RPM for 15 minutes. (1: test line 1, 2: test line 2, and c: control line) (Reprinted with permission from Ref. [20]. Copyright (2015) Royal Society of Chemistry.)

Firstly, the LAMP-lateral flow strip chip, a novel invention by Jung et al., showcases an intricate process for detecting Influenza A virus, integrating the precision of RT-LAMP with the simplicity of an ICS (Fig. 6). This operation starts with a meticulously prepared RT-LAMP cocktail, which is mixed with a viral RNA template, and an ICS running buffer. The RT-LAMP cocktail and the ICS running buffer are introduced into the chip, initiating the process by filling the designated sample dispensing microchannel through capillary force. To maintain the integrity of the process, the chip is then carefully sealed, safeguarding against any potential contamination, and preventing the evaporation of reagents.

The heart of the operation lies in the use of a custom-made centrifugal system which is responsible for dividing the RT-LAMP sample and transporting it efficiently to the reaction chambers. These chambers, containing freeze-dried primer sets, are then activated as the RT-LAMP solution rehydrates the primers. This step is important as it kick-starts the RT-LAMP reaction. The reaction itself is conducted at a controlled temperature of 66 °C, maintained for 40 minutes. This precision in temperature control is vital for the accuracy and efficiency of the RT-LAMP reaction. Following this, the RT-LAMP products, now labeled with haptens and biotins, are directed towards the ICS.

Simultaneously, a specially formulated running buffer is released, controlled by sequential siphon valves, crucial for the correct flow and interaction of reagents within the chip. After loading the running buffer, the detection phase begins. The culmination of the operation is seen in the colorimetric ICS detection. This stage provides a visual representation of gene expression, with specific color changes on the test lines indicating the presence of Influenza A virus. The test lines are designed to respond to different gene expressions, with one indicating the expression of the M gene and the other the HA gene.

In the second paper by Park et al. [20], the chip's design, which includes a more complex sample processing and flow control mechanism, reflects an advanced level of microfluidic integration compared to the earlier model. This progression not only demonstrates technical refinement but also highlights the trend towards more comprehensive, automated solutions in pathogen detection. This new chip integrates sample preparation with RT-LAMP and ICS, allowing direct DNA/RNA extraction on the chip with only centrifugal force control (Fig. 7). This eliminates the need for pre-processing samples, streamlining the entire diagnostic process. The process begins with the preparation of a cell lysate, which is then introduced into the microfluidic chip, filling a microbead-bed channel by capillary force. Various reagents, including a washing buffer and LAMP reaction cocktail, are strategically dispensed into designated reservoirs. The chip's complex design utilizes a centrifugal system and triple siphon channels to precisely control the movement of the running buffer. This ensures the effective adsorption of DNA on the microbeads, followed by sequential washing, elution, and transfer of the DNA to the LAMP chamber. The LAMP reaction is then performed at 66 °C for 50 minutes. After the LAMP reaction, the amplified products and running buffer are carefully loaded onto a lateral flow strip for detection by precise centrifugal force control. This process involves quantifying the color density in the detection zone, a critical step in determining the presence and concentration of pathogens.

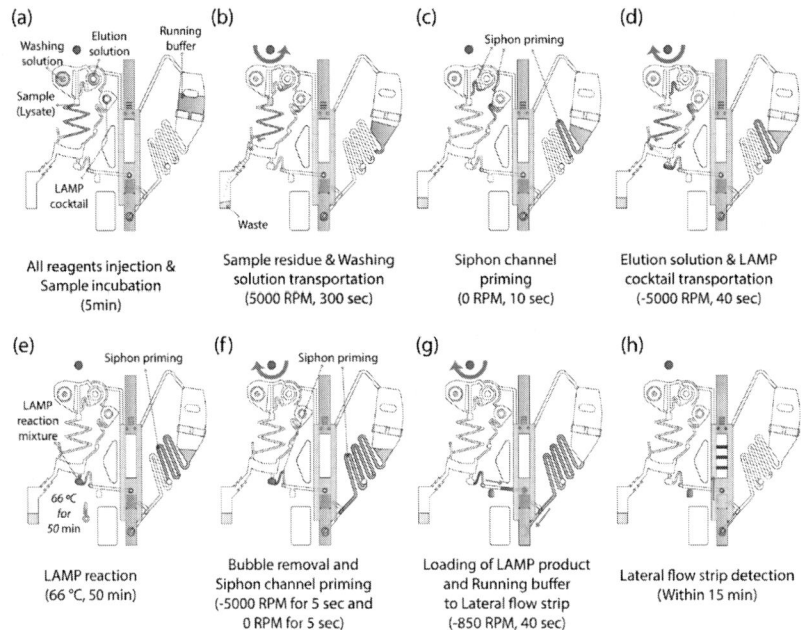

Fig. 7. The whole procedure for chip operation. (a) All reagents were injected into the designated reservoirs on the integrated centrifugal microdevice. After loading the reagents, the sample was incubated for 5 minutes for genomic DNA to be bound on the glass microbeads. (b) Residual impurities in the glass bead bed were transported into the waste chamber #1, followed by the washing solution transportation at 5000 RPM for 300 seconds. (c) Siphon channels which are linked to the elution solution, the LAMP cocktail and the running buffer reservoir, respectively, were primed at 0 RPM for 10 seconds. (d) The elution solution and the LAMP cocktail were transported into the LAMP chamber at -5000 RPM for 40 seconds. (e) For the DNA amplification, the LAMP reaction was performed at 66 °C for 50 minutes, while the running buffer primed over the second crest of the siphon channel. (f) At -5000 RPM for 5 seconds, bubbles in the LAMP chamber caused by heating were removed. Then, the siphon channel connected with the LAMP chamber and the third crest of the triple siphon channel of the running buffer were primed at 0 RPM for 5 seconds. (g) At -850 RPM for 40 seconds, the LAMP product and the running buffer were loaded into the lateral flow strip located in the third PC layer via the connecting channel patterned in the first PC layer. (h) After 15 minutes, the colorimetric signal on the strip was observed by naked eyes. (Reprinted with permission from Ref. [29]. Copyright (2017) Elsevier.)

4. Application of a pathogen detection

The advent of LFA-incorporated centrifugal microfluidic devices has emerged as a highly effective approach for the detection of viral and bacterial pathogens, as well as other infectious diseases. This integration represents a significant advancement in diagnostic technology, combining the rapid and user-friendly interface of LFA with the enhanced LAMP precision and automation offered by centrifugal microfluidics.

Such devices are particularly effective in swiftly identifying a wide range of infectious agents, making them invaluable in various healthcare scenarios— from routine clinical diagnostics to urgent outbreak response. The ability to rapidly detect and differentiate between viral and bacterial infections is crucial for appropriate treatment decisions, especially in environments where laboratory resources are limited. Moreover, the application of these advanced devices extends beyond traditional healthcare settings, offering potential for field-based diagnostics and global health surveillance. This technology provides a promising solution for early detection and management of infectious diseases, contributing significantly to public health preparedness and response strategies.

In this chapter, we focus on two representative studies that exemplify the advancements in detecting viruses and bacteria using LFA-incorporated centrifugal microfluidic devices. These studies are not only instrumental in illustrating the state-of-the-art technology in pathogen detection but also in highlighting the diverse applications of this technology in diagnosing both viral and bacterial infections.

4.1. Virus detection on a LAMP-lateral flow strip chip

The need for rapid, accurate viral detection is paramount, especially for Influenza viruses, due to their high mutation rates and potential for causing pandemics. Traditional diagnostic methods often fall short in terms of speed and sensitivity, necessitating more advanced solutions. The innovative research conducted by Jung et al. on a RT-LAMP-ICS microdevice provides a groundbreaking approach to the detection of Influenza A H1N1 virus. Their study focused on the monoplex detection of H1 and M genes, for identifying this virus strain (Fig. 8). In their experimental setup, primer sets targeting these genes were separately dried in RT-LAMP chambers, with an additional chamber serving as a negative control. This setup facilitated simultaneous processing of the RT-LAMP solution, containing viral templates, at 66 °C for a duration of 40 minutes.

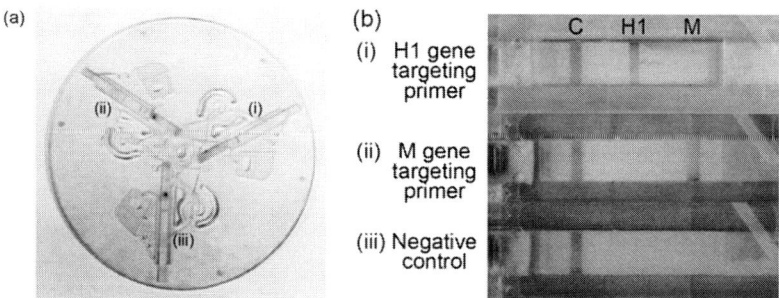

Fig. 8. (a) A digital image of the RT-LAMP—ICS microdevice after monoplex RT-LAMP reaction with influenza A H1N1 virus. (b) Digital images of the detection regions of the ICS for confirming gene expression of (i) H1 gene, (ii) M gene, and (iii) a negative control. (Reprinted with permission from Ref. [20]. Copyright (2015) Royal Society of Chemistry.)

The results of this experiment were significant. The detection of the H1 gene was confirmed by the emergence of a violet color in the test line dedicated to the H1 gene, alongside a control line. The M gene's expression was similarly verified by observing positive signals in its respective test line and the control line. In the negative control, the presence of color only in the control line validated the specificity of the experiment.

In a parallel experiment, the research team explored the time control of the multiplex RT-LAMP reaction (Fig. 9). This study aimed to assess the speed and effectiveness of the microsystem in detecting influenza A. By varying the reaction time from 60 to 20 minutes and using H1N1 viral RNA as a template, the researchers conducted a series of tests. These tests, which involved loading H1 and M gene primer sets together in a single RT-LAMP chamber, were repeated multiple times for consistency. The findings from this time-controlled experiment were illuminating. Successful detection of Influenza A H1N1 virus was indicated by three distinct violet lines, representing the expression of both the H1 and M genes. Notably, the intensity and clarity of these signal bands enhanced with longer reaction times. A 30-minute RT-LAMP reaction was determined to be sufficient for effective multiplex detection. However, reducing the reaction time further to 20 minutes resulted in faint band signals, which posed challenges for accurate interpretation.

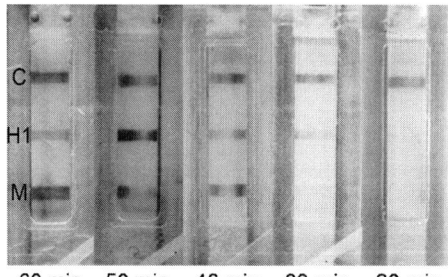

Fig. 9. The ICS results to identify influenza A H1N1 virus on the RT-LAMP—ICS microdevice with variation of multiplex RT-LAMP time from 60 minutes to 20 minutes. (Reprinted with permission from Ref. [20]. Copyright (2015) Royal Society of Chemistry.)

The study next focused on the evaluation of the LAMP-lateral flow strip chip's specificity and sensitivity for detecting Influenza A virus. These tests are integral to establishing the reliability and efficacy of diagnostic tools, especially in the context of managing potential pandemics. The specificity test conducted by the researchers was designed to assess the microdevice's ability to accurately identify specific strains of Influenza A virus (Fig. 10). This aspect of diagnostic testing is significant to ensure that positive results are attributed to the target pathogen and not other similar organisms. In this experiment, the team used three different RT-LAMP reactions on a single RT-LAMP-ICS microdevice, with each reaction targeting a different combination of genes (H1 & M, H3 & M, and H5 & M) to subtype Influenza A virus into H1N1, H3N2, and H5N1, respectively.

The results of this specificity test were quite revealing. When influenza A H1N1 viral RNA was introduced to the RT-LAMP chambers, only the chamber with H1 & M gene targeting primer sets produced the three bands on the ICS, indicative of a positive result. This outcome demonstrated the chip's precision in identifying the specific strain of Influenza A virus, confirming its capability to distinguish between closely related viral subtypes. The successful differentiation of these strains underlines the potential of this microdevice in various clinical settings, where accurate strain identification is essential.

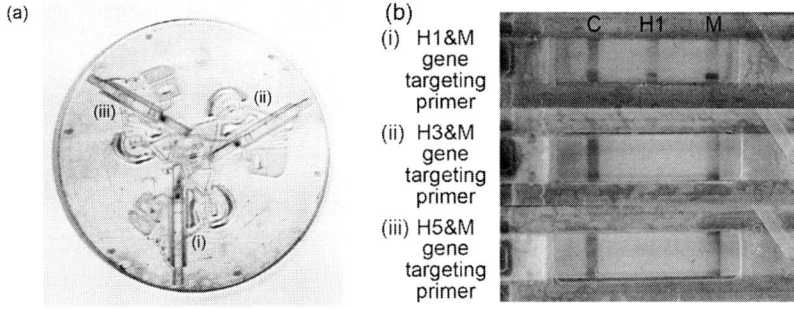

Fig. 10. (a) A digital image of the RT-LAMP—ICS microdevice after multiplex RT-LAMP reaction for subtyping influenza A virus. (b) The ICS results when the used primer sets targeted (i) H1 & M gene, (ii) H3 & M gene, and (iii) H5 & M gene. (Reprinted with permission from Ref. [20]. Copyright (2015) Royal Society of Chemistry.)

Following the specificity test, the sensitivity test was conducted to evaluate the chip's effectiveness in detecting low quantities of Influenza virus (Fig. 11). Given the virus's potential for rapid spread and pandemic development, the ability of a diagnostic tool to detect even minimal amounts of the virus is vital.

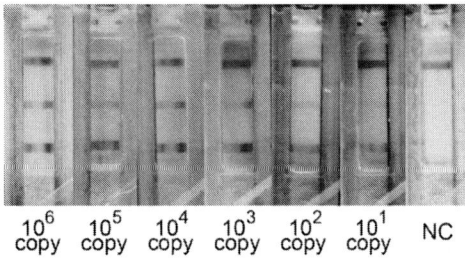

Fig. 11. LOD test of the multiplex RT-LAMP reaction in the RT-LAMP—ICS microdevice by using Influenza A/H1N1 viral RNA templates ranging from 10^6 to 10 copies. (Reprinted with permission from Ref. [20]. Copyright (2015) Royal Society of Chemistry.)

The researchers performed a limit of detection (LOD) study using 10-fold serial dilutions of influenza A H1N1 viral RNAs, ranging from 10^6 to 10 copies. These diluted samples underwent a multiplex RT-LAMP reaction at 66°C for 40 minutes, a duration selected to optimize the detection of low RNA copy numbers.

The chip successfully amplified and detected the multiplex targets of H1 and M genes, with even the lowest quantity of 10 copies of H1N1 viral RNAs producing positive signals on the ICS. This remarkable sensitivity, reflected in the detection of minimal viral quantities, highlights the chip's potential as an early-stage diagnostic tool, particularly crucial in managing and controlling flu outbreaks.

The integration of RT-LAMP with ICS on a centrifugal microdevice, as demonstrated in this study, marks a significant leap in the field of viral detection technologies. Such a compact and automated system capable of conducting both monoplex and multiplex detection has wide-ranging implications, especially in clinical diagnostics and resource-limited environments. The study underscores the critical balance between rapidity and accuracy in diagnostic processes, highlighting the need for optimized reaction times to ensure precise and reliable detection.

4.2. Bacterial detection on a sample pretreatment-LAMP-lateral flow strip chip

In their innovative study, Park et al. placed considerable emphasis on optimizing the bacterial detection capabilities of their integrated centrifugal microdevice. *Salmonella enterica subsp. enterica serotype Typhimurium* (*S. Typhimurium*) and *Vibrio parahaemolyticus* (*V. parahaemolyticus*) were used as model organisms to demonstrate integrated genetic analysis.

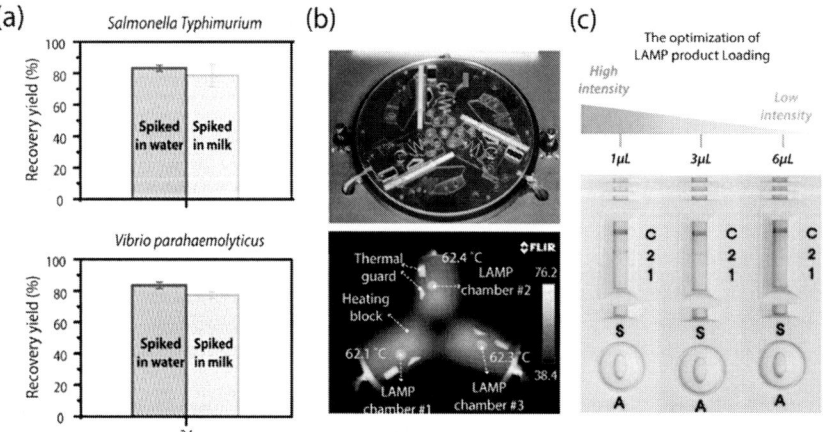

Fig. 12. (a) Evaluation of DNA recovery yield on glass microbeads depending on the type of bacterial cell and sample. (b) Temperature measurement of the LAMP chamber of the microdevice mounted on top of three heating blocks of a rotary system. (c) Optimization of loading amount of the LAMP product into the lateral flow strip. (Reprinted with permission from Ref. [22]. Copyright (2016) Elsevier.)

As mentioned, the sample pretreatment-LAMP-lateral flow strip chip represents an integrated approach that combines sample pretreatment, LAMP amplification, and lateral flow assay into a cohesive unit. The team's approach was rooted in ensuring that each component of the device functioned at its highest capacity, directly impacting the device's ability to accurately and efficiently detect bacterial pathogens. Recognizing that the successful integration of these modules hinges on their individual performance, the team focused on three critical components: the DNA extraction efficiency of the microbead bed, temperature control within the loop-mediated isothermal amplification (LAMP) chamber, and the optimization of the LAMP product amount for effective strip detection. (Fig. 12)

DNA Extraction Efficiency

The study began with an assessment of DNA extraction efficiency. The researchers utilized a method involving the spiking of pure genomic DNA from bacteria into water and milk samples. These samples were then treated with 6M Gu-HCl and ethanol. This step was critical as it mirrored the complex conditions under which the microdevice would be expected to operate in real-world scenarios. Upon completion of the extraction process in the microfluidic chip, the purified DNA was recovered. Remarkably, the results displayed an extraction efficiency of about 80% across various bacterial cells (Fig. 12A). However, a observation was made regarding the samples in milk; the DNA recovery yield was relatively lower compared to that in water. This discrepancy was attributed to milk's inherent properties, such as the presence of fats, proteinases, and calcium ions, which potentially interfered with the DNA extraction process. Despite these challenges, the method exhibited high performance, comparable to commercial silica-membrane kits, and was effective in purifying genomic DNA from substances in milk that could inhibit PCR.

Temperature Control in the LAMP Chamber

The next focal point was the temperature control within the LAMP chamber. The team used an infrared camera to capture the temperature distribution of the microdevice, particularly focusing on three heating blocks that applied heat to the LAMP chamber. This setup was governed by a proportional–integral–derivative controller. The results showed that the temperature on the top of the LAMP chambers was consistently maintained at around 62.0 ± 0.4 °C, with the heat block set at 69.0 °C. This careful management of temperature was important, as it ensured that the heat diffusion to the strip was minimized, thus preserving the integrity of the sample and the reagents 5[40]. The temperature of the LAMP mixture in the chamber, crucial for the activation of the Bst polymerase, was calculated to be around 66.0 °C. This temperature is optimal for the enzyme, facilitating efficient DNA amplification (Fig. 12B).

Optimizing LAMP Product Loading for Strip Detection

Finally, the study addressed the importance of the LAMP product loading amount for strip detection. This aspect was important because the accuracy and reliability of the lateral flow strip results depended on this parameter. The researchers found that the loading amount of the LAMP product was inversely proportional to the signal intensity on the lateral flow strip. They discovered that when the amount of LAMP product exceeded that of the streptavidin-coated gold nanoparticles, unlabelled LAMP products migrated more rapidly to the anti-hapten than the Au NP labelled LAMP product. This led to potential false-negative results due to the saturation of the binding sites of the anti-hapten.

Based on these findings, the optimal loading amount of the LAMP product was determined to be 1 µL. To facilitate this precise loading, a 4 µL waste chamber was designed into the microdevice (Fig. 12C).

Monoplex Bacterial Detection and Transition to Multiplex Detection

The sample pretreatment-LAMP-lateral flow strip chip demonstrated remarkable capabilities in both monoplex and multiplex bacterial detection within a single, integrated microfluidic platform.

In the domain of monoplex bacterial detection, the study focused on *S. Typhimurium*, a bacterium commonly associated with food-borne illnesses. The experiment was meticulously designed to test the device's detection range and sensitivity. Samples were prepared by spiking *S. Typhimurium* into two different mediums – tap water and milk – with bacterial concentrations varying from 50 to 50,000 colony-forming units (CFU) per 5 µL. This wide range in bacterial concentration was significant for assessing the device's ability to accurately detect varying levels of bacterial presence.

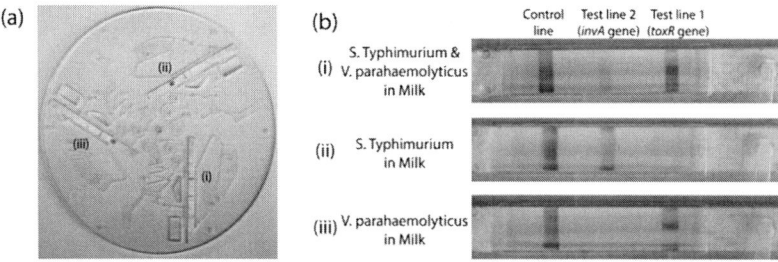

Fig. 13. (a) Digital image of the integrated centrifugal microdevice after simultaneous analysis of three samples; (i) *S. Typhimurium* and *V. parahaemolyticus* spiked in milk, (ii) *S. Typhimurium* spiked in milk, (iii) *V. parahaemolyticus* spiked in milk, (b) The lateral flow strip results when each sample was used. (Reprinted with permission from Ref. [22]. Copyright (2016) Elsevier.)

Within the microdevice, each sample was subjected to a systematic process that began with DNA extraction, followed by the LAMP reaction for amplification, and culminated in the detection phase using a lateral flow strip. The notable outcome of the study was the successful identification of *S. Typhimurium* in both water and milk samples within 80 minutes. This rapid detection time is a key attribute for any on-site diagnostic tool. The *invA* gene of *S. Typhimurium* was specifically targeted and amplified using Texas Red hapten-labelled loop primers and biotin-dUTP. The resultant LAMP amplicons were detectable through the streptavidin-biotin interaction with streptavidin-coated gold nanoparticles and the hapten-anti-hapten interaction on the lateral flow strip.

To quantify the detection results, densitograms of the violet signals on the lateral flow strips were analyzed, and the optical densities of the test lines were measured with high reproducibility. The color intensity of the detection signal decreased as the input bacterial number reduced, indicating a direct correlation between the bacterial concentration and the signal intensity. Additionally, higher signal intensities were observed in samples spiked with tap water compared to milk, which aligns with the earlier observation of lower DNA recovery yield in milk. The device demonstrated the capability to detect as low as 10 CFU/μL, equivalent to 10^4 CFU/mL in the original spiked sample. This level of sensitivity is particularly significant considering the infective doses of pathogenic bacteria in real samples, which typically range from 10^2 to 10^6 CFU/mL. This finding underscores the microdevice's applicability for on-site pathogen detection, meeting the required thresholds for detecting infectious doses in real-world scenarios.

In the sphere of multiplex bacterial detection, the ingenuity of the microdevice's lateral flow strip was highlighted by its design, which incorporated two distinct test lines for anti-Digoxigenin and anti-Texas Red (Fig. 13). This unique feature facilitated the colorimetric detection of multiple pathogens on a single strip. Specifically, LAMP reactions were designed to produce two different types of amplicons: one that binds to anti-Digoxigenin, using the *invA* gene of *S. Typhimurium*, and another that binds to anti-Texas Red, using the *toxR* gene of *V. parahaemolyticus*. This approach enabled the simultaneous detection of these two distinct bacterial pathogens, marking a significant enhancement in the microdevice's multiplex pathogen detection capabilities. To validate this technology, the research team conducted experiments with various samples, including milk spiked with either *S. Typhimurium* or V. parahaemolyticus, as well as milk containing both.

The integrated centrifugal microdevice's design, featuring three identical units, was instrumental in enabling the simultaneous analysis of three different samples (Fig. 13A). The bacterial loads in the samples were meticulously adjusted, and the device successfully performed DNA extraction and multiplex LAMP reactions. The results were compelling. For the sample containing both *S. Typhimurium* and *V. parahaemolyticus*, the lateral flow strip displayed three positive bands – one for each pathogen and a control line (Fig.13B-i). This outcome was a definitive demonstration of the device's capability for multiplex pathogen detection.

Furthermore, the utilization of loop primers tagged with Texas Red and Digoxigenin hapten for *S. Typhimurium* and V. *parahaemolyticus*, respectively, facilitated distinct analyses on the lateral flow strip. The type of bacterial pathogen in the sample determined the position of the positive band on the test line (test line 1 or test line 2, Fig. 13B-ii/iii). Such precision in pathogen differentiation and the ability to conduct these analyses simultaneously on a single chip highlight the potential of microfluidic devices in complex diagnostic scenarios. This advancement in multiplex colorimetric detection marks a significant progression in the field, paving the way for more efficient, reliable, and versatile diagnostic solutions, particularly relevant in clinical settings and food safety testing.

5. Conclusion

This chapter has highlighted the significant advancements and innovative strides in genetic analysis, particularly focusing on the integration of lateral flow strip assays with centrifugal microfluidic technologies. This fusion represents a new era in diagnostic and analytical capabilities. The centrifugal microfluidic device, as a prominent example, stands as a testament to the evolution of diagnostic tools, especially in genetic analysis. Its ability to combine the simplicity and effectiveness of lateral flow assays with the precision and automation of centrifugal microfluidics signifies a substantial advancement. This integration not only improves the efficiency and accuracy of genetic analyses but also introduces a level of convenience and accessibility that was previously unattainable in conventional laboratory settings.

The innovative device has been adeptly applied in various scenarios, ranging from the detection of specific pathogens to broader applications in clinical diagnostics and foodborne pathogen detection. Its versatility in handling different sample types, along with the capability for both monoplex and multiplex detection, emphasizes its potential in addressing a diverse range of diagnostic challenges.

Furthermore, the case studies and research findings presented in this chapter underscore the device's ability to deliver rapid, reliable results. The integration of lateral flow strip assays into the centrifugal microfluidic platform has notably streamlined the genetic analysis process, making it more accessible and user-friendly, which is essential for on-site and point-of-care applications. The insights and innovations discussed here offer a promising outlook on the future of medical diagnostics. This technology, with its remarkable capabilities and potential for further development, is well-positioned to transform the approach to genetic analysis, leading to more effective, efficient, and accessible healthcare solutions.

References

1. Y. Ge, B. Wu, X. Qi, K. Zhao, X. Guo, Y. Zhu, Y. Qi, Z. Shi, M. Zhou, H. Wang, H. Wang and L. Cui, *PLoS One*, 2013, 8(8), e69941.

2. M. Ito, M. Watanabe, N. Nakagawa, T. Ihara and Y. Okuno, *J Virol Methods*, 2006, 135, 272–275.

3. X. Chen, Y. Xu, J. Yu, J. Li, X. Zhou, C. Wu, Q. Ji, Y. Ren, L. Wang, Z. Huang, Y. Wang and J. Lou, *Anal Chim Acta*, 2014, 841, 44–50.

4. H. Gao, J. Han, S. Yang, Z. Wang, L. Wang and Z. Fu, *Anal Chim Acta*, 2014, 839, 91–96.

5. A. K. Yetisen, M. S. Akram and C. R. Lowe, *Lab Chip*, 2013, 13, 2210–2251.

6. B. Ngom, Y. Guo, X. Wang and D. Bi, *Anal Bioanal Chem*, 2010, 397, 1113–1135.

7. G. A. Posthuma-Trumpie, J. Korf and A. Van Amerongen, *Anal Bioanal Chem*, 2009, 393, 569–582.

8   J. H. Jung, S. J. Oh, Y. T. Kim, S. Y. Kim, W.-J. Kim, J. Jung and T. S. Seo, *Anal Chim Acta*, 2015, 853, 541–547.

9   Y. T. Kim, J. H. Jung, Y. K. Choi and T. S. Seo, *Biosens Bioelectron*, 2014, 61, 485–490.

10 Y. Mori and T. Notomi, *Journal of Infection and Chemotherapy*, 2009, 15, 62–69.

11 H. T. C. Thai, M. Q. Le, C. D. Vuong, M. Parida, H. Minekawa, T. Notomi, F. Hasebe and K. Morita, *J Clin Microbiol*, 2004, 42, 1956–1961.

12 J. P. Dukes, D. P. King and S. Alexandersen, *Arch Virol*, 2006, 151, 1093–1106.

13 T. Notomi, H. Okayama, H. Masubuchi, T. Yonekawa, K. Watanabe, N. Amino and T. Hase, *Nucleic Acids Res*, 2000, 28(12), e63.

14 X. Fang, H. Chen, S. Yu, X. Jiang and J. Kong, *Anal Chem*, 2011, 83, 690–695.

15 Y. Sun, R. Dhumpa, D. D. Bang, J. Høgberg, K. Handberg and A. Wolff, *Lab Chip*, 2011, 11, 1457–1463.

16 M. Schulze, A. Nitsche, B. Schweiger and B. Biere, *PLoS One*, 2010, 5, e9966.

17 J. H. Jung, B. H. Park, Y. K. Choi and T. S. Seo, *Lab Chip*, 2013, 13, 3383–3388.

18 J. H. Jung, S. J. Choi, B. H. Park, Y. K. Choi and T. S. Seo, *Lab Chip*, 2012, 12, 1598–1600.

19 J. H. Jung, B. H. Park, S. J. Oh, G. Choi and T. S. Seo, *Biosens Bioelectron*, 2015, 68, 218–224.

20 J. H. Jung, B. H. Park, S. J. Oh, G. Choi and T. S. Seo, *Lab Chip*, 2015, 15, 718–725.

21 B. H. Park, J. H. Jung, H. Zhang, N. Y. Lee and T. S. Seo, *Lab Chip*, 2012, 12, 3875–3881.

22 S. J. Oh, B. H. Park, J. H. Jung, G. Choi, D. C. Lee, D. H. Kim and T. S. Seo, *Biosens Bioelectron*, 2016, 75, 293–300.

23 Y. H. Kim, J. S. Kim, J. H. Joo and J. W. Park, *J Arthroplasty*, 2012, 27, 88–94.

24 G. P. Zhang, J. Q. Guo, X. N. Wang, J. X. Yang, Y. Y. Yang, Q. M. Li, X. W. Li, R. G. Deng, Z. J. Xiao, J. F. Yang, G. X. Xing and D. Zhao, *Vet Parasitol*, 2006, 137, 286–293.

25 K. Glynou, P. C. Ioannou, T. K. Christopoulos and V. Syriopoulou, *Anal Chem*, 2003, 75, 4155–4160.

26 S. R. Farrah, D. O. Shah and L. O. Ingram, *Proc Natl Acad Sci U S A*, 1981, 78, 1229–1232.

27 W. L. Hoffman, A. A. Jump, P. J. Kelly and A. O. Ruggles, *Anal*

Biochem, 1991, 198, 112–118.

28 M. Focke, F. Stumpf, B. Faltin, P. Reith, D. Bamarni, S. Wadle, C. Müller, H. Reinecke, J. Schrenzel, P. Francois, R. Zengerle and F. Von Stetten, *Lab Chip*, 2010, 10, 2519–2526.

29 B. H. Park, S. J. Oh, J. H. Jung, G. Choi, J. H. Seo, D. H. Kim, E. Y. Lee and T. S. Seo, *Biosens Bioelectron*, 2017, 91, 334–340.

Chapter 6
Combination of a Solution-loading Cartridge with a Centrifugal Microfluidic Device

1. Background

In recent years, point-of-care testing (POCT) has become increasingly important in the healthcare industry, offering rapid and on-site diagnostic results that support immediate treatment decisions. Traditional central laboratories, with their high-cost and automatic diagnostic platforms, offer sensitive, precise, and accurate analyses but are typically bulky and not suited for on-site diagnostics [1]. In contrast, POCT devices provide user-friendly, fully integrated, and automatic operation for fast analyses, minimizing human interference and contamination risks [2–14]. These devices are especially valuable in resource-limited settings where they offer a cost-effective alternative to traditional lab-based diagnostics.

Centrifugal microfluidics has garnered considerable attention as a powerful platform for microfluidic operations, including liquid pumping, valving, metering, and aliquoting, all controlled by the rotational speed of a single device [15–26]. This simplicity and portability make centrifugal microfluidic devices highly suitable for POCT applications across various fields, such as immunoassays, cell analysis, nanoparticle synthesis, and molecular diagnostics [27–35]. Despite the proven capabilities of integrated centrifugal microfluidics in biological applications, a major challenge has been the treatment of the designated solutions into the chip [36]. The sequential loading of the designated solutions to the chip without manual steps is of importance to realize the POCT platform in a user-friendly way.

Addressing this challenge, the integration of a solution-loading cartridge with the centrifugal microdevice has been a significant step forward. This combination has enabled the automation of complex diagnostic procedures and has simplified the operation process by eliminating labor-intensive steps such as manual pipetting for sample loading and device sealing. The solution-loading cartridge stores and sequentially releases essential solutions into the centrifugal microdevice by a rotation program, which has significantly improved sample handling capacity and reduced the likelihood of human error [37–40]. A portable genetic analyzer with an integrated centrifugal disc, equipped with a glass-filter extraction column for nucleic acid purification and multiple reaction chambers for analysis, exemplifies the practical application of this technology [39, 40]. This setup allows for the automation of solution loading, nucleic acid purification, and both RT-LAMP reactions and RT-PCR, culminating in a compact and efficient system suitable for POCT. Furthermore, the combination of solution-loading cartridges with centrifugal microfluidic devices has been applied to detect a wide range of pathogens, including foodborne bacteria and respiratory viruses, enhancing the capacity for multiplex detection. The integrated system not only automates the entire diagnostic process but also provides rapid results with high sensitivity and specificity.

In this chapter, we will explore various studies that demonstrate the application and efficacy of integrating solution-loading cartridges with centrifugal microfluidic devices. This technology has shown immense potential in detecting a wide range of pathogens, from food-borne bacteria to respiratory viruses, including COVID-19. The advancements highlighted in these studies illustrate the increasing importance of POCT and centrifugal microfluidic devices in the healthcare industry. By combining a solution-loading cartridge with a centrifugal microfluidic device, researchers have created a powerful tool that bridges the gap between the need for rapid, on-site diagnostics and the complexities associated with processing a sample solution.

In their groundbreaking study, Nguyen et al. (2019) tackled the critical public health issue of rapidly detecting food-borne pathogens [37]. Recognizing the limitations of traditional pathogen detection methods, which often require time-consuming procedures and bulky, expensive equipment, they developed a portable, efficient solution. The study introduced an integrated centrifugal microsystem, comprising a portable genetic analyzer and a centrifugal microdevice. This innovative device consisted of a 3D-printed solution-loading cartridge and a centrifugal microfluidic disc, designed to handle large sample volumes (up to 1 mL). The device streamlined various processes, including solution release, bead-based DNA extraction, isothermal gene amplification, and detection of pathogens, thereby automating the entire procedure. This system was capable of multiplex detection, identifying several bacteria such as *E. coli* O157:H7, *Salmonella Typhimurium*, and *Vibrio parahaemolyticus*, in a rapid and sensitive manner. The detection was achieved through Eriochrome Black T (EBT)-mediated LAMP reaction, facilitating both colorimetric and UV-visible detection. This advancement in centrifugal microfluidics demonstrated its potential in POCT, enabling rapid on-site detection of food pathogens, reducing manual handling, and minimizing contamination risks.

Oh et al. (2019)'s research contributed significantly to lab-on-a-chip technology, focusing on the automated detection of foodborne pathogens [38]. Their device, comprising a centrifugal microfluidic disc and a 3D-printed solution-loading cartridge, marked a leap in fully automated molecular diagnostics. The device was adept at conducting bead-based DNA extraction, isothermal DNA amplification (LAMP), and colorimetric detection without manual intervention. It was designed to detect 18 different pathogens simultaneously, each device housing 20 reaction chambers. The study successfully demonstrated the device's effectiveness in identifying pathogens such as *E. coli* O157:H7, *Salmonella typhimurium*, *Vibrio parahaemolyticus*, and *Listeria monocytogenes*. The entire diagnostic process could be completed in just 65 minutes, with a detection limit of 100 bacterial cells, thus representing a substantial advancement towards true automation in molecular diagnostics and expanding the capabilities of POCT.

In response to the growing need for advanced tools in veterinary diagnostics, Nguyen et al. (2021) developed a portable genetic analyzer targeting feline upper respiratory tract diseases (FURTD) [39]. This integrated device combined a centrifugal disc with a glass-filter extraction column to facilitate nucleic acid purification and analysis. It was specifically designed to identify pathogens such as *Feline herpesvirus* 1 (FHV), *Mycoplasma felis* (MPF), *Bordetella bronchiseptica* (BDB), and *Chlamydophila felis* (CDF) [41, 42]. The system included components such as a spinning motor, heaters, a fluorescence detector, and a touch screen for real-time data analysis. Its compact and lightweight design made it ideal for POCT in veterinary settings. The device automated the entire process, from sample injection and DNA extraction to multiplex gene amplification and result display, completing the procedure within 1.5 hours. It was capable of performing both RT-LAMP and RT-PCR assays, offering a versatile and practical solution for rapid, accurate on-site diagnostics of common feline diseases and thereby enhancing veterinary care. Similarly, Phan et al. (2023)'s study focused on developing a comprehensive diagnostic system for multiple respiratory viruses, including the novel coronavirus (COVID-19), underlining the critical need for rapid diagnostics in managing pandemics [40]. Their research was instrumental in creating an integrated genetic analyzer capable of performing reverse transcription-loop-mediated isothermal amplification (RT-LAMP) assays for the simultaneous detection of various respiratory viruses. The study was significant in demonstrating the practical application and effectiveness of the integrated system in real medical settings. It highlighted the device's capability to streamline the diagnostic process, from viral RNA extraction to gene amplification and fluorescence detection, completing the procedure in just 80 minutes.

The integration of a solution-loading cartridge with a centrifugal microfluidic device represents a significant leap in molecular diagnostics. This technology has proven its versatility and effectiveness in various fields, from food safety to human and veterinary medicine. As we advance, it is expected that such technologies will become more accessible, further transforming the landscape of POCT and molecular diagnostics.

2. Chip design

2.1. Design and fabrication of a solution-loading cartridge

In the realm of microfluidic technology, particularly within the sphere of molecular diagnostics and POCT, the design and implementation of solution-loading cartridges have emerged as a critical element. These cartridges play a pivotal role in centrifugal microfluidic devices, facilitating automated and precise management of various reagents crucial for pathogen detection. The studies conducted by Nguyen et al., Oh et al., Nguyen et al., and Phan et al. exemplify the innovative approaches taken in designing these cartridges to cater to specific diagnostic needs while maintaining a set of core features that highlight their utility in POCT settings [37–40]. Central to the operational efficacy of these devices, the cartridges serve as a sophisticated means for storing and methodically dispensing various solutions essential for the diagnostic process.

A common thread running through all these designs is the utilization of 3D printing technology. This method allows for meticulous control over the geometry and internal structure of the cartridges, ensuring a perfect fit and functionality within the microfluidic systems. Each cartridge, crafted with multiple chambers, is designated for storing distinct solutions—ranging from sample solutions and washing buffers to elution solutions and various diagnostic reagents like LAMP cocktails or lysis buffers. The designs are planned to enable automated release of the stored solutions, leveraging the centrifugal forces generated by the microfluidic device's rotation. This strategic design choice not only streamlines the diagnostic process but also significantly minimizes human error by reducing manual intervention. Compactness and integration are key in these designs. The cartridges are dimensioned to seamlessly integrate with the centrifugal microfluidic devices, aligning perfectly with the device's injection holes for efficient solution transfer. In some instances, as seen in Phan et al.'s work [40], the cartridges undergo additional post-processing steps like sonication in isopropanol and UV irradiation for sterilization and solidification of the 3D printed structure, ensuring their readiness for immediate use in diagnostics. The scalability and potential for mass production of these cartridges are also considered in these studies. Nguyen et al. [39], for instance, explored hot injection molding as a technique for mass-producing the cartridges,

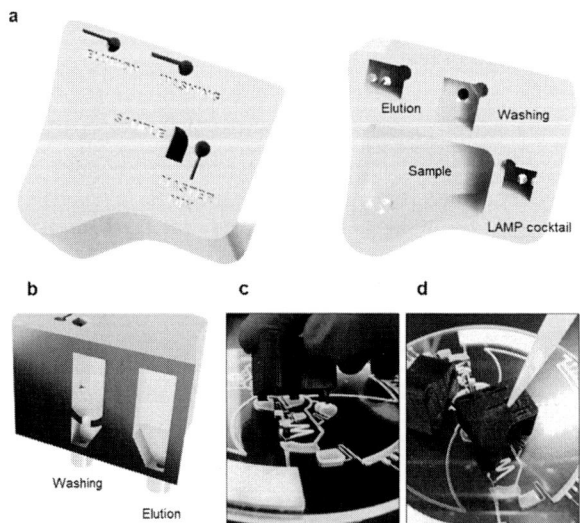

Fig. 1. (a) Design of the 3D printed solution loading cartridge with internal chambers for containing the sample, elution, washing, and LAMP cocktail solution (top and bottom view). (b) Design of the washing chamber with double rooms, and the elution chamber with a single room. (c) Attachment of the sample loading cartridge on the centrifugal microdevice. (d) Injection of the solution into the cartridge through the injection hole. (Reprinted with permission from Ref. [37]. Copyright (2019) Elsevier.)

indicating the practicality and scalability of these designs for broader applications.

Nguyen et al. engineered a solution-loading cartridge for their centrifugal microsystem using 3D printing technology [37]. This cartridge was specifically designed to store the sample solution, washing solution, elution solution, and the LAMP cocktail. Each compartment within the cartridge was allocated a specific volume, with 1 mL for the sample and 50 µL each for the other solutions (Fig. 1). The cartridge was designed using Fusion 360 software and printed using a DLP 3D printer, using ABS-like resin. Its compact dimensions (3.0 cm × 2.6 cm × 1.7 cm) allowed for easy integration with the microdevice. The design ensured that each solution could be ejected separately at different RPMs, optimizing the process based on the fluid dynamics within the microfluidic system.

The solution-loading cartridge developed by Oh et al. aimed to boost the sample handling capacity of the microdevice and automate the operation process [38]. Their design stored solutions like a sample lysis buffer, a washing solution, an elution solution, and a LAMP cocktail/EBT solution. The cartridges, manufactured using FDM 3D printing technology, were made of poly-lactic acid (PLA) with dimensions of 3.2 cm × 2.0 cm × 1.0 cm. A critical feature of this design was the way the cartridge could release solutions in an ordered manner

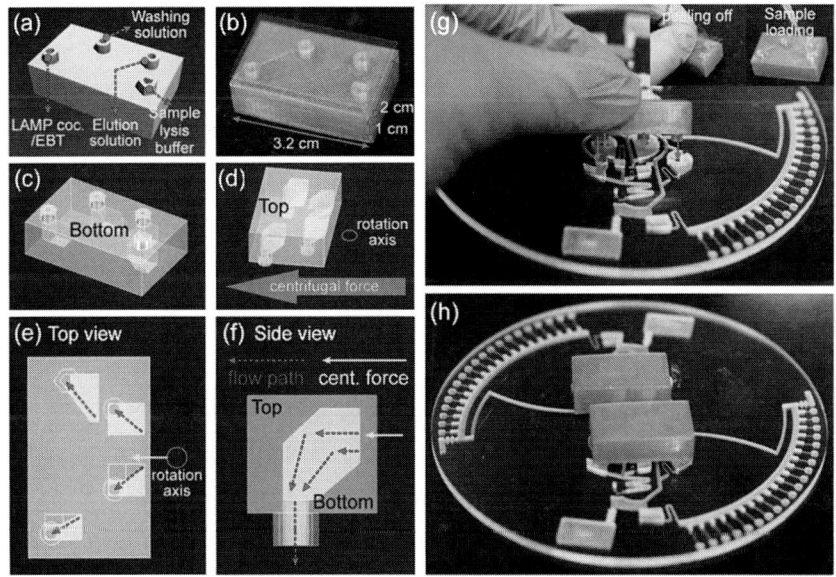

Fig. 2. Design, principle and usage of solution-loading cartridge in combination with the microdevice. (a) An image of the solution loading cartridge that stores a sample lysis buffer, a washing solution, an elution solution and a LAMP cocktail/EBT solution. (b) A photograph of the sealed cartridge before use. (c) A perspective view of the cartridge. (d) An image of the cartridge when it is used on the microdevice. (e) A top view of the cartridge and the flow path of the solutions. (f) A side view of the cartridge and the flow path of the solutions. (g) The instructions for the cartridge: i) peeling off the tape; ii) sample injection; and iii) link on the microdevice. (h) An assembled microdevice. (Reprinted with permission from Ref. [38]. Copyright (2019) Royal Society of Chemistry.)

under centrifugal force when combined with the microdevice (Fig. 2). The cartridge also included a unique structure with four spouts aligning with the injection holes of the microdevice. The cartridges were sealed with adhesive before use to prevent spillage or evaporation.

For mass production of their solution-loading cartridge, Nguyen et al. employed a hot injection molding technique [39]. Due to the internal complexity of having four chambers, the cartridge was divided into upper and lower parts (Fig. 3). These parts were made through metal molding and then bonded using ultrasonic welding and gluing. This approach highlighted the scalability and practicality of producing such devices for broader applications.

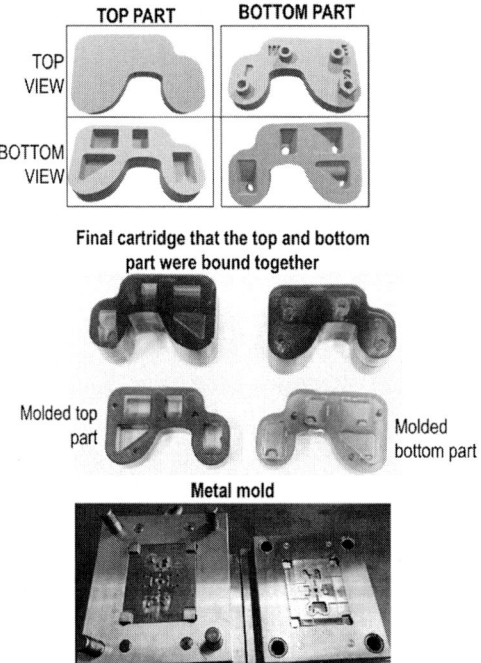

Fig. 3. Illustration of the mold cartridge. (Reprinted with permission from Ref. [39]. Copyright (2021) Elsevier.)

Phan et al. designed their solution-loading cartridge using Fusion 360 and fabricated it via the stereolithography method [40]. This cartridge was composed of four chambers, each with a capacity of 170 μL, and included a separator wall to prevent overflow during solution injection (Fig. 6C). Post-fabrication, the cartridge underwent a cleaning process involving isopropanol and sonication, followed by UV irradiation for solidification and sterilization. The cartridge was then attached to the centrifugal microfluidic device using double-sided adhesive tape and secured with a metal lid (Fig. 6F). Each chamber was equipped with a connection tube at the bottom for seamless integration with the microdevice, and a 3D printed lid was added to prevent backflow.

Together, these designs not only share common principles of automation and fluid control but also exhibit unique features that cater to the specific diagnostic requirements of varied applications. These solution-loading cartridges, with their ability to automate and simplify complex diagnostic tasks, are emblematic of the advancements in the field and highlight the potential for further innovation in centrifugal microfluidic technologies.

2.2. Design and fabrication of the integrated centrifugal microdevice

In the evolving landscape of microfluidic technology for molecular diagnostics, particularly for POCT, the design and fabrication of centrifugal microfluidic chips stand out for their complexity and precision. These chips, as detailed in the studies by Nguyen et al., Oh et al., Nguyen et al., and Phan et al., demonstrate a remarkable integration of microengineering and microfluidic principles, tailored to address specific diagnostic needs [37–40]. Across these designs, a few common themes emerge. Firstly, the use of poly(methylmethacrylate) (PMMA) as a material choice is consistent, underscoring its suitability for biomedical applications due to its durability and compatibility with biological samples. Secondly, each design features innovative mechanisms for fluid control, such as delay chambers and siphon valves, which are crucial for ensuring precision and efficiency in diagnostics. Thirdly, the integration of these chips with solution-loading cartridges is a key aspect, highlighting the synergy between different components of the diagnostic system. Moreover, these designs illustrate the importance of specific areas dedicated to sample preparation and analysis, such as areas for DNA extraction, sample washing, and reaction chambers. This compartmentalization within the chips ensures that each step of the diagnostic process is conducted efficiently and accurately.

The centrifugal microfluidic chip designed by Nguyen et al. represents a significant innovation in the field of molecular diagnostics, particularly for POCT [37]. Their design demonstrates a thoughtful integration of intricate microengineering with functional practicality, aiming to facilitate rapid and accurate pathogen detection. In their design, Nguyen et al. conceptualized a centrifugal disc with a 13 cm diameter, ingeniously structured to incorporate two distinct units on a single platform. This dual-functionality was crucial, as it allowed for simultaneous solid-phase DNA extraction and multiplex LAMP assay. The chip included four specifically designed reservoirs, each designated for a particular type of solution—sample, washing, elution, and LAMP cocktail (Fig. 4). These reservoirs were intricately connected to a 3D-printed solution-loading cartridge, ensuring seamless integration and fluid transfer for the diagnostic process. A key feature of this design was the bead bed channel, connected to the sample injection hole and packed with acid-treated silica beads in a zigzag pattern. This arrangement was pivotal in enhancing the DNA extraction process. Additionally, the chip incorporated a delay chamber strategically placed after the washing solution reservoir. This chamber played a role in managing the flow and timing of the solution release. Another innovative aspect of the design was the waste chamber, integrated with super absorbent polymer (SAP). This chamber absorbed the sample and washing solutions, an important step in maintaining the integrity and cleanliness of the assay process. The device also featured a collection chamber where the LAMP cocktail and elution solution with purified genomic DNA were gathered, and then aliquoted into reaction chambers for the assay.

The fabrication process of the chip reflected Nguyen et al.'s commitment to precision engineering and adaptability to biomedical needs. The chip was etched from a 3.0 mm thick PMMA plate using CNC machining, a technique chosen for its precision and compatibility with medical applications. All siphon channels on the chip were coated with a hydrophobic reagent, Vistex 111-50, to facilitate efficient fluid movement, and the reaction chambers were primed with specific primer sets targeting the bacteria of interest. The integration of SAP from baby diapers into the waste chamber to absorb up to 1 mL of sample solution showcased an inventive approach to material utilization (Fig. 4C). Finally, the chip was sealed with a pressure-sensitive adhesive (PSA) foil layer, and the acid-washed glass beads were packed into the zigzag channel and treated to enhance DNA capture capacity.

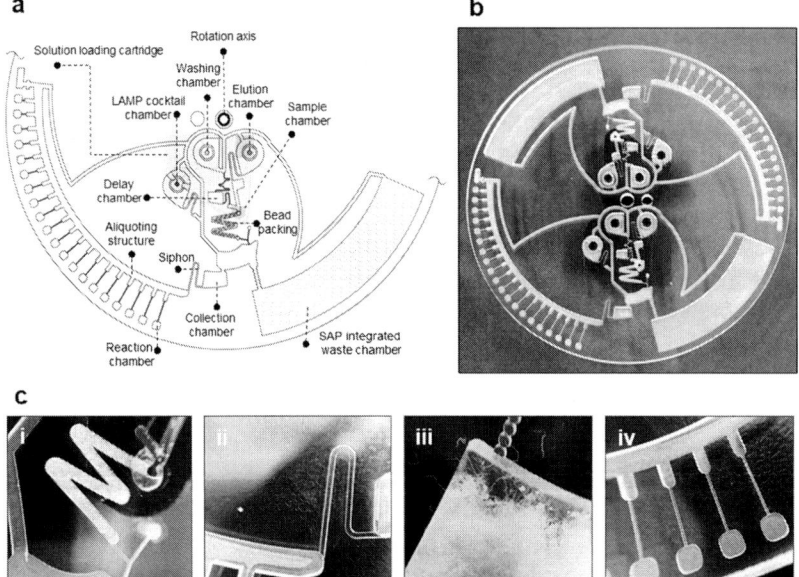

Fig. 4. (a) Schematic illustration of the integrated centrifugal disc. (b) A digital image of the disc. (c) Components of the centrifugal microdevice. (i) A glass bead-packed microchannel for DNA extraction, (ii) a siphon channel coated with Vistex, (iii) a SAP incorporated waste chamber, and (iv) aliquoting structure and the LAMP reaction chambers. (Reprinted with permission from Ref. [37]. Copyright (2019) Elsevier.)

In Oh et al. innovative design, the device was structured to facilitate the automated handling of various solutions essential for pathogen detection (Fig. 5) [38]. The inclusion of four injection holes was a critical aspect of this design. These holes were meticulously aligned with the corresponding solutions from the solution-loading cartridge, ensuring a streamlined and automated transfer of solutions into the microdevice. The device featured a bead-packed channel, which included a unique weir structure at the end (Fig. 5A). This structure was specifically designed for the efficient packing of silica microbeads, essential for DNA extraction and purification. The microbeads were introduced into the channel via vacuum-suction through the sample injection hole [43]. This method of bead introduction was not only efficient but also allowed for long-term storage of the beads on the device. One of the standout parameters in their design was the delay chamber, strategically placed between the washing solution injection hole and the bead-packed channel. The exact dimensions and volume of this delay chamber were tailored to ensure the proper timing and sequencing of the washing solution injection into the bead-packed channel. The precise control of fluid flow and timing within this chamber was essential for the accuracy and reliability of the diagnostic process. Additionally, the device incorporated siphon valves, which were crucial for the sequential release of the LAMP cocktail/EBT and the elution solution. The dimensions and design of these siphon valves were specifically optimized to control the flow of solutions within the microdevice. The waste chamber and the collection chamber were other significant components of the device. The waste chamber was designed to store waste materials such as sample solution, cell debris, and washing solution, while the collection chamber received the LAMP cocktail/EBT and the eluted DNA solution. The exact volumes of these chambers were important to ensure that they could accommodate the necessary amounts of solutions and waste materials without spillage or overflow.

In the fabrication process, Oh et al.'s device was crafted from a PMMA sheet using a CNC milling machine, demonstrating the precision of this fabrication technique. The thickness of the PMMA sheet, typically around 3.0 mm, was chosen to provide the necessary durability and stability for the device (Fig. 5B). The hydrophilic coating on the siphon valves, a key feature for enhancing fluid movement, was applied in a controlled manner to ensure uniform coverage and effectiveness.

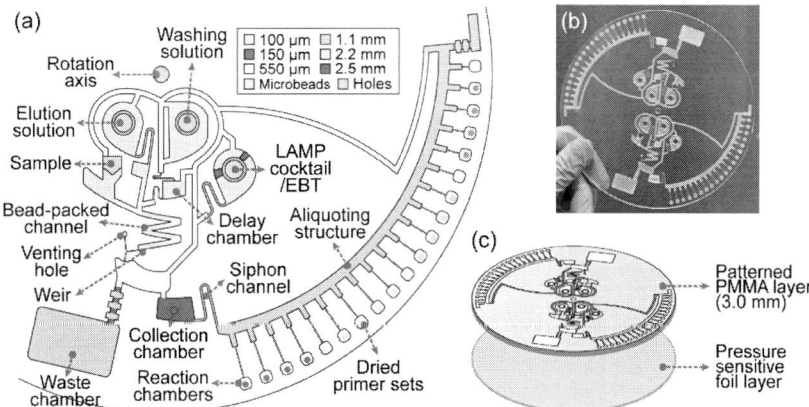

Fig. 5. (a) Design and components of the microdevice: i) Four holes (gray color) for loading of a sample, a washing solution, an elution solution and a LAMP cocktail/EBT; ii) a bead-packed channel for DNA extraction filled with silica micro-beads; iii) a delay chamber, a waste chamber and siphon channels for sequential fluid manipulation; iv) a collection chamber, an aliquoting structure and 20 reaction chambers for multiplexed LAMP. (b) A real photograph of the microdevice. (c) Assembly of a patterned PMMA layer and a pressure sensitive foil layer to form an integrated microdevice. (Reprinted with permission from Ref. [38]. Copyright (2019) Royal Society of Chemistry.)

The centrifugal microfluidic device developed by Nguyen et al. and Phan et al. is similar for efficient and simultaneous nucleic acid (NA) purification and multiplex reaction. Nguyen et al.'s chip is a centrifugal disc with a diameter of 13 cm, ingeniously segmented into two operational units [39]. Each of these units is dedicated to specific functions: NA purification and multiplex reaction for RT-LAMP or RT-PCR. This dual-functionality approach facilitates the simultaneous processing of different samples or targets, an essential feature for versatile and comprehensive diagnostics. The micropattern of the disc, crafted with precision using Cut2D software, reflects a design that integrates various critical operational units. These include siphon valves, passive valves, transfer chambers, a metering structure, and solution reservoirs. Each component plays a crucial role in the device's functionality, ensuring efficient fluid movement and accurate processing of the samples [14, 38, 44, 45]. Phan et al. incorporated a

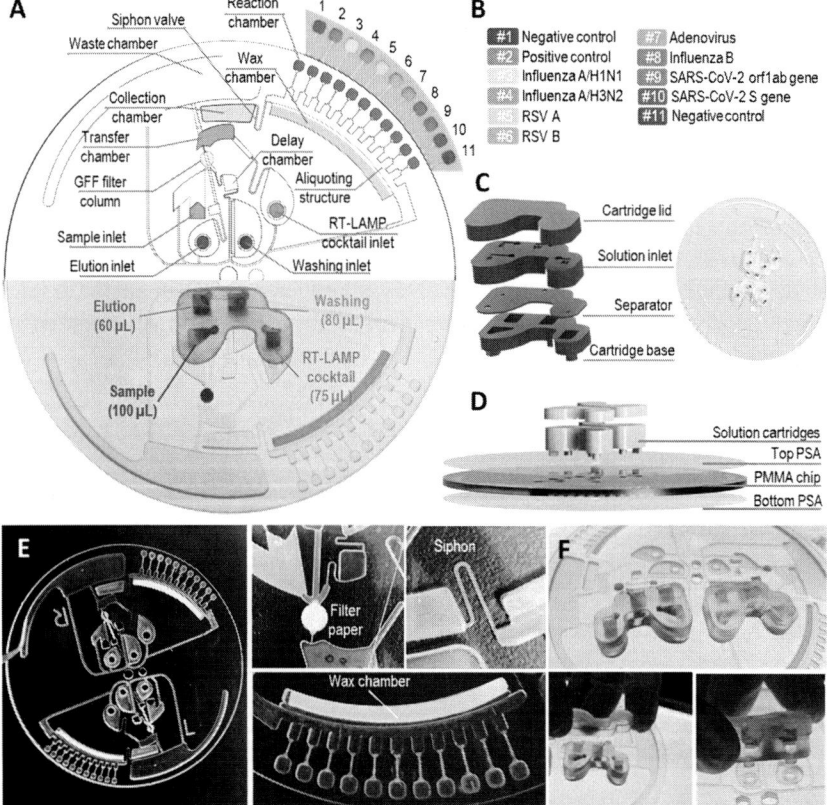

Fig. 6. The proposed centrifugal microfluidic device and the solution loading cartridge. (A) Schematics (top half of a device) and the digital image for the solution loading cartridge combined centrifugal microfluidic device (bottom half of a device). (B) Position order of the primers on the 11 reaction chambers. (C) Design of the solution loading cartridge, consisting of a lid, a solution inlet layer, a thin separator, and a base. (D) Assembly of the centrifugal microfluidic device with the cartridge. (E) Digital images for the whole chip, the filter column part, a siphon valve and a wax chamber. (F) Digital images for the cartridge attachment on a chip. (Reprinted with permission from Ref. [40]. Copyright (2023) Elsevier.)

highly efficient feature that significantly enhances its diagnostic capabilities: the inclusion of 11 reaction chambers capable of simultaneously detecting 11 different targets on a single unit (Fig. 6A) [40]. This device exemplifies the integration of sophisticated design with practical functionality, tailored to meet the challenges of diagnosing respiratory illnesses.

Fabrication of the device involved the use of a 3 mm thick PMMA sheet, processed using a conventional CNC machine. This method allowed for the creation of microfluidic structures on both the bottom and top surfaces of the disc. The double-sided etching technique employed in the fabrication process created a comprehensive and integrated microfluidic network. Key functional components of the device include siphon valves, which were coated with a 6% Vistex 111-50 reagent in 33 % isopropanol. This coating was critical for controlling the fluid flow within the device, ensuring that the reactions occurred under optimal conditions. After the coating was applied, the device was dried at 80 °C for 30 minutes to cure the layer, ensuring its durability and effectiveness. The extraction column included GF/F glass filter paper, shaped into a circular form, facilitating efficient NA extraction. Additionally, a paraffin wax chamber was incorporated to actively seal the reaction chambers during amplification processes, thus preventing evaporation and contamination. The reaction chambers were meticulously coated with specific primer mixes, tailored for either RT-LAMP or RT-PCR assays. The device was sealed using PSA films, and specific chambers were designated as negative controls (Fig. 6D). This arrangement was integral to the device's functionality, allowing for the simultaneous investigation of multiple targets on one side of the disc.

Collectively, these integrated centrifugal microdevices exhibit a remarkable level of design sophistication, combining programmable solution loading, efficient DNA/RNA extraction, RT-LAMP reaction, and data display into compact, high-performance systems. The meticulous detailing extends to the microfabrication of the devices, where siphon valves are treated with anti-fog reagents, and primer mixes are pipetted with exactitude into reaction chambers for precise gene amplification. These systems prove to be highly efficient and portable, capable of providing rapid, real-time diagnostic results, a feature that proves invaluable, especially in the face of pandemic threats.

3. Development of a portable genetic analyzer

In the dynamic landscape of molecular diagnostics, the evolution of technology has led to the development of sophisticated devices that are POCT. Four notable papers, each presenting a breakthrough in this field, demonstrate the breadth and depth of innovation currently shaping molecular diagnostics. These studies, while distinct in their focus and outcomes, collectively represent the cutting edge of diagnostic technology, addressing critical needs in healthcare and beyond. Each of these studies not only contributes to the field of molecular diagnostics through innovative designs and practical applications but also sets a precedent for future advancements. They collectively represent a shift towards more accessible, efficient, and accurate diagnostic methods, crucial in both clinical and non-clinical settings.

Nguyen et al. introduced a significant innovation in the field of molecular diagnostics with the development of a portable genetic analyzer designed for POC DNA testing [37]. This device represents a major advancement in the accessibility and efficiency of DNA testing outside traditional laboratory environments, marking a pivotal shift in the realm of molecular diagnostics. The portable genetic analyzer is a compact and sophisticated tool, specifically designed to operate in conjunction with an integrated centrifugal microdevice. This synergy between the analyzer and the microdevice is critical for enhancing the capabilities of DNA testing in various settings, ranging from fieldwork to clinical environments. The compact design of the analyzer, with dimensions of 20 cm in length, 22 cm in width, and 20 cm in height, underscores its portability, making it an ideal solution for settings where conventional laboratory equipment is impractical (Fig. 7). Key components of the analyzer include a spindle motor, which is essential for the precise rotation of the centrifugal microdevice. This controlled rotation is crucial for effective DNA sample processing. The analyzer also features a couple of Minco heaters, providing accurate temperature control necessary for DNA amplification processes. The ramping rate of these heaters, at 0.43 °C/s for heating and 0.09 °C/s for natural cooling, ensures rapid and precise thermal cycling, which is vital for efficient molecular diagnostics.

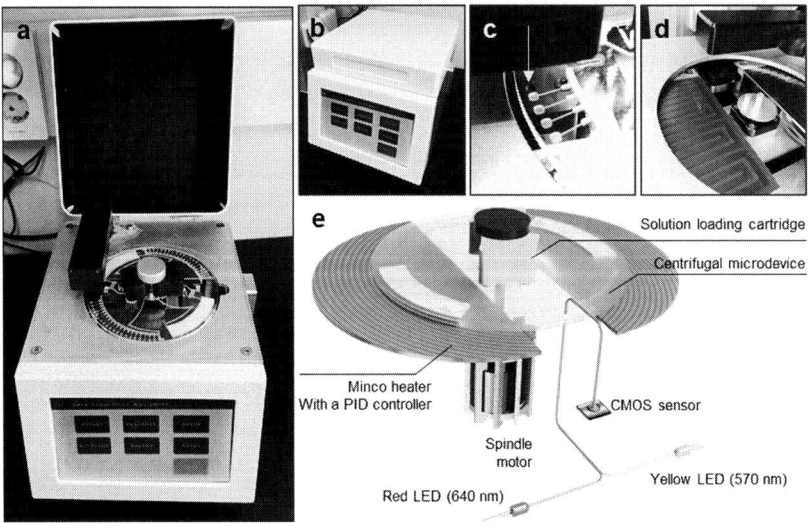

Fig. 7. (a, b) Digital images of the portable genetic analyzer. (c) A UV-vis detector for measuring the absorbance of the reaction chamber. (d) A couple of Minco heaters. (e) Schematic illustration of the integrated genetic analyzer platform with a motor, a couple of Minco heaters, and a UV-vis detector with two LED light sources at 640 and 570 nm. (Reprinted with permission from Ref. [37]. Copyright (2019) Elsevier.)

For UV-visible absorbance detection, the analyzer employs a yellow LED (570 nm) and a red LED (650 nm). These LEDs are strategically positioned to direct light toward the LAMP reaction chamber via an optical fiber, a setup that is critical for accurate DNA amplification measurement. The system's optical precision is further enhanced by the inclusion of a filter and an aspheric lens, used to eliminate interference from excited light and reduce optical aberrations. The analyzer's optical readings are captured by a CMOS camera sensor, which converts the transmitted light intensity into an absorbance ratio of Abs_{640}/Abs_{570}. This conversion is key to quantifying DNA amplification in the samples. Furthermore, the front side of the analyzer features an LCD panel with touchscreen functionality, allowing users to easily input experimental protocols. This user-friendly interface makes the device accessible to operators with varying levels of technical expertise in molecular diagnostics.

In their study, Oh et al. developed a centrifugal microfluidic device, operated by a customized rotary system, marking a significant advancement in the field of molecular diagnostics for POCT [38]. This system plays a crucial role in automating and enhancing the efficiency of diagnostics, particularly for foodborne pathogens. Central to the functionality of their microfluidic device is the rotary system, which incorporates a servo motor for precise device rotation [23]. This controlled rotation is essential for effective sample processing within the microfluidic device. Alongside the servo motor, the system includes aluminum heating blocks designed specifically for LAMP. These heating blocks provide necessary temperature control for DNA amplification processes, ensuring the accuracy and efficiency of molecular diagnostics. A unique feature of the rotary system is the integration of a CCD camera combined with a stroboscope; a setup employed for monitoring fluidic movement within the device. This real-time monitoring is vital for verifying the reliability and accuracy of the diagnostic process, allowing for adjustments and ensuring consistent results. The operation of the rotary system, including device rotation and temperature control, is managed by optimized software, which ensures precise and sequential control of these crucial parameters. This level of precision is particularly important in the complex processes involved in molecular diagnostics. One of the significant advantages of Oh et al.'s centrifugal system over traditional pressure-driven microfluidic systems is the elimination of the need for external pumps or active valves. This simplification in design and operation not only makes the system more user-friendly but also enhances its adaptability for POCT applications. The miniaturized and streamlined nature of this rotary system augments the potential of the centrifugal microfluidic device as a powerful diagnostic tool, particularly suitable for diverse environments, including those that are remote, or resource limited.

Similarly, Nguyen et al. and Phan et al. made a significant advancement in the field of POCT with the development of a portable genetic analyzer (Fig. 8) [39, 40]. This device is a testament to the growing sophistication in molecular diagnostics, particularly in creating accessible and rapid genetic analysis tools. The design of the genetic analyzer prioritizes compactness and user-friendliness. The device features a chip mounting base similar to a CD hard drive, which is secured with an M3 thread screw, ensuring stability and ease of use during operations. Additionally, the device's dimensions, measuring 28 cm in width, 26 cm in length, and 28 cm in height, and weighing approximately 10 kg, make it highly portable. This portability is a crucial attribute for POCT applications, allowing the device to be used in various settings outside traditional laboratories.

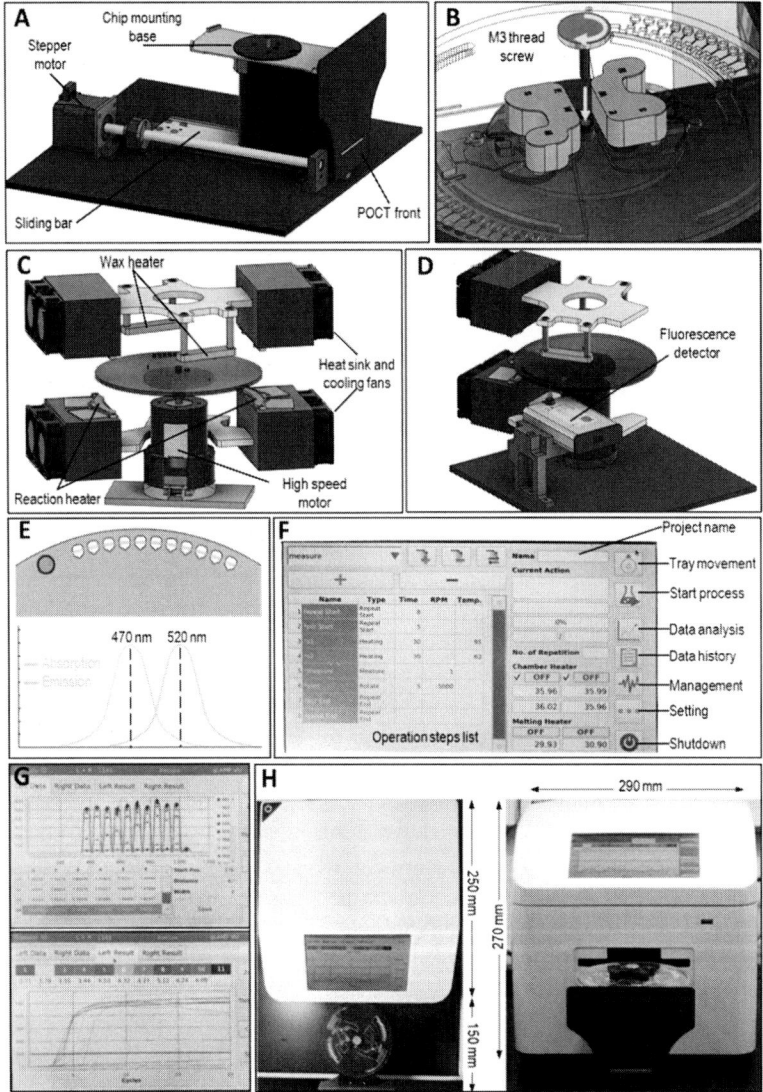

Fig. 8. The schematics and digital images of the portable POC genetic analyzer. (A) A chip loading part composed of a chip mounting base and a stepper motor. (B) A M3 thread screw to secure the chip to the chip mounting base. (C) Two wax melting heaters and two sets of the reaction heaters for the RT-LAMP reaction. (D) A miniaturized fluorescence detector positioned underneath of a chip. (E) Scanning mode of the optics with the excitation/emission wavelength of 470/520 nm. (F) Graphic user interface of the POC genetic analyzer. (G) Data analysis window for the raw data (top) and the amplification curve (bottom) on the touch screen. (H) The real images and the dimension of the POC genetic analyzer. (Reprinted with permission from Ref. [40]. Copyright (2023) Elsevier.)

In terms of functionality, the analyzer is equipped with six heaters—two for melting wax and four for heating the reaction chambers. This setup provides precise temperature control, essential for accurate molecular reactions in diagnostics. Notably, the device includes two up-and-down Peltier heaters, complete with heat sinks and cooling fans, designed for rapid thermal cycling. The inclusion of a stepper motor, connecting the top and bottom heating units, ensures that the heaters are in firm contact with the chip, thus maintaining consistent temperatures during diagnostic processes. Peltier heaters, a fluorescence detector, and a touchscreen interface for data input and analysis. These components facilitate various diagnostic processes such as RT-LAMP or RT-PCR, allowing for real-time detection and monitoring of pathogen presence. The analyzer also features a state-of-the-art Fluo Sens Integrated fluorescence detector. Capable of scanning up to 1500 points in a single measurement, the detector is tailored for detecting intercalating dyes such as SYTO 9, with specific excitation and emission wavelengths. This advanced optical detection capability is key to the analyzer's effectiveness in molecular diagnostics. User interaction with the device is facilitated through a touch screen interface, allowing for the easy creation of experiment protocols and access to multiple functions. The screen's data analysis tab displays real-time amplification data during optical detection in scan mode. This feature not only enhances the user experience but also provides critical information for accurate diagnostics. Users can select data points from peaks revealed from each reaction chamber to plot amplification curves, further adding to the device's analytical capabilities.

4. Chip operation

The operation protocols of the centrifugal microfluidic devices developed by Nguyen et al., Oh et al., Nguyen et al., and Phan et al. showcase an impressive array of similarities and sophisticated mechanisms, underscoring the advancements in molecular diagnostics [37–40]. These devices, while distinct in their specific applications, share core operational principles and utilize forces to achieve precise fluidic control, essential for accurate and efficient pathogen detection. Each of these devices leverages centrifugal force as the primary mechanism for fluid movement. By spinning the discs at high speeds, fluids are directed through microchannels, enabling the sequential processing of samples, reagents, and washing solutions. This force is controlled to ensure precise timing and routing of fluids within the devices. In addition to centrifugal force, these devices also employ Coriolis and Euler forces to enhance their functionality [46]. The Coriolis force, arising due to the rotation of the disc, is utilized to direct fluids along specific paths, aiding in the separation of samples and waste [45, 47]. The Euler force is employed for mixing solutions, crucial for ensuring uniform reaction conditions within the devices [48]. Capillary force plays a significant role in these devices, particularly in controlling the movement of fluids through narrow channels and valves [49]. This force is used to prime siphon valves, facilitating the release of solutions at the desired stages of the operation, thus ensuring that reagents and samples interact at optimal times for the reactions.

All four studies use customized rotary systems to control the spinning of the microfluidic discs. These systems allow for precise control over rotational speed, which is critical for managing the timing and direction of fluid flow. Additionally, temperature control, achieved through heaters and Peltier elements, is a crucial aspect of these devices, ensuring that the conditions for reactions like LAMP and PCR are consistently maintained. Despite their varied applications, from foodborne pathogen detection to respiratory virus diagnosis, these devices share a common operational approach. They start with the strategic placement of samples and reagents in different chambers or on a cartridge. The spinning protocol is then initiated, guiding the solutions through a series of steps involving washing, mixing, and eventual distribution into reaction chambers for analysis. The precision in this process is key to the devices' efficacy, ensuring that each step occurs sequentially and accurately for optimal diagnostic results.

Nguyen et al. employed a detailed operational protocol for their centrifugal microdevice, crucial for the accurate detection of foodborne pathogens [37]. The device's operation begins with the orderly flushing of the lysed sample and washing solution into a bead-packed channel at +3500 rpm, with the sample absorbed by SAP in the waste chamber (Fig. 9). Bacteria DNA is captured and washed on the surface of the glass beads. After solid phase extraction in the bead-packed channel [43, 50], the disc is stopped to initiate siphon priming of the elution solution and the LAMP cocktail, which is driven by capillary force. A speed of -3500 rpm is used to eject the purified DNA and LAMP cocktail/EBT solution into the collection chamber. To enhance the mixing of these solutions, the disc undergoes swirling motions. An additional siphon priming step follows, and then the disc is spun at -1000 rpm to separate the LAMP mixture into 20 aliquots. The final step involves spinning at -3500 rpm to transfer these aliquots into individual reaction chambers. For the LAMP reaction, the Minco heater maintains a temperature of 63 °C, with absorbance at 640 nm and 570 nm measured continuously over 60 minutes. The ratio of Abs_{640}/Abs_{570} is calculated, and a real-time curve is plotted to monitor the reaction.

Oh et al.'s microdevice operation had similar approach to Nguyen et al. Four essential solutions for molecular diagnostics are stored in a solution-loading cartridge and connected with four holes in the device [38]. The operational protocol involves rotating the microdevice counterclockwise at 5000 RPM for 60 seconds, allowing the sample lysate to flow through the bead-packed channel and bind genomic DNA to the microbeads. Following this, washing solution passes through the microbeads, purifying the bound genomic DNA. Siphon valves are then primed to allow the flow of the elution solution and the LAMP cocktail/EBT, gathering in the collection chamber. The device then undergoes a shake-mode mixing to homogenize the LAMP mixture [51]. The homogenized mixture is split into 20 aliquots and transferred into reaction chambers for pathogen detection. LAMP occurs at 63 °C for 1 hour, with results interpreted by colorimetric detection [43].

Nguyen et al. [39] utilize a centrifugal microfluidic disc, where fluidic control is mediated by forces such as centrifugal force, Euler force, Coriolis force, and capillary force [44, 52, 53]. The operation begins with the counter-clockwise rotation of the disc at +5000 rpm, releasing the sample and washing solution into the waste chamber (Fig. 10). The elution solution and master mix are then released in the right direction, gathering in the collection chamber. The solutions are mixed using a shaking mode and then distributed into aliquoting chambers.

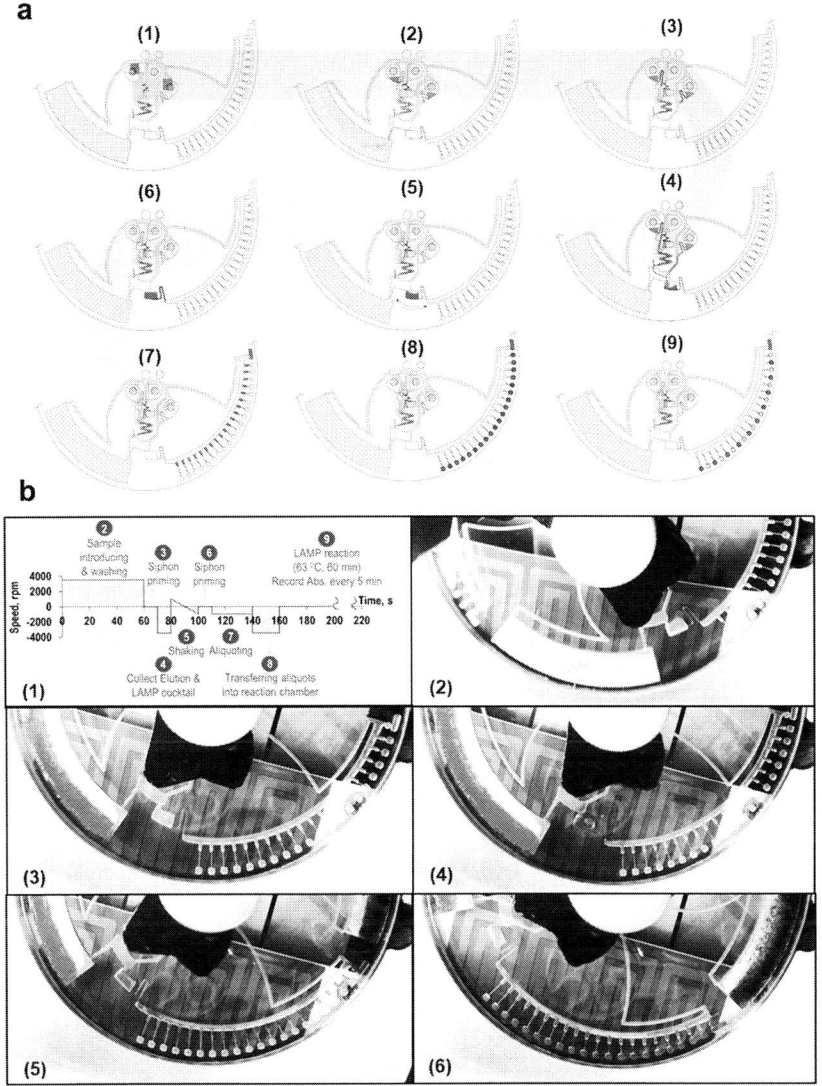

Fig. 9. (a) The overall schematics for the chip operation: (1) Injection of the solution into the solution cartridge. (2) Flushing of the lysed sample and the washing solution into the bead-packed channel at +3500 rpm. (3) Starting the siphon priming of the elution solution and the LAMP cocktail solution. (4) Ejecting the purified captured DNA with the elution solution and the LAMP cocktail/EBT solution into the collection chamber at −3500 rpm. (5) Mixing the LAMP mixture by swirling counter-clockwise and clockwise repeatedly. (6) Siphon priming of the LAMP mixture. (7) Rotation at −1000 rpm for separating the LAMP mixture into 20 aliquots. (8) Transferring the 20 aliquots into the individual reaction chamber at −3500 rpm. (9) Initiation of the LAMP reaction at 63 °C. (b) The rotation profile and representative real images of the chip operation. (Reprinted with permission from Ref. [37]. Copyright (2019) Elsevier.)

High RPM is required to transfer the solution from the aliquoting chamber to the reaction chamber [54]. The RT-LAMP or RT-PCR mixture is prevented from evaporating during the amplification reaction by filling the chambers with melted wax. The reaction proceeds at a controlled temperature, with fluorescence signals measured at regular intervals. Phan et al.'s centrifugal microfluidic device also operates with similar mechanism [40]. The device rotation begins at +5000 rpm, directing solutions through the glass filter column to the waste chamber. The elution buffer and RT-LAMP cocktail pass through the siphon valve when the device stops, then are released at -5000 rpm into the collection chamber. The device then undergoes shaking at ±1000 rpm to mix the eluted RNAs with the RT-LAMP cocktail. After siphon priming, the RT-LAMP mixture is divided into 11 chambers and transferred to the reaction chambers at −5000 RPM. The wax is then melted and filled into the aliquoting structure, and the RT-LAMP reaction is initiated at a constant temperature. The fluorescence signal from the reaction chambers is continuously monitored, providing insights into the amplification of target genes.

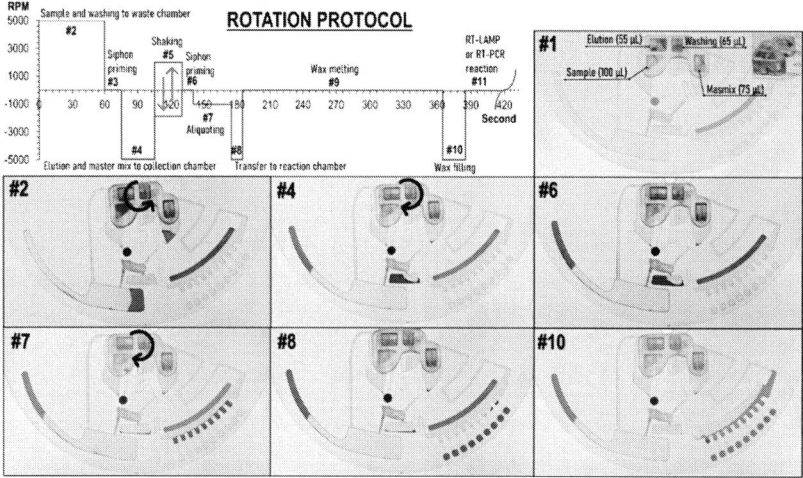

Fig. 10. The digital images of the disc at each operation step, and the rotational protocol. (Reprinted with permission from Ref. [39]. Copyright (2021) Elsevier.)

5. Application of multiplex pathogen detection

The results obtained from the studies of Nguyen et al., Oh et al., Nguyen et al. and Phan et al. demonstrate the effectiveness and precision of their respective centrifugal microfluidic devices and portable genetic analyzers in molecular diagnostics. These studies successfully employed advanced microfluidic technology for the detection of various pathogens, showcasing their potential in POCT [37–40].

In their 2019 study, Nguyen et al. demonstrated the effectiveness of their portable genetic analyzer and centrifugal microfluidic device for detecting foodborne pathogens [37]. Their approach was characterized by the use of UV-vis absorption measurements and the execution of rigorous limit of detection (LOD) tests. A key aspect of Nguyen et al.'s method involved utilizing UV-vis absorption spectroscopy to validate the success of the LAMP reactions. Initially, the LAMP mixture appeared violet due to the presence of the EBT and Mg^{2+} complex. Upon successful completion of the LAMP reaction, this complex was disrupted by the binding of pyrophosphate ions ($P_2O_7^{4-}$) to the Mg ions, leading to the formation of EBT− ions and causing a color change to sky blue. This colorimetric shift was a crucial indicator of the reaction's success [43, 55]. Correspondingly, the absorption peak shifted from 570 nm to 640 nm before and after the LAMP reaction. The device's UV-vis detector, equipped with two LED light sources at these wavelengths, was designed to calculate and display the relative absorbance intensities (Abs_{640}/Abs_{570}) on the touchscreen, providing a clear and quantifiable measure of the reaction outcome.

The LOD tests conducted by Nguyen et al. were pivotal in establishing the device's sensitivity and specificity. Focusing on the detection of *E. coli* O157:H7, they preloaded the device's 20 reaction chambers with specific primer sets for this pathogen. The tests involved using cultured *E. coli* O157:H7 cells at concentrations ranging from 10^2 to 10^5 cells/mL. The sensitivity of the device was evident as even at the lowest concentration of 10^2 cells/mL, the reaction chambers targeting *E. coli* O157:H7 displayed a color change to sky blue (Fig. 11A, B and C). This change indicated the high specificity of the primers and the absence of cross-contamination between chambers. To bolster these visual results, the portable genetic analyzer performed real-time UV-vis absorption measurements for each chamber, producing Abs_{640}/Abs_{570} ratios. Chambers showing a ratio higher than 1 were indicative of positive reactions, correlating with the observed color change.

Additionally, Nguyen et al. demonstrated the robust capabilities of their centrifugal microfluidic device for monoplex and multiplex analysis (Fig. 11D and E). This part of their study highlights the device's potential for comprehensive pathogen screening, crucial in food safety and public health applications. The device's design enabled parallel processing of two samples simultaneously, each capable of undergoing 20 multiplex reactions. This innovative approach allowed for the potential simultaneous monitoring of up to 20 different foodborne pathogens per sample. In their multiplex experiment, Nguyen et al. targeted three bacteria as a proof of concept: *E. coli* O157:H7, *Salmonella Typhimurium*, and *Vibrio parahaemolyticus*. The experiment was conducted with a sample solution containing 10^5 cells/mL of each of these pathogens. A key aspect of the analysis was the detection of color changes in the reaction chambers. Chambers with air-dried primers specifically targeting the *fliC* gene of *E. coli* O157:H7, the *invA* gene of *S. Typhimurium*, and the *toxR* gene of *V. parahaemolyticus* exhibited a color shift from purple to sky blue, indicative of positive reactions. This change provided a visual confirmation of the presence of these bacteria. Importantly, no color change was observed in the negative control chambers, underscoring the specificity of the reactions and the absence of cross-contamination. The results were further validated through the measurement of the absorbance ratio Abs_{640}/Abs_{570}. Chambers showing a positive reaction (with a color change to sky blue) displayed an Abs_{640}/Abs_{570} ratio higher than 1.0, while those with negative results (no color change) had ratios below 1. This method of using the absorbance ratio as determined by the optical sensor based on Abs_{640}/Abs_{570} measurements allowed for the accurate and simultaneous detection of multiple pathogens.

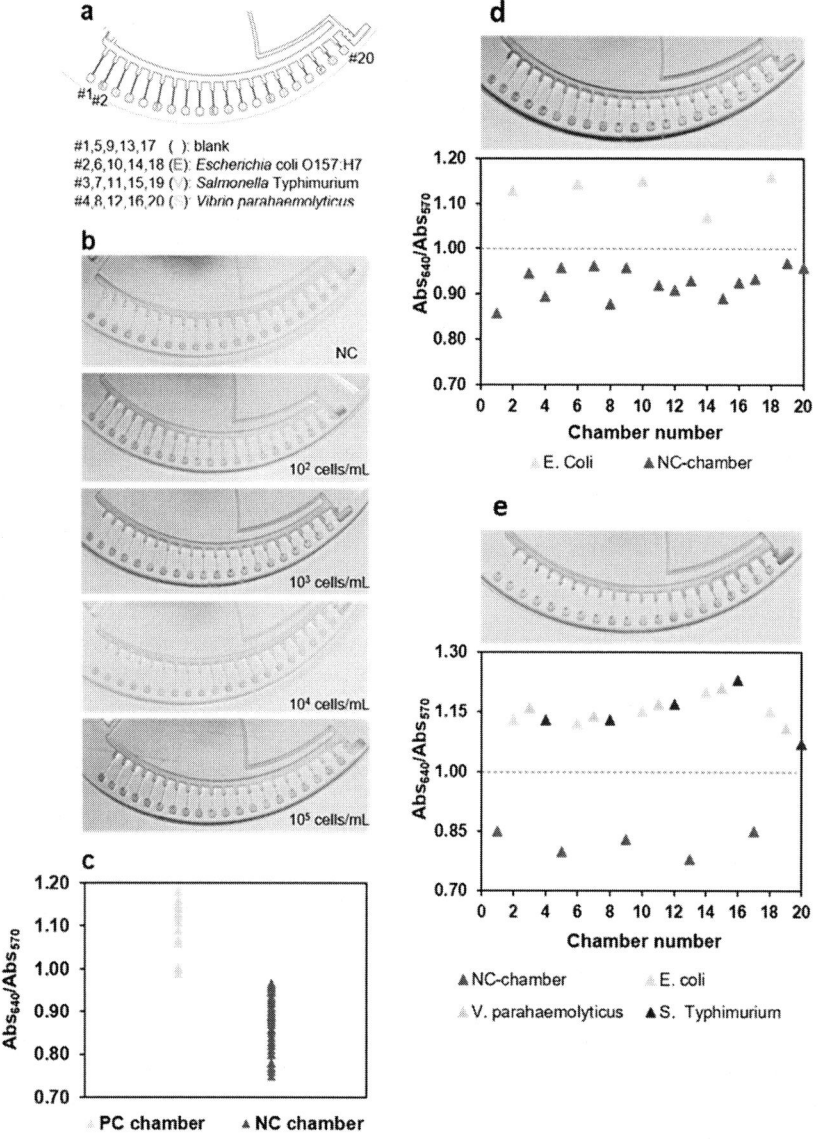

Fig. 11. (a) 20 reaction chambers were numbered from left to right, and specific primer sets were coated in each reaction chamber. (b) Digital images of the reaction chambers after amplifying *fliC* gene of *E. coli* O157:H7 with concentrations ranging from 10^2 to 10^5 cells/mL. (c) Analysis of the Abs_{640}/Abs_{570} value for the positive and negative chambers. (d) (Top) Colorimetric detection for identifying single pathogen (*E. coli* O157:H7) and (Bottom) the graph of the Abs_{640}/Abs_{570} ratio of the 15 negative chambers and 5 positive chambers on the portable genetic analyzer. (e) (Top) Colorimetric detection for identifying three pathogens (*E. coli* O157:H7, *S. Typhimurium* and *V. parahaemolyticus*) and (Bottom) the graph of the Abs_{640}/Abs_{570} ratio of the 5 negative chambers and 15 positive chambers on the portable genetic analyzer. (Reprinted with permission from Ref. [37]. Copyright (2019) Elsevier.

In their study, Oh et al. demonstrated the efficacy of their microfluidic device, equipped with a solution-loading cartridge, for the detection of foodborne pathogens [38]. They conducted detailed experiments on both singleplex and multiplex detection, as well as limit of detection (LOD) tests, showcasing the device's versatility and precision in molecular diagnostics.

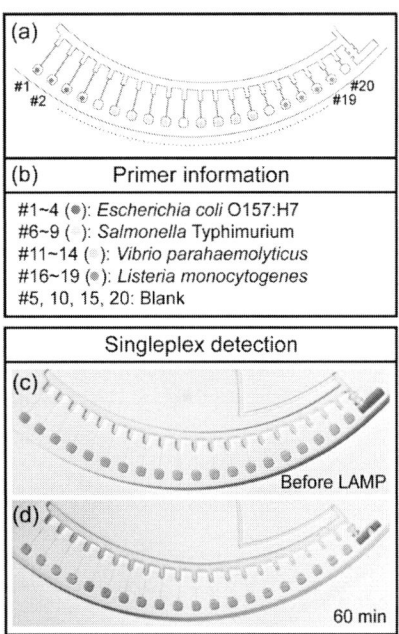

Fig. 12. Singleplex detection of *E. coli* O157:H7 in real samples on the microdevice. (a) 20 reaction chambers were numbered as #1 to #20. (b) Four kinds of LAMP primer sets were pre-stored in the reaction chambers. (c) Before LAMP, the color of the LAMP mixtures was purple due to EBT. (d) After LAMP for 60 minutes, the color of the LAMP mixtures in reaction chamber #1~#4 turned into sky blue because of the amplification of target gene *E. coli* O157:H7 DNA. (Reprinted with permission from Ref. [38]. Copyright (2019) Royal Society of Chemistry.)

The focus of the singleplex detection was *E. coli* O157:H7 in real milk samples. A milk sample contaminated with *E. coli* O157:H7 at a concentration of 4×10^3 cells/µL was prepared, introducing a total of 10^5 cells into the solution-loading cartridge mixed with a lysis buffer. After bacterial cell lysis within the cartridge, it was inserted into the microdevice, which then underwent automatic operation based on a predefined rotational speed control protocol. Post-LAMP reaction, a distinct color change was observed in the reaction chambers #1 to #4, shifting from purple to sky blue (Fig. 12). This color change indicated the successful amplification of the gene *of E. coli* O157:H7, corresponding to the specific primers in these chambers. Importantly, no color change was noted in the other reaction chambers (#5 to #20), affirming the high specificity of the primers and ruling out cross-reactivity among different pathogens. Oh et al. also extended their analysis to multiplex detection, leveraging the capacity of their microdevice to incorporate 20 reaction chambers. This design theoretically allows for the simultaneous detection of up to 18 different foodborne pathogens, excluding controls. In their experiment, real milk samples spiked with 10^5 cells each of *E. coli* O157:H7 and *Salmonella Typhimurium* were analyzed. The results showed color changes in chambers designated for these two bacteria (#1 to #4 for *E. coli* O157:H7 and #6 to #9 for *Salmonella Typhimurium*), confirming their presence in the sample. No color change in other chambers underscored the specificity of the detection, and the device successfully demonstrated its capability to identify multiple pathogens in a single test. Expanding the detection scope, the device was then used to analyze samples spiked with *E. coli* O157:H7, *S. Typhimurium*, and *V. parahaemolyticus*. Again, the device accurately identified all three pathogens, as evidenced by the color changes in the respective chambers. In a more complex setup, milk samples were spiked with four pathogens: *E. coli* O157:H7, *S. Typhimurium*, *V. parahaemolyticus*, and *L. monocytogenes*. The device successfully detected all four, with the correct chambers displaying the indicative color changes (Fig. 13). These results showcased the device's exceptional capability for comprehensive multiplex testing, essential for broad-spectrum pathogen detection and food safety analysis.

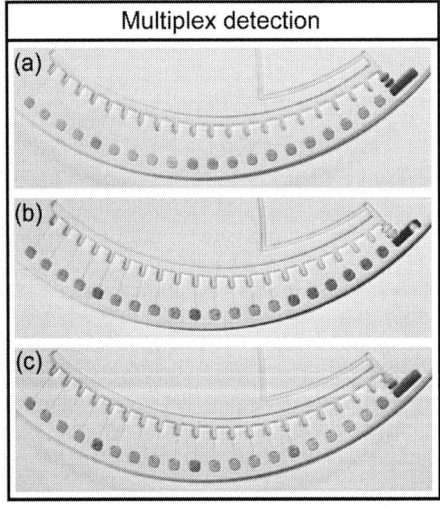

Fig. 13. Multiplex detection of foodborne pathogens contaminated in real samples on the microdevice. (a) Colorimetric detection of two pathogens *(E. coli* O157:H7 and *S. typhimurium)*. (b) Colorimetric detection of three pathogens (*E. coli* O157:H7, *S. typhimurium* and *V. parahaemolyticus*). (c) Colorimetric detection of four pathogens (*E. coli* O157:H7, *S. typhimurium*, *V. parahaemolyticus* and *L. monocytogenes*). (Reprinted with permission from Ref. [38]. Copyright (2019) Royal Society of Chemistry.)

The LOD test conducted by Oh et al. was crucial in evaluating the sensitivity of their microfluidic device. They prepared milk samples with varying concentrations of pathogenic bacteria, including *E. coli* O157:H7, *S. Typhimurium*, *V. parahaemolyticus*, and *L. monocytogenes*. The device could detect these pathogens even at a low cell concentration of 10^2 per reaction, equivalent to about 4.5 cells per chamber, considering the aliquoting process (Fig. 14). The color change in specific reaction chambers validated the presence of the respective pathogens, while the absence of color change in the negative control confirmed the absence of contamination. The LOD of 100 cells, corresponding to 4×10^3 cells/mL, was competitive compared to other devices, showcasing the device's practical applicability in molecular diagnostics [43, 56].

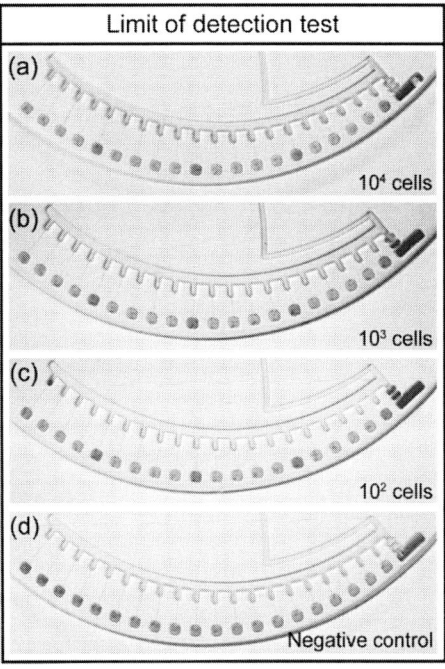

Fig. 14. Limit of detection test of the microsystem. (a) Multiplex detection of four pathogens with the input cell number of 10^4. (b) Multiplex detection of four pathogens with the input cell number of 10^3. (c) Multiplex detection of four pathogens with the input cell number of 10^2. (d) Result of a negative control experiment.

Nguyen et al. presented extensive results from their advanced study using a centrifugal microfluidic device and a portable genetic analyzer, offering a comprehensive look into their capabilities in molecular diagnostics [39]. Their approach incorporated scanning mode-based fluorescence detection, where the detector was precisely aligned to capture fluorescence intensity from the reaction chambers. During RT LAMP and RT-PCR reactions, the device rotated under the detector at a constant speed, enabling continuous recording of fluorescence intensity. The data obtained provided 1200 fluorescence intensity values per measurement, which were plotted against the position to produce a fluorescence spectrum graph. This graph was used to monitor the amplification of target genes, with increasing fluorescence signals indicating successful amplification. The raw data were also converted into real-time amplification curves, depicting the progression of the reactions. They optimized the number of glass filter papers used for nucleic acid extraction. Nguyen et al. experimented with varying numbers of GF/F filter papers, comparing their efficiency against a commercial Qiagen extraction kit. They found that using four GF/F papers, totaling a thickness comparable to the commercial kit's membranes [57], yielded similar Ct values, indicating efficient nucleic acid capture [58, 59]. For the FHV target, the Ct value of the Qiagen and the 4 GF/F papers was 20.0 and 21.6, respectively. For the BDB target, the Ct value of the Qiagen and the 4 GF/F papers was similar at 17.3. For the MPF target, the Ct value of the Qiagen and the 4 GF/F papers was 19.4 and 20.8, respectively. Five layers of the GF/F paper was too thick to recover the captured NA, while one or two layers of the GF/F paper were relatively thin to capture enough NA. This optimization was crucial for the device's ability to perform successful genetic diagnostics.

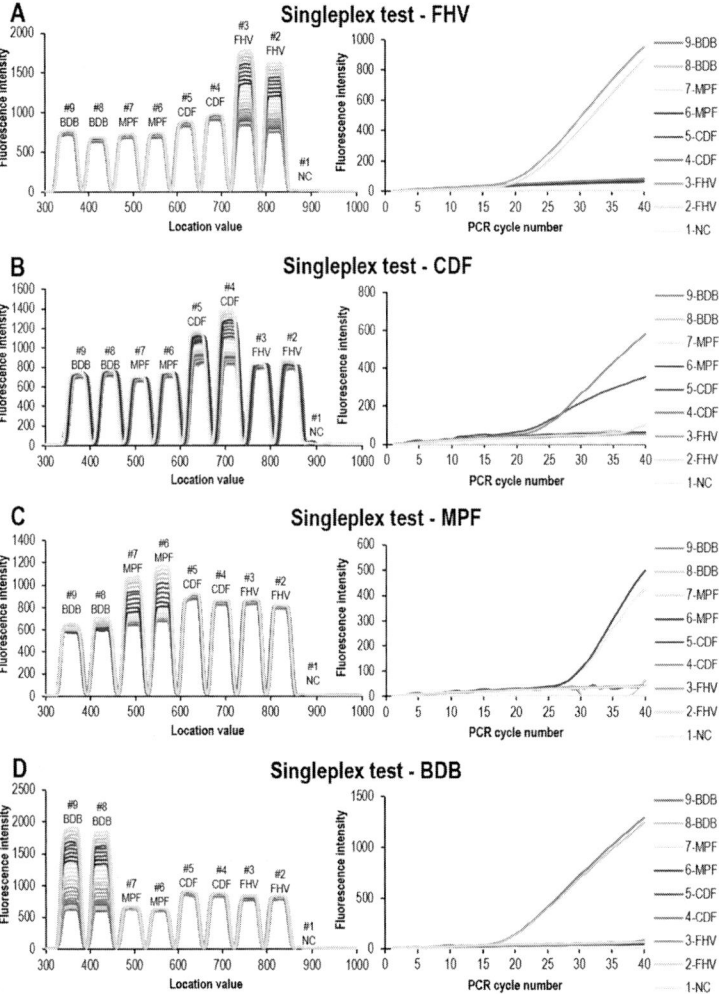

Fig. 15. Singleplex test on the proposed centrifugal microsystem with RT-PCR. The raw data of the fluorescence intensity for each chamber (left panel) and the amplification profile by plotting the fluorescence intensity vs. the PCR cycle number (right panel) for targeting (A) FHV, (B) CDF, (C) MPF, and (D) BDB. (Reprinted with permission from Ref. [39]. Copyright (2021) Elsevier.)

The device's capability for singleplex testing was clearly illustrated through experiments involving samples infected with individual FURTD pathogens (Fig. 15). In these tests, amplification was observed only in the chambers that were coated with the corresponding primers and probes. This selective amplification validated the device's precision in targeting and identifying specific pathogens. Moreover, the device's versatility was further evidenced in duplex and triplex tests. Here, samples infected with multiple FURTD pathogens were processed, and the device accurately detected each pathogen. The results showed amplification in respective chambers designated for each pathogen, demonstrating the device's ability to handle complex samples with multiple infections (Fig. 16). Complementing the RT-PCR tests, Nguyen et al. also conducted RT-LAMP reaction tests. These tests followed similar protocols to the RT-PCR experiments and served to confirm the device's efficacy in detecting a variety of FURTD pathogens. The RT-LAMP tests, both singleplex and multiplex, consistently aligned with the results obtained from the RT-PCR tests.

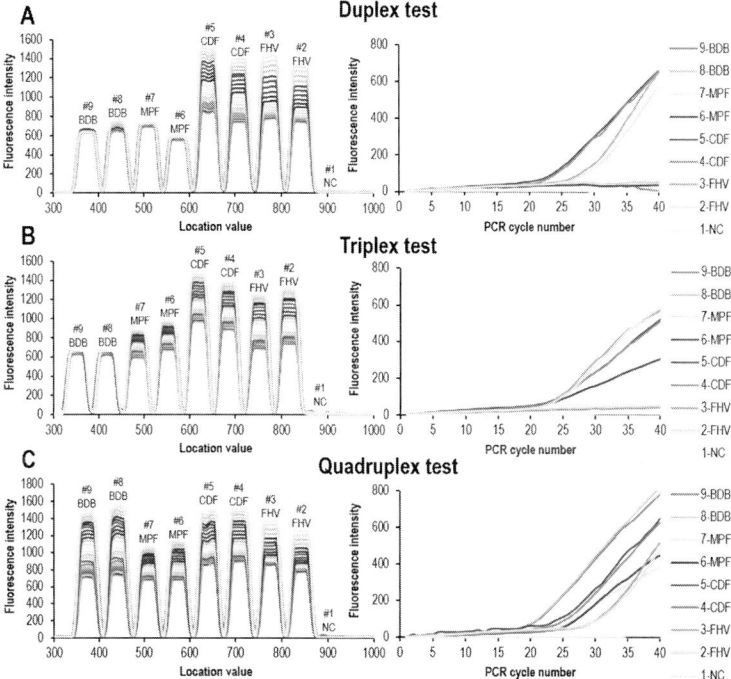

Fig. 16. Multiplex test on the proposed centrifugal microsystem with RT-PCR. (A) Duplex test for the FHV and MPF infected samples. (B) Triplex test for the FHV, MPF and BDB infected samples. (C) Quadruplex test for the FHV, MPF, BDB, and CDF infected samples. (Reprinted with permission from Ref. [39]. Copyright (2021) Elsevier.)

This consistency underscored the device's reliability and accuracy in simultaneous pathogen detection, irrespective of the molecular diagnostic method employed. When compared with the data of the RT-PCR, the RT-LAMP gproduced the same results for the singleplex and multiplex experiments. The sample pretreatment step was completed in 10 minutes in both the RT-PCR and the RT-LAMP, and the amplification time for the RT-PCR and the RT-LAMP was 50 minutes and 60 minutes, respectively. The whole process was finished within 1.5 hours. A pivotal part of the study was the LOD test, which evaluated the device's sensitivity in detecting FURTD pathogens (Fig. 17). LOD of LAMP reaction was 10^3 copies/µL for MPF, BDB and CDF, and 10^4 copies/µL for FHV. The LOD of PCR was 10^3 copies/µL for FHV, MPF, and CDF, and 10^4 copies/µL for BDB.

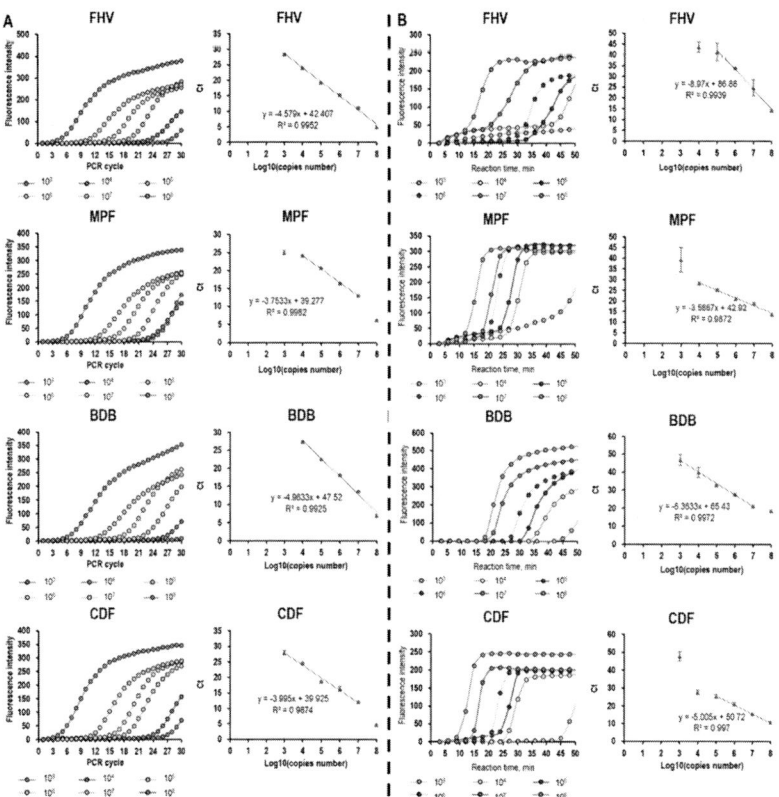

Fig. 17. (Right) Amplification profiles and the correspondent threshold times of the LAMP reaction for the synthesis gene of the four targets with serially diluted from 10^8 to 10^3 DNA copies/µL. (Left) Amplification profiles and the correspondent threshold times of the PCR reaction for the synthesis gene of the four targets with serially diluted from 10^8 to 10^3 DNA copies/µL. (Reprinted with permission from Ref. [39]. Copyright (2021) Elsevier.)

Phan et al. demonstrated the remarkable capabilities of their centrifugal microfluidic device and portable genetic analyzer, focusing on the detection of a range of respiratory viruses, including the novel SARS-CoV-2 [40]. The study encompassed an extensive evaluation through clinical sample tests and limit-of-detection assessments, providing a comprehensive view of the device's efficacy in real-world scenarios. The clinical sample testing involved 21 samples with unknown viral concentrations, offering a robust framework for assessing the device's diagnostic accuracy. This broad testing spectrum was crucial,

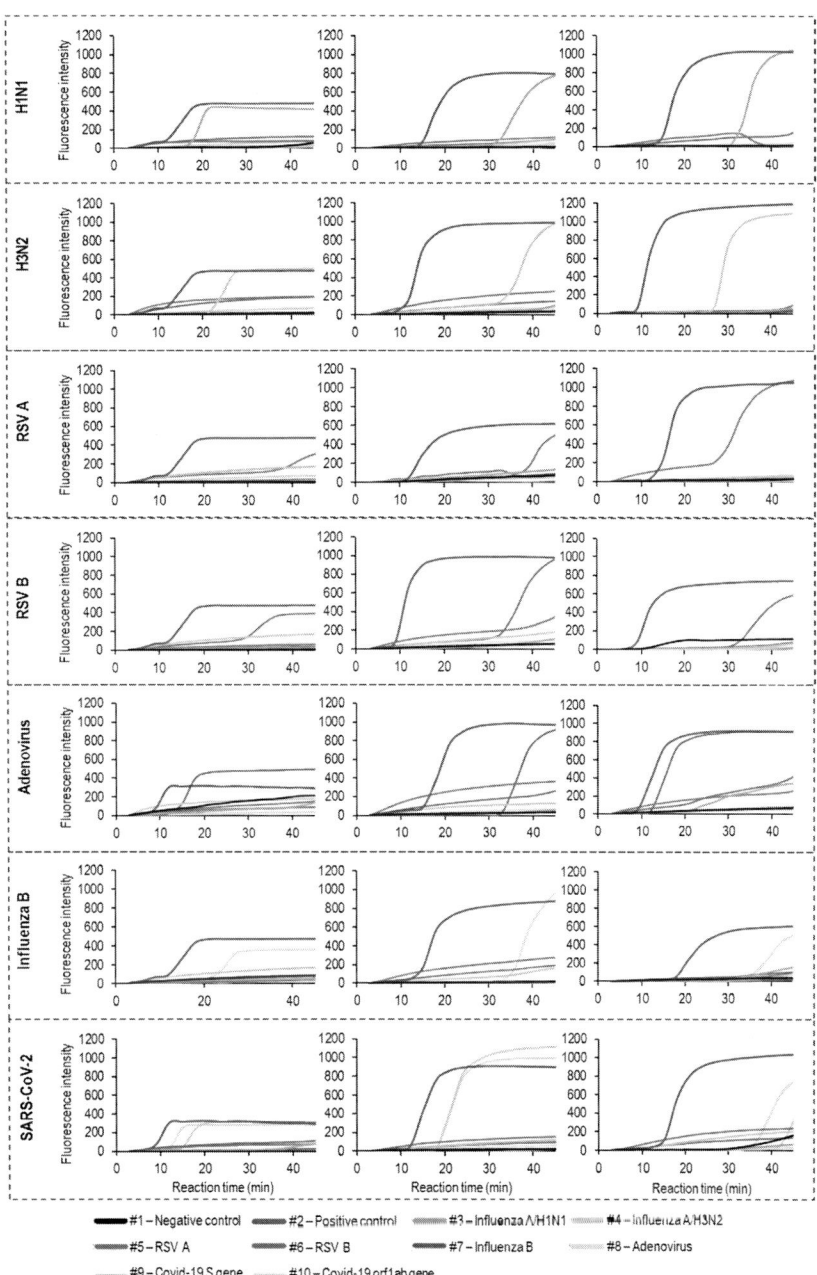

Fig. 18. Results of the clinical sample testing with the integrated POC genetic analyzer for detecting 7 respiratory viruses with an 18S rRNA as a positive internal control. (Reprinted with permission from Ref. [40]. Copyright (2023) Elsevier.)

particularly in the context of a pandemic, where rapid and accurate diagnostics are essential. The testing process was impressively efficient, with the complete cycle from solution injection to results generation being completed in just 1 hour and 20 minutes. This rapid turnaround is vital in situations where timely diagnosis significantly impacts patient care and infection control. Using the portable genetic analyzer, the team generated amplification curves for each target virus in the samples. These curves were instrumental in accurately identifying the presence of specific respiratory viruses, including influenza subtypes H1N1 and H3N2, RSV A/B, influenza B, adenovirus, and COVID-19 (Fig. 18). Crucially, there were no false-positive results, as indicated by the absence of amplification in the negative control chambers. The inclusion of internal positive controls in the tests consistently yielded early amplification signals, ensuring the reliability of the results and effectively ruling out the possibility of false negatives.

6. Conclusion

The collective research of Nguyen et al., Oh et al., Nguyen et al., and Phan et al. has significantly advanced the field of molecular diagnostics through the development of centrifugal microfluidic devices by combining with a solution-loading cartridge. These studies not only demonstrate the technical feasibility of these devices but also highlight their automatic capabilities, crucial for rapid, accurate, and comprehensive pathogen detection.

Nguyen et al. [37] effectively demonstrates the use of a fully integrated centrifugal microdevice for multiplex food poisoning bacteria screening. Their system is notable for its ability to handle large volume samples, achieving a low limit of detection with a rapid turnaround. This study exemplifies the shift towards more integrated, automated, and user-friendly devices in molecular diagnostics, paving the way for real-time, point-of-care testing scenarios.

Oh et al. further advance the field with a device that integrates a solution-loading cartridge to facilitate fully automated pathogen diagnostic detection. The implementation of silica bead-based DNA extraction and LAMP in their device demonstrates a remarkable leap in simplifying the operational process while enhancing the device's sample handling capacity. Their device's ability to detect multiple pathogens within a condensed timeframe exemplifies the high-throughput potential of these microfluidic systems.

Nguyen et al. [39] present a sophisticated system for diagnosing feline upper respiratory tract diseases. Their approach, integrating a centrifugal disc with a solution-loading cartridge in a portable genetic analyzer, automates the entire process of genetic diagnostics. This system's efficiency in nucleic acid extraction and rapid diagnostic capabilities illustrates the potential of such devices in diverse fields, including veterinary medicine, showcasing the broader applicability of these high-throughput systems.

Phan et al.'s work on diagnosing respiratory viruses, including COVID-19, using a centrifugal microsystem, underscores the critical role of these devices in pandemic response. Their system's high sensitivity and specificity, coupled with rapid processing times, demonstrate its suitability for in-field diagnostics in medical centers and screening clinics. The ability to analyze multiple clinical samples accurately and quickly is particularly crucial in managing infectious diseases and preventing their spread.

In summary, these studies collectively reinforce the importance of automatic properties in centrifugal microfluidic devices for molecular diagnostics. The advancements in these devices—characterized by their automation, integration, and rapid processing capabilities by a solution loading cartridge—offer a paradigm shift in how molecular diagnostics are approached. This shift is particularly significant in the context of infectious diseases, where timely and accurate detection is paramount. As the demand for rapid, in-field diagnostics continues to grow, these centrifugal microfluidic devices are poised to become indispensable tools in clinical settings, public health surveillance, and global disease management efforts.

References

1. A. Bradley, J. Kinyon, T. Frana, D. Bolte, D. R. Hyatt and M. R. Lappin, *J Vet Intern Med*, 2012, 26, 1121–1125.

2. J. R. Choi, R. Tang, S. Wang, W. A. B. Wan Abas, B. Pingguan-Murphy and F. Xu, *Biosens Bioelectron*, 2015, 74, 427–439.

3. X. Ye, J. Xu, L. Lu, X. Li, X. Fang and J. Kong, *Anal Chim Acta*, 2018, 1018, 78–85.

4. S. Choi, J. Hwang, S. Lee, D. W. Lim, H. Joo and J. Choo, *Sens Actuators B Chem*, 2017, 240, 358–364.

5. J. R. Choi, K. W. Yong, R. Tang, Y. Gong, T. Wen, F. Li, B. Pingguan-Murphy, D. Bai and F. Xu, *TrAC - Trends in Analytical Chemistry*, 2017, 93, 37–50.

6. X. Deng, C. Wang, Y. Gao, J. Li, W. Wen, X. Zhang and S. Wang, *Biosens Bioelectron*, 2018, 105, 211–217.

7. S. Takalkar, K. Baryeh and G. Liu, *Biosens Bioelectron*, 2017, 98, 147–154.

8. J. A. DuVall, D. Le Roux, B. L. Thompson, C. Birch, D. A. Nelson, J. Li, D. L. Mills, A.-C. Tsuei, M. G. Ensenberger, C. Sprecher, B. E. Root and J. P. Landers, *Anal Chim Acta*, 2017, 980, 41–49.

9. L. Zhang, B. Ding, Q. Chen, Q. Feng, L. Lin and J. Sun, *TrAC - Trends in Analytical Chemistry*, 2017, 94, 106–116.

10. B. Berg, B. Cortazar, D. Tseng, H. Ozkan, S. Feng, Q. Wei, R. Y.-L. Chan, J. Burbano, Q. Farooqui, M. Lewinski, O. B. Garner and A. Ozcan, *ACS Nano*, 2015, 9, 7857–7866.

11. A. Priye, S. Wong, Y. Bi, M. Carpio, J. Chang, M. Coen, D. Cope, J. Harris, J. Johnson, A. Keller, K. Chan and V. M. Ugaz, *Anal Chem*, 2016, 88, 4651–4660.

12. R. D. Stedtfeld, D. M. Tourlousse, G. Seyrig, T. M. Stedtfeld, M.

Kronlein, S. Price, F. Ahmad, E. Gulari, J. M. Tiedje and S. A. Hashsham, *Lab Chip*, 2012, 12, 1454–1462.

13 L.-J. Wang, Y.-C. Chang, R. Sun and L. Li, *Biosens Bioelectron*, 2017, 87, 686–692.

14 O. Strohmeier, M. Keller, F. Schwemmer, S. Zehnle, D. Mark, F. Von Stetten, R. Zengerle and N. Paust, *Chem Soc Rev*, 2015, 44, 6187–6229.

15 J. Hoffmann, D. Mark, S. Lutz, R. Zengerle and F. Von Stetten, *Lab Chip*, 2010, 10, 1480–1484.

16 F. Stumpf, F. Schwemmer, T. Hutzenlaub, D. Baumann, O. Strohmeier, G. Dingemanns, G. Simons, C. Sager, L. Plobner, F. Von Stetten, R. Zengerle and D. Mark, *Lab Chip*, 2016, 16, 199–207.

17 T. Van Oordt, Y. Barb, J. Smetana, R. Zengerle and F. Von Stetten, *Lab Chip*, 2013, 13, 2888–2892.

18 Y. Zhao, G. Czilwik, V. Klein, K. Mitsakakis, R. Zengerle and N. Paust, *Lab Chip*, 2017, 17, 1666–1677.

19 T.-H. Kim, C.-J. Kim, Y. Kim and Y.-K. Cho, *Sens Actuators B Chem*, 2018, 256, 310–317.

20 G. Choi, J. H. Jung, B. H. Park, S. J. Oh, J. H. Seo, J. S. Choi, D. H. Kim and T. S. Seo, *Lab Chip*, 2016, 16, 2309–2316.

21 J. H. Jung, B. H. Park, S. J. Oh, G. Choi and T. S. Seo, *Biosens Bioelectron*, 2015, 68, 218–224.

22 S. J. Oh, B. H. Park, J. H. Jung, G. Choi, D. C. Lee, D. H. Kim and T. S. Seo, *Biosens Bioelectron*, 2016, 75, 293–300.

23 B. H. Park, S. J. Oh, J. H. Jung, G. Choi, J. H. Seo, D. H. Kim, E. Y. Lee and T. S. Seo, *Biosens Bioelectron*, 2017, 91, 334–340.

24 S. Z. Andreasen, D. Kwasny, L. Amato, A. L. Brøgger, F. G. Bosco, K. B. Andersen, W. E. Svendsen and A. Boisen, *RSC Adv*, 2015, 5, 17187–17193.

25 J. W. Martin, M. K. Nieuwoudt, M. J. T. Vargas, O. L. C. Bodley, T. S. Yohendiran, R. N. Oosterbeek, D. E. Williams and M. C. Simpson, *Analyst*, 2017, 142, 1682–1688.

26 F. Schwemmer, C. E. Blanchet, A. Spilotros, D. Kosse, S. Zehnle, H. D. T. Mertens, M. A. Graewert, M. Rössle, N. Paust, D. I. Svergun, R. Zengerle and D. Mark, *Lab Chip*, 2016, 16, 1161–1170.

27 J. Park, V. Sunkara, T.-H. Kim, H. Hwang and Y.-K. Cho, *Anal Chem*, 2012, 84, 2133–2140.

28 R. Burger, D. Kurzbuch, R. Gorkin, G. Kijanka, M. Glynn, C. McDonagh and J. Ducrée, *Lab Chip*, 2015, 15, 378–381.

29 B. H. Park, J. H. Lee, J. H. Jung, S. J. Oh, D. C. Lee and T. S. Seo, *RSC Adv*, 2015, 5, 1846–1851.

30 B. H. Park, D. Kim, J. H. Jung, S. J. Oh, G. Choi, D. C. Lee and T. S.

Seo, *Sens Actuators B Chem*, 2015, 209, 927–933.

31 M. Focke, D. Kosse, C. Müller, H. Reinecke, R. Zengerle and F. Von Stetten, *Lab Chip*, 2010, 10, 1365–1386.

32 J. H. Jung, S. J. Choi, B. H. Park, Y. K. Choi and T. S. Seo, *Lab Chip*, 2012, 12, 1598–1600.

33 B. H. Park, J. H. Jung, H. Zhang, N. Y. Lee and T. S. Seo, *Lab Chip*, 2012, 12, 3875–3881.

34 O. Strohmeier, N. Marquart, D. Mark, G. Roth, R. Zengerle and F. Von Stetten, *Analytical Methods*, 2014, 6, 2038–2046.

35 M. Keller, S. Wadle, N. Paust, L. Dreesen, C. Nuese, O. Strohmeier, R. Zengerle and F. Von Stetten, *RSC Adv*, 2015, 5, 89603–89611.

36 M. Antillon, N. J. Saad, S. Baker, A. J. Pollard and V. E. Pitzer, *Journal of Infectious Diseases*, 2018, 218, S255–S267.

37 H. Van Nguyen, V. D. Nguyen, E. Y. Lee and T. S. Seo, *Biosens Bioelectron*, 2019, 136, 132–139.

38 S. J. Oh and T. S. Seo, *Analyst*, 2019, 144, 5766–5774.

39 H. Van Nguyen, V. M. Phan and T. S. Seo, *Biosens Bioelectron*, 2021, 193, 113546.

40 V. M. Phan, S. W. Kang, Y. H. Kim, M. Y. Lee, H. Van Nguyen, Y. La Jeon, W. I. Lee and T. S. Seo, *Sens Actuators B Chem*, 2023, 390, 133962.

41 M. Fernandez, E. G. Manzanilla, A. Lloret, M. León and J.-C. Thibault, *J Feline Med Surg*, 2017, 19, 461–469.

42 D. Nguyen, V. R. Barrs, M. Kelman and M. P. Ward, *J Feline Med Surg*, 2019, 21, 973–978.

43 S. J. Oh, B. H. Park, G. Choi, J. H. Seo, J. H. Jung, J. S. Choi, D. H. Kim and T. S. Seo, *Lab Chip*, 2016, 16, 1917–1926.

44 M. Madadelahi, L. F. Acosta-Soto, S. Hosseini, S. O. Martinez-Chapa and M. J. Madou, *Lab Chip*, 2020, 20, 1318–1357.

45 H. Van Nguyen, V. D. Nguyen, H. Q. Nguyen, T. H. T. Chau, E. Y. Lee and T. S. Seo, *Biosens Bioelectron*, 2019, 141, 111466.

46 M. Zarei, *Biosens Bioelectron*, 2017, 98, 494–506.

47 R. Agarwal, A. Sarkar and S. Chakraborty, *Analyst*, 2019, 144, 3782–3789.

48 S. Burger, M. Schulz, F. Von Stetten, R. Zengerle and N. Paust, *Lab Chip*, 2016, 16, 261–268.

49 F. Schwemmer, S. Zehnle, D. Mark, F. Von Stetten, R. Zengerle and N. Paust, *Lab Chip*, 2015, 15, 1545–1553.

50 J. H. Jung, B. H. Park, Y. K. Choi and T. S. Seo, *Lab Chip*, 2013, 13, 3383–3388.

51 M. Grumann, A. Geipel, L. Riegger, R. Zengerle and J. Ducrée, *Lab Chip*, 2005, 5, 560–565.

52 J. F. Hess, S. Zehnle, P. Juelg, T. Hutzenlaub, R. Zengerle and N. Paust, *Lab Chip*, 2019, 19, 3745–3770.

53 G. Wang, J. Tan, M. Tang, C. Zhang, D. Zhang, W. Ji, J. Chen, H.-P. Ho and X. Zhang, *Lab Chip*, 2018, 18, 1197–1206.

54 D. Mark, T. Metz, S. Haeberle, S. Lutz, J. Ducrée, R. Zengerle and F. Von Stetten, *Lab Chip*, 2009, 9, 3599–3603.

55 J. Rodriguez-Manzano, M. A. Karymov, S. Begolo, D. A. Selck, D. V. Zhukov, E. Jue and R. F. Ismagilov, *ACS Nano*, 2016, 10, 3102–3113.

56 Y. Sun, T. L. Quyen, T. Q. Hung, W. H. Chin, A. Wolff and D. D. Bang, *Lab Chip*, 2015, 15, 1898–1904.

57 D. Brassard, M. Geissler, M. Descarreaux, D. Tremblay, J. Daoud, L. Clime, M. Mounier, D. Charlebois and T. Veres, *Lab Chip*, 2019, 19, 1941–1952.

58 J. Hui, Y. Gu, Y. Zhu, Y. Chen, S.-J. Guo, S.-C. Tao, Y. Zhang and P. Liu, *Lab Chip*, 2018, 18, 2854–2864.

59 Q. Liu, X. Zhang, L. Chen, Y. Yao, S. Ke, W. Zhao, Z. Yang and G. Sui, *Sens Actuators B Chem*, 2018, 270, 371–381.

Chapter 7
Centrifugal Microfluidic Device for High-throughput Genetic Analysis

1. Background

The onset of the COVID-19 pandemic and the perennial threat of influenza have underscored the limitations of conventional diagnostic methods. Quantitative polymerase chain reaction (qPCR) and serological tests, while widely adopted [1–4], are limited by high costs, prolonged turnaround times, and a heavy reliance on established laboratory infrastructure [5–8]. This has catalyzed the demand for diagnostic tools that are not only fast and accurate but also cost-effective and high-throughput (HTP) to manage the diagnostic demands during peak pandemic periods. The global impact of respiratory viruses, which account for substantial morbidity and mortality annually, has been a driving force behind the development of innovative diagnostic solutions. The existing commercial point-of-care (POC) platforms, despite their rapid detection capabilities, face challenges in throughput and affordability, limiting their accessibility, particularly in underdeveloped regions [9–15].

On the other hands, the advances in isothermal amplification, particularly loop-mediated isothermal amplification (LAMP), have improved upon the constraints of qPCR, offering faster reactions under mild conditions, simpler operation, and colorimetric result determination [16]. Thus, the combined HTP capability with the simple LAMP reaction on the centrifugal microfluidic device is very unique and is highly amenable to POC genetic diagnostic contexts [17–19].

The centrifugal microfluidic platform has recently demonstrated its capability for HTP genetic analysis, enabling processing multiple samples simultaneously and integrating multiple diagnostic steps, thus reducing costs and simplifying the diagnostic process [20–26]. Centrifugal microfluidics, commonly referred to apart of lab-on-a-disc technology, stands at the forefront of a revolution in diagnostics and biochemical analysis [9, 27–29]. By harnessing centrifugal force for fluid manipulation in miniaturized settings, it has profoundly changed laboratory procedures. This technology has not only enabled the integration of multiple laboratory functions onto a single chip, reducing reagent consumption and waste, but it has also brought forth significant advancements in economic and environmental sustainability. The automated nature of fluidic operations, controlled by the rotation speed of the disc, simplifies complex processes, making these devices more accessible and reducing the potential for human error. Their compact and robust design, combined with low power needs, makes them especially suitable for POCT and field deployment, particularly in areas with limited resources [30–37]

The development of centrifugal microfluidics is intrinsically linked to the growing significance of POCT in modern healthcare. POCT enables healthcare providers to conduct diagnostic tests swiftly and conveniently at the patient's location, leading to immediate clinical decisions and interventions [38–42]. This rapid response is crucial in numerous healthcare scenarios, ranging from managing chronic conditions to emergency medical situations. The ability to provide on-the-spot results is particularly vital during health crises, such as pandemics. For instance, research by Nguyen et al. (2022) on a portable genetic analyzer for COVID-19 testing highlights the pivotal role of centrifugal microfluidic technology in delivering timely and accurate diagnostics during critical times. This technology's adaptability to various healthcare settings, from hospital wards to remote clinics, further underscores its value in enhancing patient care and improving health outcomes [43].

Moreover, the advancement of centrifugal microfluidics in HTP analysis has revolutionized large-scale screening and epidemiological studies. HTP capabilities, such as processing numerous samples simultaneously, are essential in situations requiring rapid and extensive testing. The development of a platform by Nguyen et al. (2024) for multiplex respiratory virus diagnostics demonstrates this capacity, showcasing the ability to analyze several samples in a single run [44]. This feature is not only a testament to the efficiency and scalability of centrifugal microfluidic devices but also underscores their importance in pandemic responses and mass screening programs, where swift and efficient processing of a large number of samples is critical.

Despite the tremendous potential of centrifugal microfluidics, challenges remain in terms of integrating complex fluidic control mechanisms and developing standardized, universally adaptable platforms. Future advancements are likely to focus on enhancing the multiplexing capabilities of these devices, improving their detection limits, and broadening the spectrum of substances that can be analyzed. As technology continues to evolve, it is poised to transform the landscape of medical diagnostics and biochemical research. The innovative designs and diverse HTP applications of centrifugal microfluidic devices in advancing medical and scientific research were demonstrated in the studies by Seo et al. (2017), Nguyen et al. (2022), and Nguyen et al. (2024) [43–45]. Offering a method for HTP genetic analysis that is rapid, cost-effective, and user-friendly, these devices are set to become an indispensable tool in the advancement of healthcare technologies and the pursuit of scientific discovery.

2. Centrifugal microfluidic chip for HTP genetic analysis

2.1. Overall design of the centrifugal HTP microfluidic chip

The design of the HTP centrifugal microfluidic chips represents a convergence of engineering, material science, and biotechnology, creating platforms that handle multiple samples simultaneously with precision and efficiency. At their core, these devices feature an intricate microfluidic layout, with networks of microscale channels, chambers, and valves intricately etched or molded onto the chip. This microfluidic architecture is critical in dictating the flow of liquids and reagents, enabling the simultaneous processing of numerous samples, a feature pivotal to HTP applications. The centrifugal HTP microfluidic chip is engineered to address the critical need for rapid, accurate, and multiple genetic analysis. The design is specifically tailored to automate and integrate multiple diagnostic processes, including the genomic purification, the LAMP reaction, and the real-time fluorescence detection, all within a compact and portable format suitable for POCT testing.

In constructing these chips, a variety of materials are employed, each selected for specific properties like optical clarity, chemical resistance, or biocompatibility. Common materials include polydimethylsiloxane (PDMS), glass, and various plastics, each offering unique advantages. Fabrication techniques such as soft lithography, injection molding, and even advanced 3D printing are utilized to create these microfluidic structures, each method offering different benefits in terms of scalability, resolution, and cost-effectiveness. More common materials for the microfluidic chip is polymethyl methacrylate (PMMA), a material chosen for its transparency, thermal stability, and cost-effectiveness, particularly for disposable applications. The PMMA substrate is intricately carved with the desired micro-pattern to create a three-layered structure that houses the necessary channels and chambers.

The integration of functional elements within these chips is a testament to the advancements in microfluidics. Elements like micro-mixers, heaters, and optical detection systems are seamlessly incorporated into the chip's design. These components are crucial for conducting complex biochemical assays, enabling processes such as gene amplification, protein analysis, and cellular studies directly on the chip. The design often includes innovative approaches to mix reagents, control temperatures, and facilitate real-time detection, all within the confined space of the microfluidic channels. Fluidic control in these chips is ingeniously managed using a combination of centrifugal forces, capillary valves, and sometimes pneumatic control systems. The centrifugal force, generated by the rotation of the chip, is a primary mechanism for moving fluids through the microchannels. This method allows for precise and programmable fluid manipulation, essential for the accuracy and repeatability of the assays. Capillary valves, which control fluid flow based on surface tension properties, and pneumatic controls, which use air pressure for fluid movement, are also integrated to achieve intricate control over fluid movement.

The design also emphasizes ease of sample loading and processing. Many HTP chips are designed with automated or semi-automated sample loading systems, allowing for quick and efficient handling of multiple samples. The capacity to process numerous samples in parallel is a defining feature of these chips, aligning with the demands of HTP analysis in various settings, from clinical laboratories to field-based studies. The underlying fluid dynamics principles are central to the design. The capillary pressure at the liquid-air interface is meticulously calculated to ensure that the solutions remain within the aliquoting chambers until precisely directed. This enables the chip to perform HTP analysis, crucial for managing large volumes of samples. The design optimizes diagnostic processes by automating sample loading, fluid partitioning, and reagent mixing, which reduces manual labor and increases reproducibility. Moreover, the chip allows for the loading and division of all necessary solutions (excluding the sample solution) in the same manner, improving operational efficiency. The aliquoting structure and passive valve system are key to the chip's scalability and multiplexing capabilities, permitting multiplex samples to be processed in parallel in a single run. This feature significantly enhances the chip's throughput, a vital attribute for responding to pandemic scenarios.

2.2. A single unit-centrifugal microfluidics chip with multiple reaction chambers

The chip designed by Seo et al. features an advanced microfluidic layout that significantly enhances the efficiency of one sample processing for multiple target detection [45]. Characterized by a sophisticated array of micro-channels and chambers, this layout is ingeniously crafted to segregate and route a sample effectively into multiple reaction chambers. Suited for rapid screening applications such as environmental monitoring and disease outbreak management, this chip streamlines complex assays into a compact and efficient format, demonstrating the potential of microfluidics in HTP target analysis.

The design of the centrifugal microdevice is a complex interplay of microfluidic engineering and molecular biology. At its core, the device uses a rotation axis to leverage centrifugal force, a method essential for moving fluids through microscale channels. This approach is particularly effective in handling small volumes of biological samples, which are common in molecular diagnostics. A critical feature of the device is its sample reservoir, designed with specific dimensions (1.2 mm deep and 5 mm radius), colored pink for visual distinction. This reservoir not only holds the sample but also includes a circular capillary valve, colored purple and sized at 200 μm deep by 1 mm wide. The capillary valve's function is to generate enough capillary force to counterbalance the pressure from the loaded sample, a delicate balance crucial in microfluidic devices. The microdevice's layout includes a spiral-shaped sample injection microchannel, colored white and measuring 550 μm in depth and 2 mm in width. The spiral design is not just aesthetic. It's a functional choice that ensures the sample can travel efficiently along the channel during centrifugation, a process vital for transferring the sample to the reaction units. In addition to these components, the device includes two holes in the sample reservoir, one for sample injection and another as an air hole. This design considers the need for air escape during sample loading, preventing potential disruptions in sample flow. These holes are sealed with an adhesive film after loading the sample, a method that uses simple yet effective materials like a double-sided adhesive and a thin polycarbonate film for sealing. The waste chamber, another key component, is situated at the lowest part of the microdevice and is designed to store the remaining sample post aliquoting. This inclusion shows a comprehensive approach to handling all aspects of the sample, from loading to waste management. Perhaps the most intricate aspect of the device is its 24 reaction units, each designed for individual reactions. These units are patterned around the perimeter of the device, allowing for simultaneous processing of multiple samples. Each unit comprises an aliquoting chamber and a reaction chamber, both colored white and having a volume of 2.5 μL. The chambers are separated by a cross capillary valve, designed to control the sample flow between the chambers during the aliquoting step (Fig. 1).

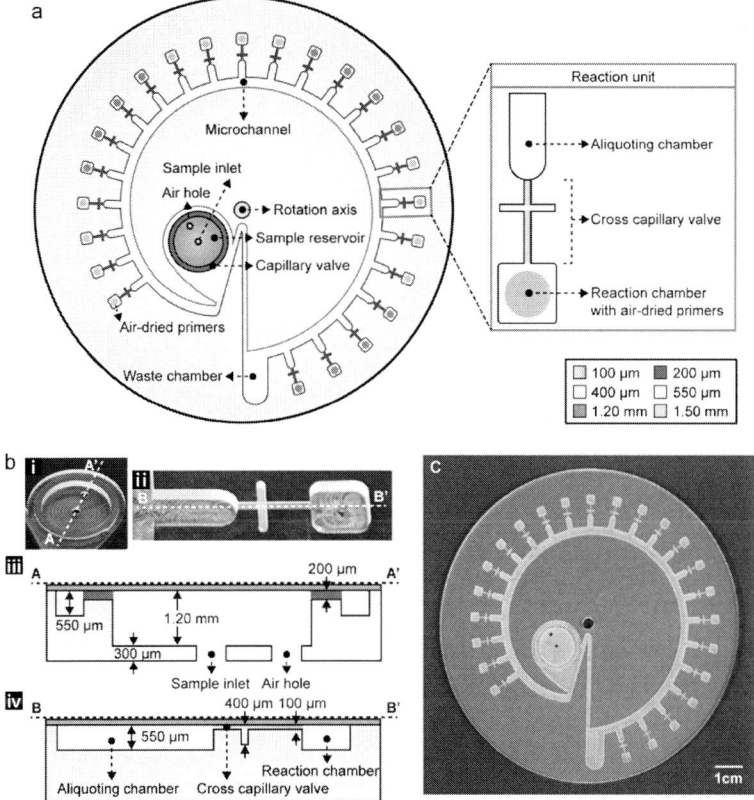

Fig. 1. (a) The design of the centrifugal LAMP microdevice of which main components are a sample reservoir, a spiral-shaped microchannel, and 24 reaction units. The enlarged image of one reaction unit containing an aliquoting chamber, a cross capillary valve, and a reaction chamber with air-dried primers (in the red box). Each color indicates the dimension of each component in the microdevice. (b) The enlarged and upside-down digital images of (i) the sample reservoir and (ii) the reaction unit. The upside-down and cross-sectional images of (iii) the sample reservoir and (iv) the reaction unit. (c) The digital image of the assembled microdevice which consists of two layers: a 1.5 mm-thick patterned PMMA layer and a PSA foil layer. (Reprinted with permission from Ref. [45]. Copyright (2017) Elsevier.)

2.3. A centrifugal HTP microfluidics chip with 10 units for COVID-19 diagnostics

In the work of Nguyen et al., the integration of real-time detection systems sets their chip apart [43]. These systems are adeptly embedded within the chip, enabling on-the-spot analysis and immediate result readout. Its design is a sophisticated integration of three layers: two pressure-sensitive adhesive (PSA) films enclosing a central PMMA layer. This layered construction ensures a hermetic environment, critical for the precise handling of biological samples. A distinctive feature of this device is the double-sided patterning on the PMMA sheet. This innovative approach allows for the segregation of functional units across the two surfaces of the disc, enhancing the device's functionality and efficiency. The top surface of the disc houses the sample chamber, aliquoting channel and chamber, and the waste chamber. This strategic placement facilitates the initial stages of sample processing. In contrast, the bottom surface is meticulously arranged with the transfer chamber, collection chamber, metering structure, wax loading chamber, and reaction chamber. This alignment is crucial for subsequent stages of sample processing and analysis.

A cornerstone of the device's design is its sophisticated aliquoting structure. It is engineered to automate the RNA purification process for up to 10 samples

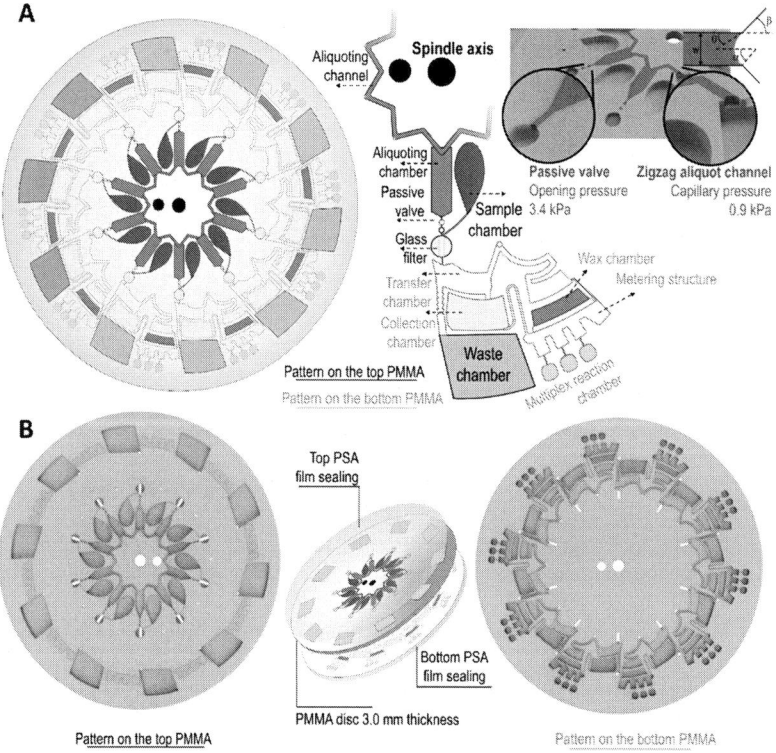

Fig. 2. A HTP centrifugal microfluidics chip with 10 units for COVID-19 diagnostics. (A) Illustration of the total integrated centrifugal disc, which can perform the genetic analysis for 10 samples in one run. The structure of the centrifugal disc is designed for the automatic 10 aliquoting process, the viral RNA extraction, the multiplex qLAMP reaction, and the fluorescence detection. (B) The 3D illustration of the micro-pattern on the top and bottom of the disc. The microfluidic channels were double-sided etched on a single PMMA layer, and then sealed with 2 PSA films. (Reprinted with permission from Ref. [43]. Copyright (2022) Elsevier.)

simultaneously, a significant advancement in terms of throughput. The aliquoting structure comprises 10 intricately designed pentagon-arrow-shaped aliquoting chambers. These chambers are interconnected by a 1 mm-wide zigzag aliquoting channel, a design choice that optimizes the fluidic path for sample distribution. Each aliquoting chamber is outfitted with passive valves. These valves are not just mere inclusions. They are precision-engineered with a straight channel measuring 0.2 mm in width and depth and supplemented with two circular patterns each 1.0 mm in diameter and 0.5 mm deep. The role of these passive valves is that they provide the necessary backpressure to keep the solution inside the aliquoting chamber, preventing it from bleeding outwards during the aliquoting process. The functioning of these passive valves is governed by the principles of capillary pressure. This pressure occurs at the liquid-air interface at the point of sudden expansion from the narrow 0.2 mm channel to the 1.0 mm diameter circle. The design leverages liquid surface tension and exploits the concept of capillary action at microscales (Fig. 2).

The opening pressure of these passive valves is mathematically calculated using a specific formula that takes into account the liquid surface tension, contact angles on various sides of the solution, and the dimensions of the channel. An intriguing aspect of the design is the differential pressure system. The capillary pressure in the zigzag aliquoting channel is carefully calibrated to be lower than the opening pressure of the passive valves. This calibration ensures that the solution remains within the aliquoting chamber, preventing premature leakage into the glass filter column. This feature is critical for transferring the solution from chamber to chamber through the zigzag channel, ultimately resulting in the formation of 10 precise aliquots. This automatic aliquoting process represents a leap forward in microfluidic technology, allowing for a single injection of the solution to fill the aliquoting chambers and divide into 10 aliquots for simultaneous reactions. The opening pressure of the passive valve (P) is calculated by the following equation [46]:

$$P = -\gamma \left(\frac{\cos\theta_t + \cos\theta_b}{h} + \frac{\cos\theta_l + \cos\theta_r}{w} \right)$$

where γ is the liquid surface tension, θ_t, θ_b, θ_l, θ_r is the contact angle of the solution on the top, bottom, left, and right, h is the height of the channel, and w is the width of the channel. The capillary pressure of the zigzag aliquoting channel (C_p) was calculated as follows [47]:

$$C_p = \frac{C\gamma \sin\theta}{h \times w}$$

where C is the length of the channel from the pentagon arrow chamber to the passive valve, γ is the solution surface tension, θ is the contact angle, h is the height of the channel, and w is the width of the channel.

2.4. A centrifugal HTP microfluidics chip with 30 units for COVID-19 diagnostics

The chip developed by Nguyen et al. showcases a remarkable HTP capability, capable of processing a large number of samples in a single run [44]. This feature is paramount in large-scale screening and epidemiological studies. The design reflects an understanding of the needs in such scenarios, setting a new benchmark for efficiency in HTP microfluidic technologies. It demonstrates the scalability and adaptability of centrifugal microfluidics in addressing large-scale health challenges.

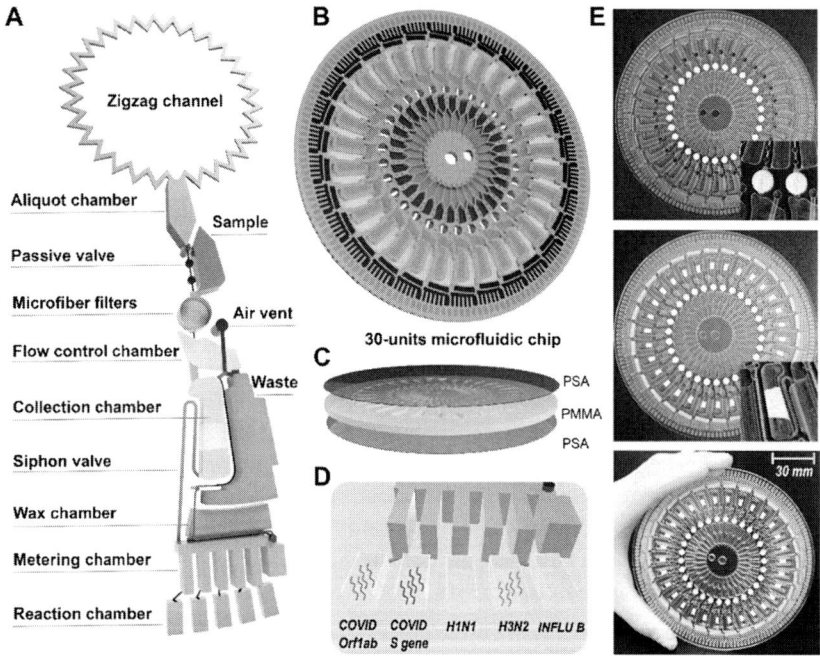

Fig. 3. Design and fabrication of the HTP centrifugal microfluidic device by Nguyen et al. (2024). (A, B) The schematics depict an HTP microfluidic device to perform the RNA extraction and the RT-LAMP reaction. The device comprises 30 parallel units, each consisting of sophisticated designs of micro-chambers and microchannels. These include a sample chamber, an aliquot chamber, a zigzag channel connecting the 30 aliquot chambers, a microfiber filter column, a flow control chamber, a waste chamber, a collection chamber, a wax chamber, a metering chamber, and five reaction chambers. (C) The device consists of a micro-patterned PMMA layer bonded with two PSA films. (D) Five reaction chambers are coated with different LAMP primer sets, targeting COVID-19 (orf1ab gene and S gene), H1N1, H3N2, and Influenza B. The simultaneous performance of 150 reactions is possible on a single microfluidic chip. (E) Digital images showcase the fabricated HTP centrifugal chip with the microfiber filter in the column chamber (top), and the lyophilized RT-LAMP master mix in the collection chambers (middle). The diameter of the chip is 130 mm (bottom). (Reprinted with permission from Ref. [44]. Copyright (2024) Elsevier.)

The chip was constructed from 3 mm thick PMMA sheets. The choice of PMMA was strategic, as it offers clarity, chemical resistance, and compatibility with biological assays – attributes that are vital for molecular diagnostic applications. To seal and protect the intricate microfluidic structures, PSA films were used. These films ensured the airtight sealing of the chip, maintaining the integrity of the internal environments which is crucial for accurate diagnostics.

The microfluidic design of the chip, etched on both the top and bottom sides of the PMMA sheet, included an array of features essential for its functionality. These features encompassed a zigzag aliquoting channel, 30 aliquoting micro-chambers, 30 sample chambers, transfer chambers, micro-channels, via holes, passive valves, siphon valves, and 150 reaction chambers. The zigzag aliquoting channel and the aliquoting micro-chambers were designed for efficient and precise distribution of samples (Fig. 3). The sample chambers and transfer chambers facilitated the introduction and movement of samples within the chip, while the micro-channels and via holes were crucial for the precise fluid transport. The inclusion of passive and siphon valves was a key aspect of the design, regulating the flow of fluids within the chip and ensuring the correct sequencing of diagnostic steps. The 150 reaction chambers significantly enhanced the throughput capacity of the chip, allowing for multiple parallel RT-LAMP reactions. Furthermore, the chip's design also took into account the integration of critical components for RNA extraction and RT-LAMP reactions. This integration included circular glass microfiber filters for RNA extraction, specific areas within the chip for drying RT-LAMP primer sets targeting COVID-19 and influenza viruses, and provisions for freeze-drying the master mix in the collection chambers.

3. Fabrication of a centrifugal HTP chip

A central material across these studies is PMMA, chosen for its optical clarity, thermal resilience, and suitability for disposable applications. This choice is consistent across the works of Seo et al., Nguyen et al., and Nguyen et al., reflecting a consensus on the material's suitability for intricate microfluidic applications [43–45]. The fabrication process, utilizing computer numerical controlled (CNC) milling machines, is a testament to the precision required in carving out the complex channel patterns essential for fluid control within these chips. The use of advanced design software facilitates the translation of intricate designs into tangible microstructures, a process that is meticulously executed in each study.

Post-fabrication, the chip undergoes a thorough cleaning process involving sonication with a detergent solution, ethanol, and distilled water. This step is crucial for removing any residues from the manufacturing process that could interfere with diagnostic assays. Following cleaning, selective surface treatment is applied: microchannels are coated with a superhydrophobic reagent to prevent unintended fluid flow, and siphon valves receive a hydrophilic coating to ensure proper liquid movement and control. These surface treatments are essential for the chip's performance, directing fluids precisely where needed during operation.

The fabrication of the microdevice, as described by Seo et al., involved a methodical and precise process, comprising three main steps [45]. Each step was essential in creating a device capable of conducting complex molecular diagnostics, particularly LAMP reactions. The device's fabrication was characterized by its meticulous attention to detail and the use of specialized materials and equipment. The initial stage of fabrication involved the precise milling of a 1.5 mm thick-PMMA layer. The microdevice, with a radius of 4.2 cm, had all its functional units meticulously patterned onto this PMMA layer. This intricate task was accomplished using a computer-controlled (CNC) milling machine, which allowed for precise and accurate creation of the microfluidic channels and chambers necessary for the device's functionality. Following the milling process, the next step involved the coating of primer sets onto the microdevice. These primer sets were essential for the LAMP reaction, as they targeted specific bacterial genes in the sample. The primers were carefully applied to the designated reaction chambers of the microdevice. This step was crucial for ensuring that the microdevice was ready for immediate use upon introduction of a sample. After the primer sets were coated, the sample injection process took place. This step involved introducing the sample into the microdevice, where it would interact with the primers in the reaction chambers for the subsequent LAMP reaction. The final step in the fabrication process was sealing the microdevice with a PSA foil layer. This PSA foil was crucial for encapsulating the milled PMMA layer, thereby creating a sealed environment for the reactions to occur. The two layers – the patterned PMMA and the PSA foil – were carefully aligned and then pressed together using a roller. This step ensured that the microdevice was airtight and that the integrity of the sample and reagents within the device was maintained.

The fabrication of HTP microfluidic chips by Nguyen et al. [43] and Nguyen et al. [44] demonstrates the evolving sophistication in microfluidic technology, particularly in the context of molecular diagnostics. Each study, while sharing a common foundation in design and fabrication techniques, introduces unique elements in the integration of diagnostic components. In both studies, the initial design of the microfluidic chips was conducted using Cut2D software and incorporated with Fusion 360 for advanced 3D modeling. These software tools enabled precise mapping of the intricate microfluidic structures needed for the chips. The actual fabrication was performed using a CNC milling machine, which precisely carved these designs into 3 mm thick PMMA sheets. Post-fabrication, each chip underwent a thorough cleaning process involving sonication in Deconex 11 Universal cleaner followed by rinsing with ethanol and further sonication in distilled water. This step was critical to ensure the removal of any residues from the fabrication process. Following cleaning, each chip was subject to specific surface treatments. The chips were coated with superhydrophobic and hydrophilic reagents in strategic areas to control the behavior of fluids within the microdevice. The super-hydrophobic coating was applied to most microchannels to prevent unintended spreading of liquids, while the siphon valves received a hydrophilic coating for efficient fluid movement.

In Nguyen et al.'s study [43], the focus was on developing a chip for COVID-19 testing. The chip included chambers for RT-LAMP primer sets specific to COVID-19, and the master mix necessary for the RT-LAMP reaction was freeze-dried in the collection chambers. Nguyen et al. further advanced the chip's design to process 30 samples at once, emphasizing its utility for multiplex pathogen detection [44]. This chip was tailored for handling a broader range of pathogens, and each reaction chamber was specified for different pathogen detection, showcasing a significant leap in the chip's diagnostic capabilities.

The final assembly of the chips across all studies involved sealing the PMMA sheets with pressure-sensitive adhesive (PSA) films. This step was crucial for creating a stacked, three-layer structure, essential for maintaining the integrity and isolation of the microfluidic pathways. The sealed chips were then ready for use in various diagnostic applications, ranging from COVID-19 testing to multiplex pathogen detection.

4. Construction of a portable HTP genetic analyzer

The portable POCT genetic analyzer is an integrated system designed to be compact and lightweight, facilitating hand delivery and use in a variety of settings. The system's portability is a critical feature, making it ideal for field diagnostics and rapid response scenarios. The analyzer comprises three main components: a heating system, a centrifugal motor, and a UV-visible or fluorescence optics unit. The heating system uses flat polyimide ring heaters or peltier heaters to cover all reaction chambers and maintain a uniform temperature on the centrifugal microdevice, essential for consistent LAMP reactions. The centrifugal motor facilitates fluid flow in the microfluidic channels, and its speed and direction are controlled to precisely manage the sample processing steps. The UV-vis or fluorescence optics, equipped with a light source and a spectrometer, are used to measure the absorbance or fluorescence during the LAMP reaction, a critical step in the diagnostic process. The analyzer's design prioritizes user-friendliness and automation. Input for the rotational program and temperature settings is managed via a touchscreen monitor, streamlining the operational workflow. The system's centrifugal microdevice can be loaded similarly to a CD-ROM, which, when combined with the automatic control of rotation and temperature, simplifies the user interaction required for genetic analysis.

Nguyen et al., [43], led the way with their compact POCT workstation, an embodiment of portability with dimensions remarkably conducive to field deployment (40 cm × 50 cm × 30 cm) and a manageable weight of 10 kg. This system amalgamates several critical components, including a spindle motor, a syringe pump with multiple ports, and a fluorescence detector. The choice of these components reflects a deep understanding of the need for robust yet flexible systems in POCT scenarios. Notably, the integration of a user-friendly touchscreen interface is a pivotal feature that simplifies the operation, making the technology accessible to a broader range of users, including those in remote or resource-limited settings (Fig. 4).

In contrast, the 2024 study by Nguyen et al. described a POC workstation designed for automatic solution injection and real-time fluorescence detection [44]. This workstation consisted of four main parts: solution storage vials, a multiple-syringe pump, a customized linear motion system, and a POC analyzer.

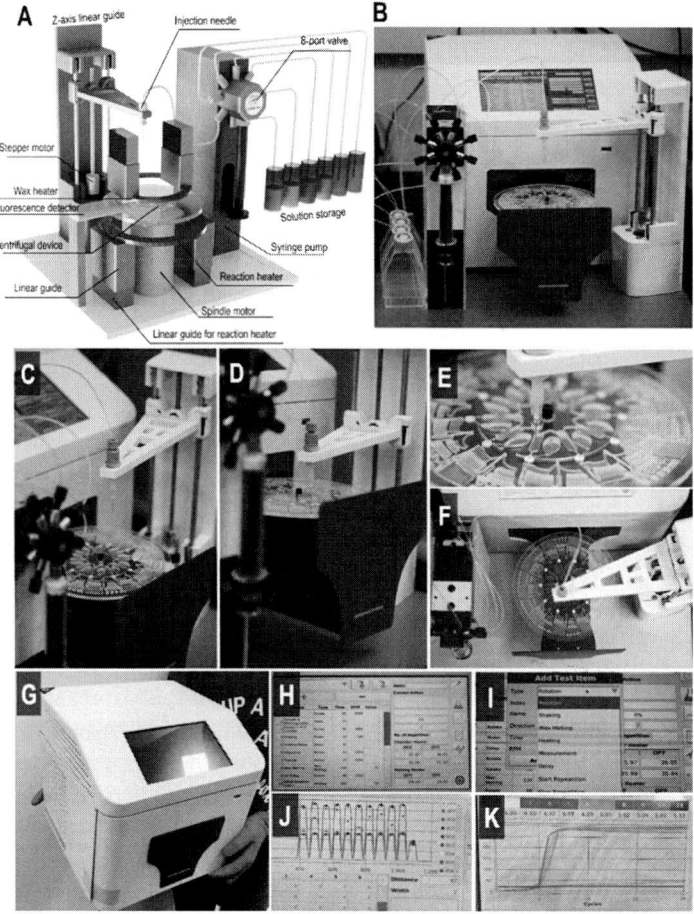

Fig. 4. A compact POCT workstation by Nguyen et al. (2022). (A) Illustration of the POCT workstation equipped with a spindle unit, a solution delivery unit, a moving unit, a heating unit, and a detection unit. (B) A digital image of the POCT workstation and a centrifugal disc. (C) The injection needle is at zero position which is 15 cm above the disc. (D-E) The injection needle moved down and penetrated into the PSA film in the middle of the one of the aliquot chambers and introduced a solution for dividing into 10 aliquoting. (F) A top view of the integrated POCT system. (G) The POCT platform is small in size and light in weight for POCT application. (H-I) The serial operation steps can be inserted to the main interface of the program. (J) Real-time monitoring of the fluorescence intensity of the RT-LAMP reaction for each reaction chamber. (K) Real-time amplification plot on the POCT as a final genetic analysis data. (Reprinted with permission from Ref. [43]. Copyright (2022) Elsevier.)

The storage vials were connected to multiple-port valves for washing and elution buffers. A linear guide system controlled the movement of an injection needle, which punctured the PSA film in the microfluidic chip's aliquot chamber for solution injection. This z-axis injecting approach was noted for its precision, simplicity, and cost-effectiveness compared to more complex commercial systems. The POC analyzer was a critical component, integrating a spindle motor, heaters, a fluorescence detector, and a screen display. This setup enabled automatic chip operation, gene amplification, real-time detection, and data analysis. The fluorescence detection was particularly notable, utilizing a commercially available detector with specific excitation and emission wavelengths to detect the intercalating dye, SYTO 9 (Fig. 4).

From the works done by Nguyen et al. [43] and Nguyen et al. [44], it becomes evident that the field of POCT is moving towards systems that are not only technically advanced but also user-centric and adaptable to diverse environments. The emphasis on compact design, automated processes, and precise control mechanisms is a clear response to the growing demand for rapid and accurate genetic analysis in various settings, from clinical laboratories to remote field locations. These advancements reflect a broader trend in medical technology, where portability, efficiency, and user-friendliness are no longer aspirational goals but essential features of cutting-edge devices.

5. Chip operation

The operation of the HTP centrifugal microfluidic chip integrates mechanical, fluidic, and biochemical principles to facilitate rapid, automated, and accurate molecular diagnostics. This process begins with sample preparation and introduction, where precision is paramount. A key aspect of the chip's functionality is the precise control of fluid movement, achieved through a combination of rotational forces and capillary action. The centrifugal microfluidic devices developed by Seo et al. [45], Nguyen et al. [43], and Nguyen et al. [44] utilize a sophisticated mechanism that leverages centrifugal force for fluid manipulation, crucial for molecular diagnostic processes. In these devices, samples are introduced into a reservoir and then propelled along microfluidic channels and into various chambers at controlled speeds and directions, dictated by a custom-built centrifugal system. This system, capable of precise rotational speed adjustments, plays a pivotal role in distributing the sample into aliquoting chambers and ensuring its interaction with pre-stored reagents like LAMP primers or RNA extraction solutions. For instance, at lower speeds (around 850 to 1000 RPM), the devices effectively split the sample into aliquoting chambers, while higher speeds (up to 5000 RPM) are employed to draw the aliquoted samples into reaction chambers or to pass them through microfiber filters for RNA purification. The direction of rotation, either clockwise or counterclockwise, is strategically altered to guide the samples towards different pathways – to waste chambers for discarding or to reaction chambers for analysis. Additionally, in some designs, specific features like passive valves and zigzag channels exploit capillary forces to aid in the precise division of solutions, enhancing the efficiency of the process. These mechanisms, combined with additional elements like temperature control for reactions and optical systems for real-time monitoring, underscore the intricate engineering and functional sophistication of these centrifugal microfluidic devices in conducting complex molecular diagnostics.

Across these studies, a common theme is the use of centrifugal force to control fluid movement within microfluidic chips. However, each study employs this mechanism differently, tailored to their specific diagnostic applications – from LAMP reactions for pathogen detection to RNA extraction for COVID-19 diagnostics. The devices' rotational speeds, fluid injection methods, and reaction monitoring techniques vary, reflecting each study's unique focus and objectives. Seo et al.'s colorimetric approach [45], Nguyen et al.'s focus on Covid-19 genes [43], and Nguyen et al.'s [44] automation and precision in fluid handling illustrate the versatility and potential of centrifugal microfluidic systems in molecular diagnostics. These studies collectively demonstrate the evolving sophistication in microfluidic technology, adapting to meet diverse diagnostic needs with precision and efficiency.

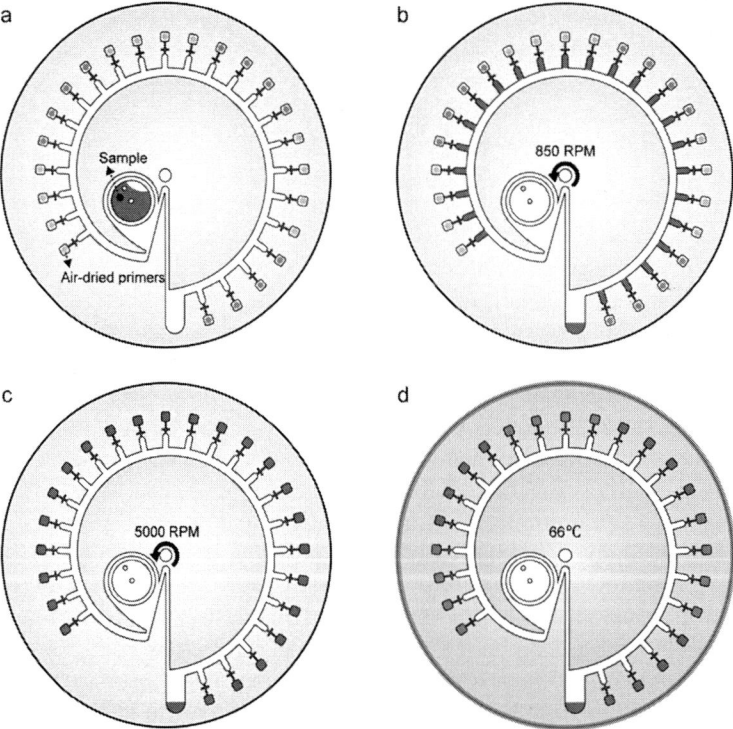

Fig. 5. The schematic images of the whole operation procedure on the microdevice. (a) The initial state of the microdevice including the sample (purple) in the reservoir and the dried primers (red and green colours mean different primer sets targeting specific foodborne pathogen) in the reaction chambers before rotational speed control process. (b) Aliqouting the loaded sample at 850 rpm (counterclockwise). (c) Transfer of the aliqouted sample to the reaction chambers at 5000 rpm (counterclockwise). (d) The LAMP isothermal amplification in the laboratory oven and the colorimetric detection. Sky blue represents a positive result and purple represents a negative one. (Reprinted with permission from Ref. [45]. Copyright (2017) Elsevier.)

Seo et al. describe a centrifugal LAMP microdevice operation beginning with the pipetting of LAMP primer mixtures into reaction chambers, followed by sample injection into the reservoir (Fig. 5) [45]. Key to this process is the use of

Ethidium Bromide Titration (EBT) for color indication of magnesium ion concentration. The device, mounted on a custom-made centrifugal system with temperature and rotational speed control, processes the sample at different RPMs. Initially, at 850 RPM (counterclockwise), the sample is divided into aliquoting chambers, and then at 5000 RPM, it's drawn into reaction chambers for mixing with primers. The LAMP reaction occurs at 66 °C in an oven, with colorimetric changes indicating pathogen presence.

The operation of the centrifugal microfluidic devices, developed by Nguyen et al. [43] and Nguyen et al. [44], illustrates a cohesive approach to molecular diagnostics, utilizing centrifugal force for precise fluid manipulation and sample processing. Despite each study targeting different applications, the core operational principles remain consistent, showcasing the versatility and efficiency of these devices in molecular diagnostics. In both studies, the initial step involves introducing the sample into the microfluidic device. In the 2022 study, the focus was on processing samples for COVID-19 detection. Here, a virus sample mixed with AVL buffer and carrier RNA was introduced into the device's sample chamber. The 2024 study also followed a similar approach, with the 2024 study focusing on multiplex pathogen detection. The process begins with the manual introduction of lysate samples into the device (Fig. 6). The lysate samples are carefully pipetted into the sample chambers of the HTP chip. This step is critical for initiating the RNA extraction process. Once the samples are in place, the chip is set into motion at a rotation speed of -5000 RPM. This rapid spinning action propels the lysates through GF/F filters, effectively capturing the RNAs and directing the remaining lysate towards waste chambers. This phase of RNA capturing is essential as it ensures the efficient separation and retention of RNA from the lysate. A key aspect of the device's functionality is the automated introduction of the washing buffer. Utilizing the POC workstation's syringe system, the washing buffer is accurately withdrawn and directed towards the needle. The linear guide then moves downward, positioning the needle to puncture the PSA layer of the aliquot chamber. This action results in the injection of the washing buffer into all aliquot chambers through a zigzag channel. The channel's design ensures an even distribution of the buffer across all chambers, streamlining the washing process. Subsequently, the centrifugal force of -5000 RPM is applied again, driving the buffer through the GF/F filters to the waste chambers and effectively purifying the captured RNAs. The elution stage follows the washing step, where the elution buffer is similarly introduced into the aliquot chambers. The rotation direction of the chip is then switched to +5000 RPM. This change in centrifugal force direction is crucial as it drives the elution buffer through the GF/F filters into the collection chambers. Here, the purified RNAs are gathered, marking a successful RNA recovery process. In preparation for the RT-LAMP reaction, the purified RNAs in the collection chambers are mixed with the lyophilized RT-LAMP master mix. The device then meters this mixture into 10 µL aliquots using a siphon valve, a step performed at +1000 RPM. The reaction mixture is then driven into the reaction chambers by increasing the centrifugal force to +5000 RPM, ensuring a precise and uniform distribution of the mixture for the upcoming reaction. The final steps involve moving the HTP chip into the POC analyzer, where the solid wax in the wax chamber is melted at 80 °C for 3 minutes. The device is then rotated at +1500 RPM to block the metering chambers with wax, a crucial step to prevent evaporation during the RT-LAMP reaction. The RT-LAMP reaction is carried out at 64 °C for one hour, with fluorescence signals being recorded every three minutes from each reaction chamber. The 2022 study utilizes fluorescence detection, detaching the heaters every three minutes to measure fluorescence intensity in each chamber. The 2024 study also records fluorescence signals to plot LAMP amplification curves, offering real-time data analysis.

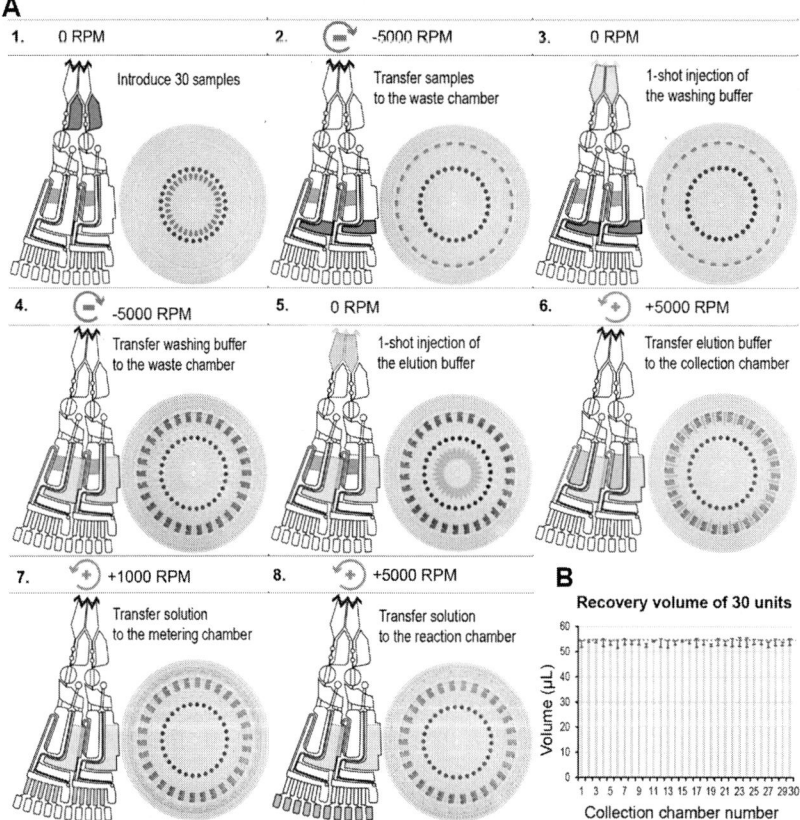

Fig. 6. Operation of the 30 units HTP centrifugal microfluidic chip. (A) (1). Thirty lysate samples were manually loaded into 30 sample chambers and (2) discharged into the waste chamber via centrifugal force at -5000 RPM. (3) The washing buffer was divided automatically into the aliquot chambers with one-shot owing to the zigzag aliquot structure. (4) The washing buffer was driven into the waste chamber at -5000 RPM. (5) In the same procedure, the elution buffer was automatically dispensed into 30 aliquot chambers with one-shot by the zigzag aliquoting structure. (6) The centrifugal rotation was switched counter-clockwise (+5000 RPM) to transfer the elution buffer into the collection chambers where the elution was mixed with the lyophilized RT-LAMP master mix. (7) At +1000 RPM, the final mixture solution was driven from the collection chambers to the metering chambers via siphon valves. (8) The centrifugal force was increased to +5000 RPM to push the mixture solution to the reaction chambers. (B) Recovered volume in the 30 collection chambers shows the high accuracy for the aliquoting process. (Reprinted with permission from Ref. [44]. Copyright (2024) Elsevier.)

6. Application of pathogen detection

6.1. A single unit-centrifugal microfluidics chip with multiple reaction chambers

The study by Seo et al. demonstrates the effective use of a centrifugal LAMP microdevice for the detection of foodborne pathogens, showcasing detailed results that emphasize the device's precision and reliability in pathogen identification [45]. The study's results focus on three representative foodborne pathogens: *E. coli* O157:H7, *Salmonella enterica*, and *Vibrio parahaemolyticus*.

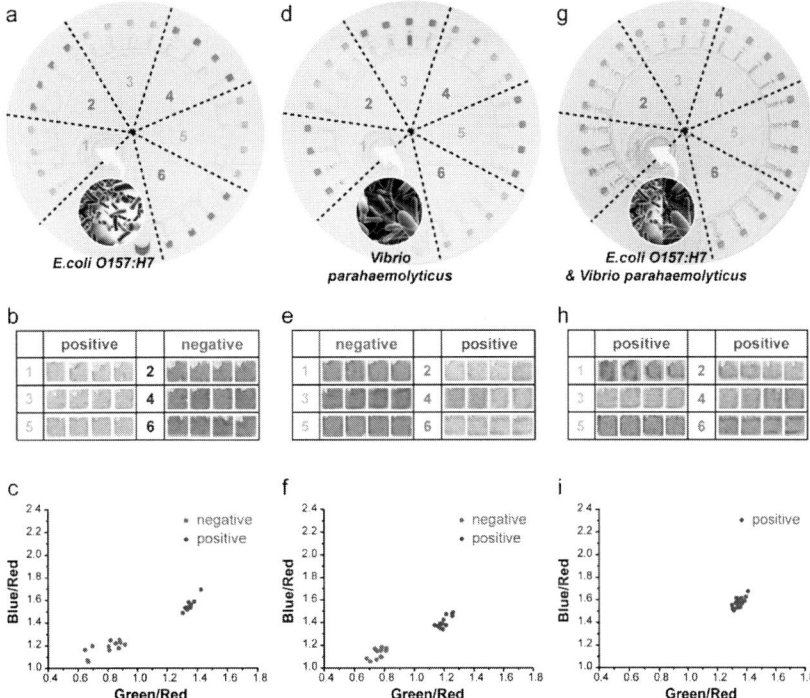

Fig. 7. Multiplex foodborne pathogen detection in the centrifugal Direct-LAMP microdevice using bacterial cells as a sample. The green and red colored bacteria images mean the cells of *E. coli* O157:H7 and *V. parahaemolyticus*. Parts 1, 3, and 5 with green color indicate the reactions with primer mixtures for targeting *E. coli* O157:H7, and parts 2, 4, and 6 with red color implies the reactions with primer mixtures for targeting *V. parahaemolyticus*. (a-c) Detection of *E. coli* O157:H7, (d-f) *V. parahaemolyticus*, and (g-i) both bacteria. (a, d, and g) The digital images of the centrifugal LAMP microdevice after 60 minutes of the LAMP reaction. (b, e, and h) Enlarged and cropped images of the positive and negative LAMP reactions. (c, f, and i) Graphs of the G/R ratio against the B/R ratio for the negative (purple circles) and positive (blue circles) LAMP reactions. (Reprinted with permission from Ref. [45]. Copyright (2017) Elsevier.)

The detection process began with the loading of the LAMP reaction mixture, containing 7500 copies of each bacterial DNA, into the sample reservoir. After distribution into the 24 reaction chambers, each chamber was estimated to contain approximately 250 copies of the bacterial DNA. The LAMP primer sets specific to each pathogen were pre-coated in designated sections of the reaction chambers. During the LAMP reaction, the color change of the mixture from purple to sky blue within 60 minutes indicated successful amplification of the target genes. This color change is due to EBT, which responds to changes in Mg^{2+} ion concentration during the LAMP reaction. The formation of insoluble $Mg_2P_2O_7$ salts due to the interaction between pyrophosphate ions and Mg^{2+} ions result in this distinct color shift.

The study's specificity in detection was underscored by the presence of blank chambers that did not contain primer mixtures and remained purple, indicating no amplification and serving as negative controls. This distinction between the reacted and non-reacted chambers validated the specificity of the primers and the absence of contamination. To enhance the reliability of the colorimetric analysis, the study utilized an RGB-based image processing method. This approach involved converting the images of reaction chambers to RGB values and calculating ratios like Green/Red (G/R) and Blue/Red (B/R). The significant difference in these ratios between positive (bacterial DNA present) and negative (no bacterial DNA) reactions provided a more accurate and objective assessment than the naked eye.

Seo et al. demonstrated the microdevice's capability for multiplex detection by simultaneously identifying *E. coli* O157:H7 and *V. parahaemolyticus* on a single device. This was achieved by directly using cultured bacterial cells without the need for DNA extraction, showcasing the device's adaptability and the robustness of the LAMP reaction even in the presence of potential inhibitors. The detailed results of the study indicated clear demarcation in the G/R and B/R ratios for negative and positive reactions. For instance, the G/R and B/R ratios for negative reactions ranged from 0.623 to 0.776 and 1.012 to 1.105, respectively. In contrast, positive reactions exhibited G/R ratios from 1.172 to 1.369 and B/R ratios from 1.273 to 1.537, demonstrating a distinct gap sufficient for reliable differentiation (Fig. 7).

6.2. A centrifugal HTP microfluidics chip with 10 units for COVID-19 diagnostics

Nguyen et al. made significant strides in optimizing RNA extraction for COVID-19 diagnostics [43]. Their study emphasized the crucial role of guanidine-HCl in binding RNA to glass fiber during the extraction process. This optimization facilitated efficient RNA capture, a critical step in the accurate detection of COVID-19. The team demonstrated the practical application of their centrifugal device using clinical COVID-19 samples, thereby highlighting its potential in pandemic response scenarios.

A significant focus of the study was the optimization of the RNA extraction process using GF/F glass filter papers (Fig. 8). The inherent challenge of RNA binding to these filters, due to the mutual negative charges of RNA and borosilicate glass microfiber, was addressed by introducing guanidine-HCl in the lysis buffer. At an acidic pH of 5, this component played a role in overcoming electrostatic repulsion and facilitating RNA's adsorption to the silica surface through intermolecular hydrogen bonding. The guanidine-HCl not only aided in reducing the Debye length, thus enhancing the electrostatic screening effect, but also hydrated the RNA, reducing the availability of free water molecules. The addition of acidic binding buffer further protonated the silanol groups, increasing the binding sites for RNA and thereby augmenting the adsorption process [48–52]. The study meticulously determined that using three layers of GF/F filter paper (Whatman, USA) yielded the most efficient RNA extraction, comparing closely with the Qiagen Viral extraction kit's performance, which had Ct values of 24.7±0.5 for the device and 23.2±0.2 for the Qiagen kit.

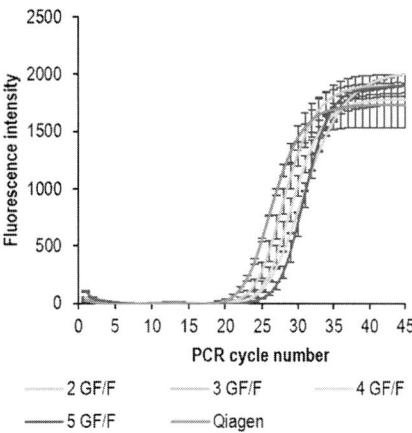

Fig. 8. RT-qPCR amplification plot using the extracted RNAs produced from different numbers of GF/F filter papers on a disc, and the comparison with the Qiagen kit. (Reprinted with permission from Ref. [43]. Copyright (2022) Elsevier.)

The sensitivity of the device was rigorously tested through a Limit of Detection (LOD) assessment using diluted heat-inactivated COVID-19 samples obtained from ATCC. The LOD for the orf1ab, S, and N genes was uniformly identified as 2×10^2 copies/µL, equivalent to 100 copies in the sample (Fig. 9). This low LOD underscores the device's capability to detect minimal viral loads, a critical feature for early disease detection. To further validate the system's applicability in a real-world scenario, clinical COVID-19 samples were processed and analyzed. These clinical samples, consisting of nasal swabs in universal transport medium, were mixed with AVL buffer and RNA carrier, and then subjected to the device for testing. The amplification profiles observed in all three reaction chambers for these clinical samples corroborated the device's feasibility and reliability for onsite COVID-19 diagnostics.

To evaluate the specificity of the device, purified RNAs from various respiratory viruses were tested using COVID-19 primers. This specificity test was crucial in ensuring that the device's detection was exclusively responsive to COVID-19 RNA and not prone to cross-reactivity with other respiratory viruses. The results were definitive: no signal was generated when using the non-COVID-19 templates, while positive amplification curves were consistently obtained with COVID-19 templates. These findings highlight the high specificity of the COVID-19 primers employed in the device.

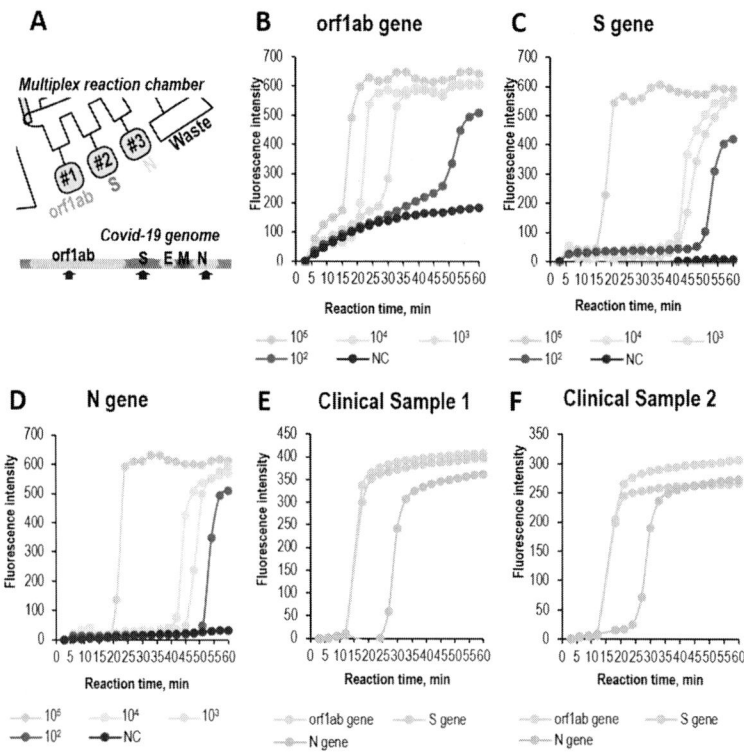

Fig. 9. (A) Position of the target gene in the reaction chambers. The LOD test of the POCT genetic analyzer to identify the COVID-19 virus (B) orf1ab gene, (C) S gene, and (D) N gene. (E-F) Amplifying profiles of two positive clinical samples. (Reprinted with permission from Ref. [43]. Copyright (2022) Elsevier.)

6.3. A centrifugal HTP microfluidics chip with 30 units for COVID-19 diagnostics

Nguyen et al. showcased the centrifugal HTP microfluidic chip, a groundbreaking device for the detection of respiratory viruses such as Influenza A, Influenza B, and SARS-CoV-2 [44]. This POC system integrates RNA purification, RT-LAMP, and real-time fluorescence detection, capable of processing up to 30 samples simultaneously. The chip's design features a zigzag aliquoting structure that facilitates ultra-rapid and uniform distribution of reagents into reaction chambers, enhancing throughput efficiency. The team developed a novel lyophilization technique for RT-LAMP master mix on a microfluidic chip to address the challenges of reagent transport and storage in areas lacking adequate infrastructure. This process involved a rapid one-pot lyophilization approach, combining primary and secondary drying into a single streamlined step. The procedure started with deep freezing at -85 °C for 12 hours, followed by a one-step freeze-drying at -110 °C for 10 hours. This innovation significantly reduced the typical duration of the lyophilization process and maintained the stability of the lyophilized master mix at 4 °C for up to 28 days. Comparative tests with RNA templates of Influenza A H3N2 showed no significant difference in effectiveness between the liquid and lyophilized forms of the RT-LAMP master mix, validating the procedure's efficacy (Fig. 10).

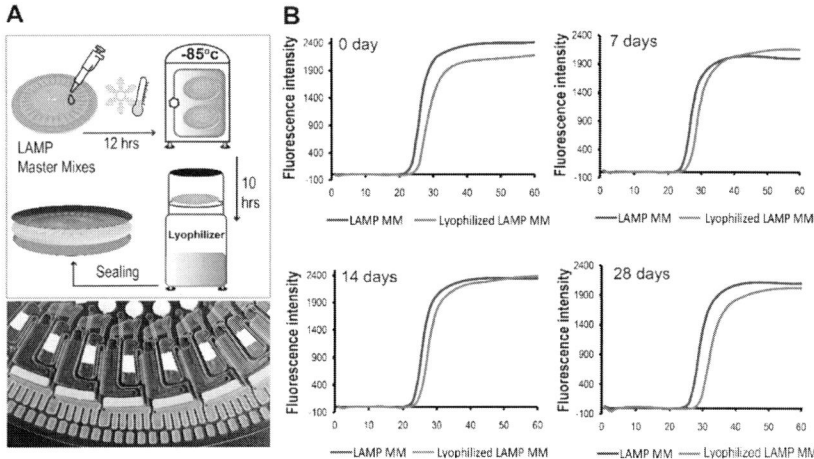

Fig. 10. Development of the lyophilization process for the RT-LAMP master mix on the HTP microfluidic chip. (A) The RT-LAMP master mix was freeze-dried in the collection chamber of the chip. After being deep-frozen at -85 °C for 12 hrs, the HTP chip was transferred to a lyophilizer for one-step lyophilization at -110 °C. (B) The lyophilized RT-LAMP master mix retains its activity even after +4 °C storage for 28 days in comparison to the liquid form of the RT-LAMP master mix, which was stored at -20 °C. (Reprinted with permission from Ref. [44]. Copyright (2024) Elsevier.)

In their specificity test for five respiratory viruses, including COVID-19 (orf1ab gene and S gene), H1N1, H3N2, and Influenza B, the chip displayed high precision in molecular diagnostics, accurately detecting two genes of SARS-CoV-2 with early emergence of fluorescence signals. The reaction chambers of the chip contained specific LAMP primer sets for each virus. During the amplification step at 64 °C, fluorescence signals from the reaction chambers were recorded every 3 minutes. The results demonstrated high specificity and reproducibility across three units, with no false positives. The system successfully detected two genes of SARS-CoV-2, as well as other respiratory viruses, with most samples showing detectable signals within the first 20 minutes (Fig. 11). This rapid detection capability underlined the system's potential for efficient diagnostics of multiple samples simultaneously.

The study also included testing 20 nasopharyngeal samples infected with SARS-CoV-2, Influenza H1N1, H3N2, or Influenza B. Additionally, five negative control samples consisting of water in a transport medium were used. All clinical samples displayed significant amplification signals with threshold times between 18 and 46 minutes, whereas the negative controls showed no amplification (Fig. 12). For the LOD test, the team selected a heat-inactivated SARS-CoV-2 sample and prepared serially diluted samples ranging from 10^5 to 10^1 copies/µL. The LODs for the orf1ab and S gene of SARS-CoV-2 were determined to be 10^2 copies/µL, highlighting the system's sensitivity and comparability to existing studies and commercial LAMP-based products.

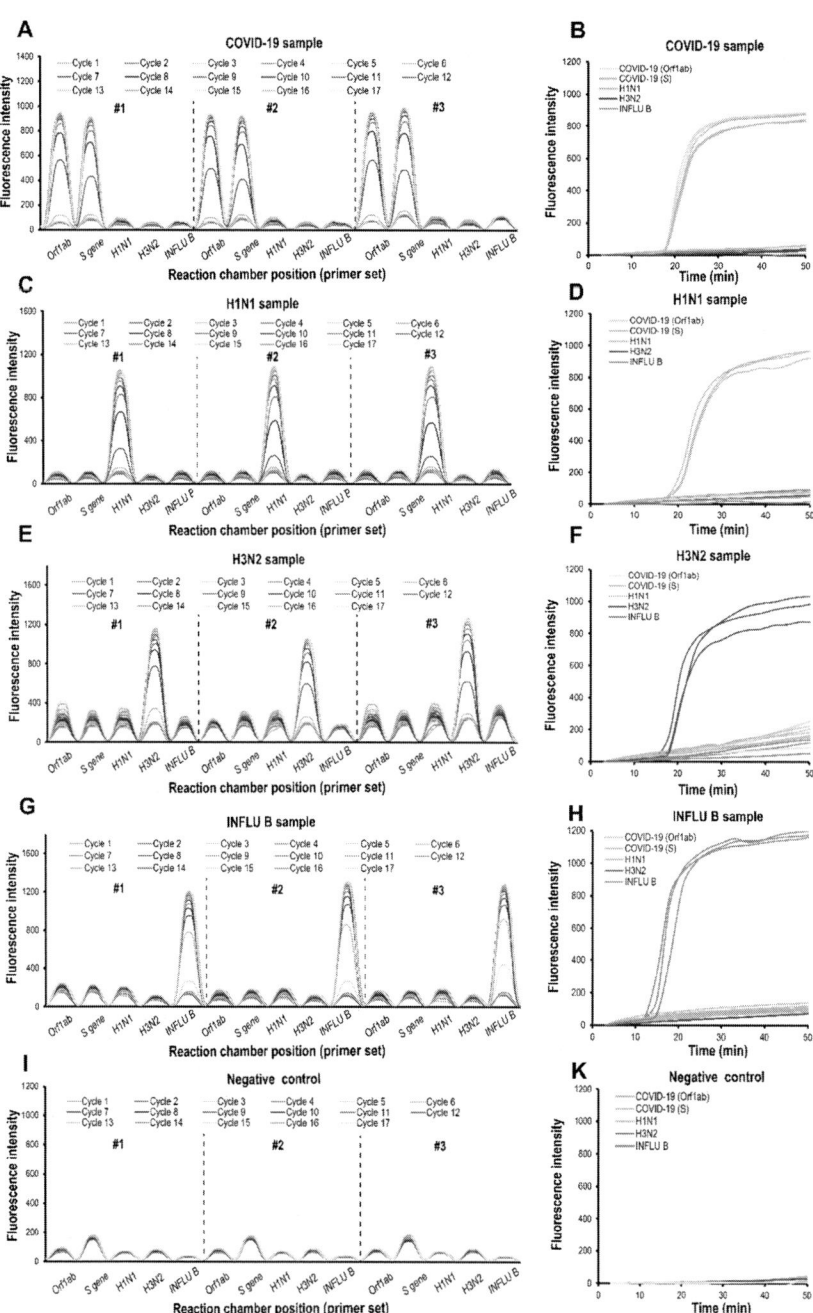

Fig. 11. The specificity and reproducibility testing of the centrifugal HTP microfluidic chip. The raw fluorescence signals obtained from the reaction chambers were scanned at 3-minute intervals (the left panel) and then used for plotting the LAMP reaction curves (the right panel). Each target was tested 3 times on the 3 separate units of the HTP chip to evaluate reproducibility. The data demonstrates the high specificity of the LAMP diagnostics for respiratory viruses without false positives and negatives. (Reprinted with permission from Ref. [44]. Copyright (2024) Elsevier.)

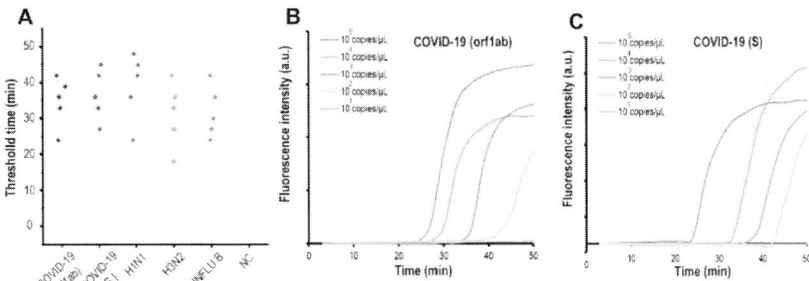

Fig. 12. The clinical sample test and LOD test for four common respiratory pathogens using the HTP centrifugal microchip. (A) Five clinical samples for each target were tested and successfully analyzed with the threshold time between 18 minutes and 46 minutes. As a negative control, water was used instead of viral templates. The LOD test for (B) orf1ab gene and (C) S gene of SARS-CoV-2. The LODs for both genes were 10^2 copies/μL. (Reprinted with permission from Ref. [44]. Copyright (2024) Elsevier.)

7. Conclusion

The emergence and development of centrifugal HTP molecular diagnostics, as exemplified by the research of Seo et al. [45], Nguyen et al. [43], and Nguyen et al. [44], represent a transformative advancement in the field of pathogen detection and molecular diagnostics. These innovations highlight a significant shift from conventional laboratory-based testing methods to more versatile, efficient, and accessible diagnostic solutions.

Seo et al.'s pioneering work in utilizing centrifugal microfluidic devices for the rapid detection of foodborne pathogens illustrates the practicality and effectiveness of this technology in public health and food safety [45]. Their approach, combining colorimetric assays with advanced image processing techniques, exemplifies how these devices can simplify and accelerate the detection process, making it more accessible for point-of-care applications.

Nguyen et al.'s development of an integrated system for COVID-19 testing further underscores the adaptability of centrifugal molecular diagnostics to respond to emergent global health crises [43]. Their work demonstrates how these devices can process multiple samples simultaneously, reducing the time and labor associated with traditional diagnostic methods. This is particularly crucial in pandemic situations, where rapid HTP testing is essential for effective disease management and control.

Lastly, Nguyen et al.'s development of a HTP platform for the detection of respiratory viruses, including SARS-CoV-2, demonstrates the critical role these devices can play in pandemic preparedness and response [44]. Their integrated system, capable of processing a large number of samples quickly, highlights how centrifugal molecular diagnostics can be pivotal in managing and controlling the spread of infectious diseases.

In conclusion, the centrifugal HTP molecular diagnostics, as evidenced by these studies, stand at the forefront of a diagnostic revolution. They offer an array of benefits, including speed, efficiency, cost-effectiveness, and adaptability, making them a cornerstone technology in modern healthcare. These advancements not only enhance our ability to respond to current and emerging health threats but also pave the way for future innovations in medical diagnostics.

References

1. M. Mustafa Hellou, A. Górska, F. Mazzaferri, E. Cremonini, E. Gentilotti, P. De Nardo, I. Poran, M. M. Leeflang, E. Tacconelli and M. Paul, *Clinical Microbiology and Infection*, 2021, 27, 341–351.

2. E. Valera, A. Jankelow, J. Lim, V. Kindratenko, A. Ganguli, K. White, J. Kumar and R. Bashir, *ACS Nano*, 2021, 15, 7899–7906.

3. W. Feng, A. M. Newbigging, C. Le, B. Pang, H. Peng, Y. Cao, J. Wu, G. Abbas, J. Song, D.-B. Wang, H. Zhang and X. C. Le, *Anal Chem*, 2020, 92, 10196–10209.

4. I. Smyrlaki, M. Ekman, A. Lentini, N. Rufino de Sousa, N. Papanicolaou, M. Vondracek, J. Aarum, H. Safari, S. Muradrasoli, A. G. Rothfuchs, B. Högberg and B. Reinius, *Nat Commun*, 2020, 11, 4812.

5. E. Nunez-Bajo, A. Silva Pinto Collins, M. Kasimatis, Y. Cotur, T. Asfour, U. Tanriverdi, M. Grell, M. Kaisti, G. Senesi, K. Stevenson, K. Stevenson and F. Güder, *Nat Commun*, 2020, 11, 6176

6. G. Liu and J. F. Rusling, *ACS Sens*, 2021, 6, 593–612.

7. N. Boonham, J. Kreuze, S. Winter, R. van der Vlugt, J. Bergervoet, J. Tomlinson and R. Mumford, *Virus Res*, 2014, 186, 20–31.

8. J. Yang, V. M. Phan, C.-K. Heo, H. V. Nguyen, W.-H. Lim, E.-W. Cho, H. Poo and T. S. Seo, *Sens Actuators B Chem*, 2023, 380, 133331.

9. H. Zhu, Z. Fohlerová, J. Pekárek, E. Basova and P. Neužil, *Biosens Bioelectron*, 2020, 153, 112041.

10. H. Zhu, H. Zhang, S. Ni, M. Korabečná, L. Yobas and P. Neuzil, *TrAC - Trends in Analytical Chemistry*, 2020, 130, 115984.

11. N. Ravi, D. L. Cortade, E. Ng and S. X. Wang, *Biosens Bioelectron*, 2020, 165, 112454.

12. M. J. Loeffelholz, D. Alland, S. M. Butler-Wu, U. Pandey, C. F. Perno, A. Nava, K. C. Carroll, H. Mostafa, E. Davies, A. McEwan, N. Zhang and D. H. Persing, *J Clin Microbiol*, 2020, 58, e00926.

13. P. M. Thwe and P. Ren, *Diagn Microbiol Infect Dis*, 2020, 98, 115123.

14. J. A. SoRelle, L. Mahimainathan, C. McCormick-Baw, D. Cavuoti, F. Lee, A. Thomas, R. Sarode, A. E. Clark and A. Muthukumar, *Clinica Chimica Acta*, 2020, 510, 685–686.

15. G. Hansen, J. Marino, Z.-X. Wang, K. G. Beavis, J. Rodrigo, K. Labog,

L. F. Westblade, R. Jin, N. Love, K. Ding, J. Sickler and N. K. Tran, *J Clin Microbiol*, 2021, 59, e02811.

16 K. Nagamine, K. Watanabe, K. Ohtsuka, T. Hase and T. Notomi, *Clin Chem*, 2001, 47, 1742–1743.

17 X. Song, F. J. Coulter, M. Yang, J. L. Smith, F. G. Tafesse, W. B. Messer and J. H. Reif, *Sci Rep*, 2022, 12, 7043.

18 B. Özay and S. E. McCalla, *Sensors and Actuators Reports*, 2021, 3, 100033.

19 B. B. Oliveira, B. Veigas and P. V. Baptista, *Frontiers in Sensors*, 2021, 2, 752600.

20 H. V. Nguyen, V. D. Nguyen, H. Q. Nguyen, T. H. T. Chau, E. Y. Lee and T. S. Seo, *Biosens Bioelectron*, 2019, 141, 11146.

21 O. Strohmeier, M. Keller, F. Schwemmer, S. Zehnle, D. Mark, F. Von Stetten, R. Zengerle and N. Paust, *Chem Soc Rev*, 2015, 44, 6187–6229.

22 H. V. Nguyen, V. D. Nguyen, E. Y. Lee and T. S. Seo, *Biosens Bioelectron*, 2019, 136, 132–139.

23 Y. Yao, X. Chen, X. Zhang, Q. Liu, J. Zhu, W. Zhao, S. Liu and G. Sui, *ACS Sens*, 2020, 5, 1354–1362.

24 M. Geissler, D. Brassard, L. Clime, A. V. C. Pilar, L. Malic, J. Daoud, V. Barrère, C. Luebbert, B. W. Blais, N. Corneau, N. Corneau and T. Veres, *Analyst*, 2020, 145, 6831–6845.

25 S. J. Oh and T. S. Seo, *Analyst*, 2019, 144, 5766–5774.

26 M. Ji, Y. Xia, F.-C. Loo, L. Li, H.-P. Ho, J. He and D. Gu, *RSC Adv*, 2020, 10, 34088–34098.

27 S. F. Berlanda, M. Breitfeld, C. L. Dietsche and P. S. Dittrich, *Anal Chem*, 2021, 93, 311–331.

28 C. Dincer, R. Bruch, A. Kling, P. S. Dittrich and G. A. Urban, *Trends Biotechnol*, 2017, 35, 728–742.

29 H. Q. Nguyen, H. K. Bui, V. M. Phan and T. S. Seo, *Biosens Bioelectron*, 2022, 195, 113655.

30 A. Lee, J. Park, M. Lim, V. Sunkara, S. Y. Kim, G. H. Kim, M.-H. Kim and Y.-K. Cho, *Anal Chem*, 2014, 86, 11349–11356.

31 J.-M. Park, M. S. Kim, H.-S. Moon, C. E. Yoo, D. Park, Y. J. Kim, K.-Y. Han, J.-Y. Lee, J. H. Oh, S. S. Kim, W.-Y. Lee and N. Huh, *Anal Chem*, 2014, 86, 3735–3742.

32 P. Arosio, T. Müller, L. Mahadevan and T. P. J. Knowles, *Nano Lett*, 2014, 14, 2365–2371.

33 B. H. Park, D. Kim, J. H. Jung, S. J. Oh, G. Choi, D. C. Lee and T. S. Seo, *Sens Actuators B Chem*, 2015, 209, 927–933.

34 U. Y. Schaff and G. J. Sommer, *Clin Chem*, 2011, 57, 753–761.

35 J. H. Jung, B. H. Park, Y. K. Choi and T. S. Seo, *Lab Chip*, 2013, 13,

3383–3388.

36 J. H. Jung, B. H. Park, S. J. Oh, G. Choi and T. S. Seo, *Biosens Bioelectron*, 2015, 68, 218–224.

37 G. Choi, J. H. Jung, B. H. Park, S. J. Oh, J. H. Seo, J. S. Choi, D. H. Kim and T. S. Seo, *Lab Chip*, 2016, 16, 2309–2316.

38 M. Safavieh, M. U. Ahmed, E. Sokullu, A. Ng and L. Braescu, *Analyst*, 2013, 139, 482–487.

39 D. Liu, G. Liang, Q. Zhang and B. Chen, *Anal Chem*, 2013, 85, 4698–4704.

40 X. Fang, Y. Liu, J. Kong and X. Jiang, *Anal Chem*, 2010, 82, 3002–3006.

41 D. Liu, H. Shen, Y. Zhang, D. Shen, M. Zhu, Y. Song, Z. Zhu and C. Yang, *Lab Chip*, 2021, 21, 2019–2026.

42 V. M. Phan, S. W. Kang, Y. H. Kim, M. Y. Lee, H. V. Nguyen, Y. L. Jeon, W. I. Lee and T. S. Seo, *Sens Actuators B Chem*, 2020, 5, 1354-1362.

43 H. Van Nguyen, V. M. Phan and T. S. Seo, *Sens Actuators B Chem*, 2022, 353, 131088.

44 H. Van Nguyen, V. M. Phan and T. S. Seo, *Sens Actuators B Chem*, 2024, 399, 134771.

45 J. H. Seo, B. H. Park, S. J. Oh, G. Choi, D. H. Kim, E. Y. Lee and T. S. Seo, *Sens Actuators B Chem*, 2017, 246, 146–153.

46 A. Olanrewaju, M. Beaugrand, M. Yafia and D. Juncker, *Lab Chip*, 2018, 18, 2323–2347.

47 S. Lai, S. Wang, J. Luo, L. J. Lee, S.-T. Yang and M. J. Madou, *Anal Chem*, 2004, 76, 1832–1837.

48 X. Li, Q. R. Xie, J. Zhang, W. Xia and H. Gu, *Biomaterials*, 2011, 32, 9546–9556.

49 U. Lehmann, C. Vandevyver, V. K. Parashar and M. A. M. Gijs, *Angewandte Chemie - International Edition*, 2006, 45, 3062–3067.

50 P. R. Nair and M. A. Alam, *Nano Lett*, 2008, 8, 1281–1285.

51 X. Li, J. Zhang and H. Gu, *Langmuir*, 2012, 28, 2827–2834.

52 W.-P. Hu, Y.-C. Chen and W.-Y. Chen, *Sci Rep*, 2020, 10, 21132.

Chapter 8
Centrifugal Microfluidic Device for High-throughput Enzyme-linked Immunosorbent Assay

1. Background

In the realm of clinical diagnostics, two categories of assays stand as the bedrock of disease detection and health monitoring: molecular assays and serological assays. Molecular assays, such as the PCR and isothermal amplification, target the genetic footprint of pathogens or genetic markers of disease. These techniques have revolutionized our ability to detect and quantify nucleic acids with unparalleled specificity and sensitivity, thereby facilitating early diagnosis and monitoring of disease progression. Serological assays, on the other hand, focus on detecting the immune response to pathogens or biomarkers within bodily fluids. Methods like lateral flow assays, Enzyme-linked Immunosorbent Assays (ELISA), Western blotting, immunofluorescence assays (IFA), and chemiluminescence immunoassays (CLIA) etc. are quintessential in this category, providing critical information on the presence and concentration of antibodies or antigens [1–5].

The previous book chapters delved into the innovation of incorporating molecular assays onto centrifugal microfluidic chips, demonstrating how this integration can streamline complex PCR or isothermal amplification processes for point-of-care applications. This technology facilitates rapid, sensitive, and specific detection of nucleic acids, thereby enhancing the capability for on-field diagnostics and personalized medicine [7–11]. Building upon that foundation, this chapter explores the integration of serological assays - especially ELISA - onto centrifugal microfluidic devices. The goal is to address how these advanced platforms can overcome the traditional limitations of ELISA, such as prolonged assay times and the need for skilled technicians, by automating the assay process on a microfluidic chip. This integration can potentially transform serological testing, making it more accessible and feasible for high-throughput screening in hospitals, medical centers, and research laboratories.

ELISA represents a cornerstone in diagnostic immunology, epitomizing the integration of enzymatic reactions with antibody specificity to detect and quantify molecular targets. Since its inception in the early 1970s, ELISA has become a fundamental tool in diagnostics, vaccine development, and quality control in various industries due to its high sensitivity and specificity [12, 13].

Fundamentally, ELISA is predicated on the antigen-antibody interaction, where the specificity of antibodies towards their unique antigens is employed to capture and quantify the antigen from complex biological mixtures [14]. An antigen is captured by an immobilized antibody on a solid phase, and a matching antibody conjugated with an enzyme is directed to bind the antigen. The interaction is then visualized through an enzymatic reaction, typically resulting in a colorimetric change, which can be quantitatively measured. This technique leverages the amplifying power of enzymes, thus allowing the detection of minute quantities of antigen with high precision (Fig. 1) [15].

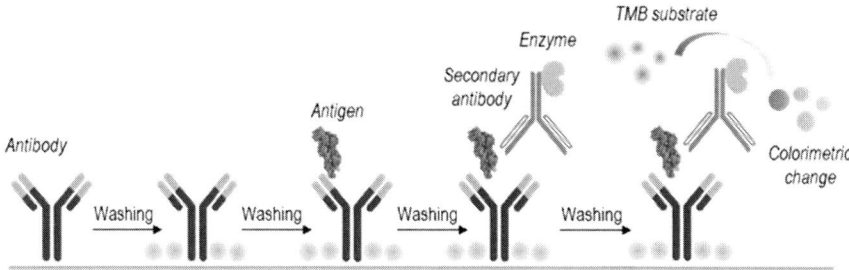

Fig. 1. Sandwich ELISA Workflow. A capture antibody is immobilized on the plate to seize the target antigen, followed by the introduction of a secondary antibody conjugated with an enzyme. Subsequent addition of a substrate catalyzes a color reaction, signaling the antigen's presence and quantity. Between each subsequent step, several washing processes are employed to remove nonspecific bindings.

Over the years, ELISA has evolved through several formats - from direct and indirect to sandwich and competitive - each tailored to meet specific experimental needs and sensitivities [1, 16–18] However, despite its versatility and robustness, conventional ELISA protocols are not without drawbacks. They often necessitate prolonged incubation times, extensive washing steps to reduce nonspecific binding, and substantial volumes of reagents and samples. The procedural complexity and manual intervention required in traditional ELISA setups have led to an inevitable quest for more streamlined, cost-effective, and high-throughput alternatives [16, 19].

The advent of microfluidics, the science of manipulating and controlling fluids at the microscale, has addressed many traditional challenges by miniaturizing bioassays and conducting them with enhanced efficiency and automation [2, 3, 5]. The integration of ELISA with microfluidics has given rise to a new paradigm in immunoassays, conferring numerous advantages including lower reagent consumption, reduced sample volume, and the parallel processing of multiple assays, thereby significantly diminishing the time and cost per analysis [5, 6, 20–25].

Centrifugal microfluidics, in particular, represents a transformative leap in this progression. It utilizes the centrifugal force generated from a spinning disc to manipulate fluids through microchannels and chambers on the disc substrate. This approach circumvents the need for external pumps and intricate channel designs inherent in other microfluidic platforms, thus simplifying the device architecture and potentially reducing the manufacturing costs. In the context of ELISA, the centrifugal microfluidic platforms enable the sequential delivery of reagents, mixing, incubation, and washing to be performed with precision and automation. By leveraging the control over fluid movement at varying rotational speeds, the centrifugal devices can finely tune the assay conditions, thereby enhancing reaction kinetics and improving the overall assay performance [26–28].

Moreover, recent advances have seen the integration of optical and electrochemical sensors within these discs, allowing for real-time detection and quantification of the ELISA reactions. This has opened avenues for point-of-care testing and on-site diagnostics, which are invaluable for rapid disease surveillance, environmental monitoring, and food safety testing.

In the research domain, high-throughput ELISA on centrifugal microfluidic devices has facilitated large-scale screening of biomarkers, expedited vaccine development, and supported the intricate study of immune responses. For instance, in the face of a global pandemic, such technologies can be pivotal for swift serological testing to assess the immune status of populations [20, 27].

The evolution of ELISA from benchtop to centrifugal microfluidic formats epitomizes the convergence of biotechnology with engineering and materials science to meet the pressing demands of modern diagnostics and analytical biochemistry. As such, the centrifugal microfluidic device for high-throughput ELISA stands not merely as a technological achievement but as a beacon of multidisciplinary innovation, heralding a future where rapid, sensitive, and accessible diagnostics are a ubiquitous reality.

2. Design of a centrifugal ELISA HTP chip

In the context of high-throughput (HTP) microfluidic designs, it provides a comprehensive understanding of their importance in modern diagnostics and research. These systems are integral for large-scale testing, crucial in areas like pandemic response, drug discovery, and personalized medicine. HTP microfluidic devices offer rapid, detailed analyses while processing numerous samples simultaneously. They significantly reduce sample and reagent volumes, thereby lowering costs and enhancing assay sensitivity and speed.

The centrifugal ELISA chip, a notable example, exemplifies the automation and miniaturization of ELISA. It features specialized microchannels, reaction chambers, and valves, with a disc-shaped design utilizing centrifugal force for efficient sample and reagent distribution, thus enabling the simultaneous processing of multiple tests.

This chapter examines an ELISA HTP chip designed by Nguyen et al. (2023) [29]. The chip's innovative design, tailored for high-throughput applications, marks a notable advancement in diagnostic technologies. It features a 6.5 cm radius and includes 40 parallel reaction units. The top side of the chip, made from PMMA, includes reaction chambers coated with antibodies, siphon valves, a common waste chamber, and aliquot chambers for distributing washing buffers and an HRP-secondary antibody solution. The bottom side contains microchannels, sample-loading holes, and air vent holes. Each unit features a sample-loading inlet, two aliquot chambers, passive valves, a capture antibody-coated reaction chamber, and a siphon valve leading to the common waste chamber (Fig. 2).

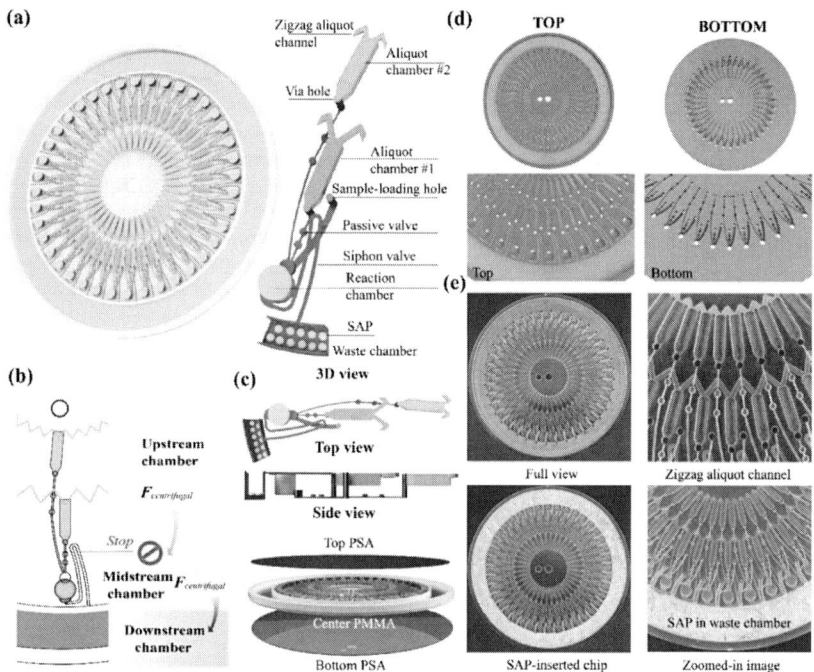

Fig. 2. Design and fabrication of the centrifugal microfluidic device for HTP ELISA. (a) Schematics illustrating the HTP microfluidic device containing 40 parallel units for ELISA. One unit includes an antibody-coated reaction chamber, a siphon valve, two aliquot chambers (aliquot chamber #1 and #2), two zigzag aliquot channels connecting the neighboring aliquot chambers, two sets of passive valves, and a sample-loading hole. (b) Aliquot chambers #1 and #2 serve as upstream chambers, responsible for dispensing reagents into the antibody-coated chamber (midstream) after rotation. The siphon valve allows for the halt of reagents at the reaction chamber during the incubation step, followed by their discharge into a waste chamber (downstream) through another centrifugal force. These steps can be repeated with high reproducibility. (c-d) CAD images for the micro-patterns on a top and a bottom side of the central PMMA layer. The micro-patterned PMMA layer was bonded with two PSA films to create the final HTP microfluidic device. (e) Digital images of the fabricated HTP centrifugal device with an SAP inserted into a waste chamber. (Reprinted with permission from Ref. [29]. Copyright (2023) Elsevier.)

In detail, a key challenge in adapting ELISA to centrifugal microfluidics is the strong centrifugal force generated during rotation, which tends to push liquids to the outermost part of the disc. This is incompatible with the ELISA process that necessitates both incubation and waste discharge steps, often requiring repetition during the assay. Traditional centrifugal microfluidic designs either employ complex structures or passive valves, which struggle to perform ELISA operations with the needed precision and reproducibility. Additionally, active valves introduce their own complexities, requiring additional equipment like wax, heaters, or lasers.

To address these challenges, the centrifugal microfluidic chip incorporates an innovative upstream-midstream-downstream model, augmented by a series of siphon structures. This design effectively halts the flow of the solution post the initial high-speed rotation, facilitating the incubation of the antigen solution, a washing solution, and the TMB substrate solution in the reaction chamber. Upon siphon priming, the valves open, allowing continuous transfer of solutions to the endpoint during the second rotation. This arrangement not only ensures repeatable steps but also allows for more complex assay performance.

A notable aspect of the design is the zigzag aliquot structure linking the aliquot chambers, enabling the division and filling of solutions in all chambers with a single injection. This automatic aliquoting process, governed by controlling the capillary pressure in the microchannels, marks a significant advancement over manual methods. Furthermore, the use of super-absorbing polymer (SAP) in the waste chamber to absorb and retain the sample and washing solutions showcases a thoughtful approach to managing the increased volume of waste typical in ELISA processes.

3. Immobilization of the antibody on the microfluidic device

The immobilization of antibodies on a microfluidic device refers to the process of firmly attaching antibodies to the surface of the microfluidic channels or chambers within the device. This attachment is typically achieved through chemical or physical means, ensuring that the antibodies remain fixed in a specific location, allowing for interactions with target molecules in a controlled and predictable manner. This process is crucial in the ELISA, ensuring accurate and reliable molecular detection. It initiates with the application of O_2 plasma to the PMMA surface, creating oxygen radicals and ions that enhance the surface's reactivity. This critical step modifies the surface, rendering it more conducive for subsequent chemical reactions.

Following plasma treatment, a solution of 3-aminopropyltriethoxysilane (APTES) is introduced into the reaction chamber. During an incubation period, the APTES molecules chemically bind to the surface, forming robust siloxane bonds. This covalent attachment results in the introduction of amino groups onto the surface, providing anchoring points for the capture antibody. The stability of this coating is normally further ensured through a carefully controlled heating process, where the surface is subjected to 80 °C for an hour. This heating step not only strengthens the siloxane bonds but also eliminates any unbound or loosely attached APTES molecules, reinforcing the reliability of the immobilization.

Moving forward, a capture antibody is prepared for immobilization by mixing it with (1-ethyl-3-(3-dimethylaminopropyl) carbodiimide)/N-hydroxysuccinimide (EDC/NHS) reagents. This preparation activates the antibody by forming an Oacylisourea intermediate that readily reacts with the amino groups present on the APTES-coated surface, ultimately forming amide bonds. These covalent bonds firmly anchor the antibody to the surface, ensuring its stability and functionality.

gTo maintain the specificity of the immobilized antibody and prevent non-specific binding interactions, the reaction chamber is meticulously washed with a blocking buffer. A typical blocking buffer used in immunochemical assays like ELISA contains proteins (e.g., BSA or non-fat dry milk) to block non-specific binding sites, detergents (e.g., Tween-20 or Triton X-100) to reduce hydrophobic interactions, salts for ionic strength, and may include stabilizers and preservatives for long-term stability (normally at 4 °C or -20 °C). This critical step effectively minimizes the risk of undesired interactions and false-positive results (Fig. 3).

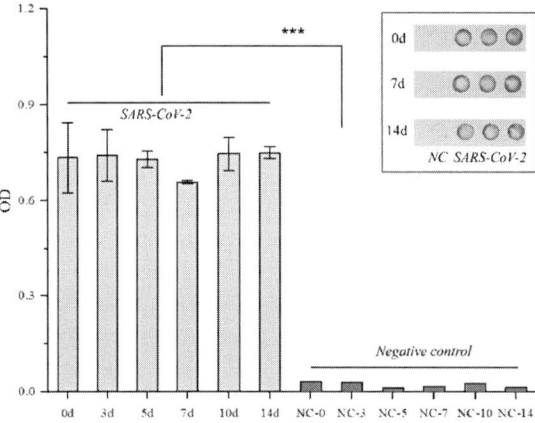

Fig. 3. Immobilization of the antibody on the PMMA sheet. SARS-CoV-2 nucleocapsid samples were used to test the reproducibility and stability of the coated antibodies on the chambers. After 2 weeks stored at 4 °C, there is no significant difference among the positive samples, while the difference between the positive groups and the negative controls is huge (***: P<0.001). (Reprinted with permission from Ref. [29]. Copyright (2023) Elsevier.)

4. Fabrication of a centrifugal ELISA HTP chip

In the process of preparing a microfluidic chip for conducting ELISA, several critical factors must be considered to ensure the accuracy and reliability of the assay results. Material selection is a foundational step in chip fabrication. Choosing the right material, such as PMMA, is pivotal. PMMA is commonly preferred due to its biocompatibility and compatibility with ELISA components. The material should not introduce interference or contaminants that could compromise the assay's integrity. Once the appropriate material is selected, the fabrication process comes into play. The design of microchannels within the chip must be meticulously executed. These microchannels play a crucial role in facilitating the precise and controlled movement of reagents and samples. Ensuring proper channel sizing and shaping is essential to confirm that the

ELISA components interact effectively within the chip [2, 30].

Following fabrication, a rigorous cleaning process is imperative to eliminate potential contaminants that could interfere with ELISA results. This process includes sonication with a detergent solution, rinsing with ethanol, and a final sonication step with distilled water. The outcome of this cleaning procedure is pristine surfaces, which are essential for obtaining accurate measurements and minimizing background noise during the assay.

Surface treatment is another vital step. In particular, applying a super hydrophobic coating to all microchannels (excluding the reaction chamber) is of utmost importance. This hydrophobic coating serves multiple purposes. It prevents nonspecific binding of biomolecules, facilitating smooth flow of reagents and samples. Furthermore, it reduces background noise, thereby enhancing the sensitivity and specificity of the ELISA assay.

Finally, pressure-sensitive adhesive (PSA) films are employed to seal both the top and bottom surfaces of the chip. A secure and tight seal ensures that there are no leaks or cross-contamination between channels, preserving the accuracy and reliability of the ELISA assay results.

5. Construction of a portable HTP genetic analyzer

In recent years, there has been a concerted effort to advance the field of diagnostics through the miniaturization and automation of complex assays like the ELISA. Conventional ELISA workflows often rely on bulky systems that include washing plate machines, microplate readers, and other equipment, making them space-consuming and labor-intensive. While existing commercial ELISA workstations, including systems like the ABBOTT PRISM System and Abbott Architect Immunology Analyzer, have undoubtedly improved assay efficiency, reduced manual labor, and enhanced precision in sample handling and processing, they are often associated with a significant drawback – a high initial cost of procurement and maintenance. These cost barriers limit their accessibility, particularly for smaller research or diagnostic facilities.

To address these challenges and pave the way for more accessible and efficient high-throughput ELISA, the construction of a portable HTP genetic analyzer has emerged as a promising solution. This analyzer combines various components, including a precision syringe pump with an 8-port valve, a spindle motor, a 3D-printed linear guidance system, a stepper motor, and a portable UV-vis absorbance spectrometer (Fig. 4). These components work in unison to automate the entire ELISA process, from reagent loading to UV-vis absorbance detection, eliminating much of the manual intervention required in traditional ELISA setups. The fully automated operation of this analyzer not only streamlines the assay workflow but also significantly reduces the potential for human error, leading to more reliable and reproducible results. This level of automation is particularly valuable in settings where rapid and high-throughput analysis is important, such as clinical diagnostics or environmental monitoring.

One of the standout features of this analyzer is its compact design, with dimensions measuring 40 cm × 35 cm × 60 cm (width × height × length). This portability allows it to be easily transported and deployed in various locations, including remote or field settings, without the need for a dedicated laboratory space. As a result, it holds great potential for point-of-care applications, where rapid on-site testing is essential for timely decision-making and patient care.

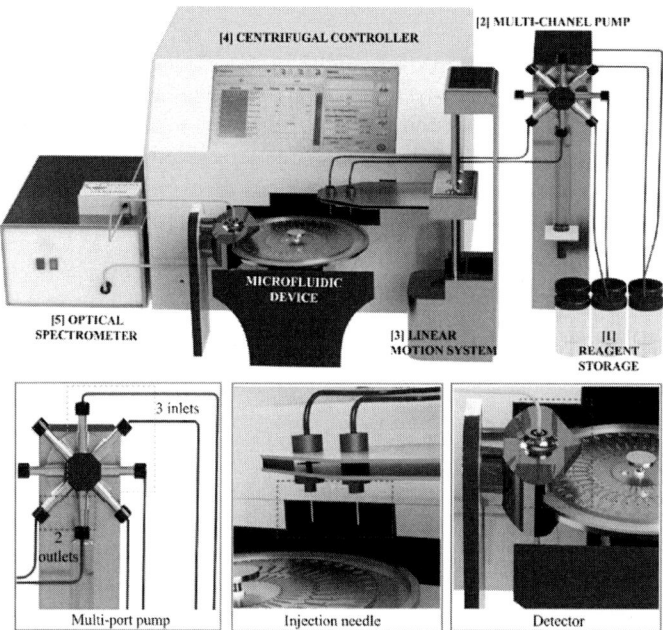

Fig. 4. The POCT ELISA analyzer for SARS-CoV-2 diagnostics. The proposed platform includes 5 parts: (1) a reagent storage, (2) an 8-port valve pump, (3) a linear guide, (4) a centrifugal controller with a touch screen, and (5) an optical spectrometer. After the microfluidic device was inserted into the POCT analyzer, all reagents of the HRP-antibody, the washing buffer, the TMB were stored and transferred into the microfluidic device via a multiple-port pump. The linear guide was responsible for controlling the position of 2 injection needles. After loading the designated solutions, centrifugal force drove the solutions into the reaction chambers or the waste chamber. The POCT ELISA analyzer was equipped with a portable optical detector, so that the optical density of each reaction chamber was measured to produce the results of ELISA. (Reprinted with permission from Ref. [29]. Copyright (2023) Elsevier.)

The sequential and automatic control capabilities of the analyzer encompass various important processes, from reagent injection to microfluidic chip control and optical detection. Reagents are withdrawn from different inlets and precisely directed to designated outlets by the syringe pump with an 8-port valve, ensuring accurate and reproducible reagent handling. The use of a linear motion system with needles enables controlled injection of reagents into the microfluidic chip's aliquot chambers, minimizing the risk of cross-contamination.

Furthermore, the analyzer's ability to rotate the chip under the portable UV-vis absorbance spectrometry at a wavelength of 640 nm for color intensity measurement offers a significant advantage in terms of throughput. This rotational movement allows multiple reaction chambers (up to 40) to be analyzed in a single run, drastically increasing the efficiency of ELISA assays.

In comparison to existing microfluidic-based immunoassay platforms, this analyzer's design stands out for its cost-effectiveness and ease of fabrication. By focusing on z-control for reagent injection, it provides an alternative to complex and expensive x-y-z control systems. This design choice simplifies the manufacturing process and makes the system more accessible to a wider range of users.

6. Chip operation

The chip operation procedure outlines the sequence of steps in the ELISA process (Fig. 5). Initially, 40 SARS-CoV-2 lysate samples are introduced into the reaction chambers through sample-loading holes coated with the capture antibody. This step ensures effective capture of the SARS-CoV-2 nucleocapsid antigen, establishing the foundation for the assay. Following a 30-minute incubation, centrifugation swiftly discharges the samples into the waste chamber.

Efficiency is a hallmark of this device's design. Washing steps, essential for removing non-specific bindings, are notably streamlined. Washing buffer 1 is introduced into 40 aliquot chamber #1 through a zigzag aliquot channel, mimicking the manual pipetting steps in traditional ELISA but with automated precision. The washing buffer is then directed to the reaction chambers. A siphon valve at the bottom of each reaction chamber controls the flow, ensuring retention of the washing buffer in the reaction chamber until subsequent centrifugation discharges it into the waste chamber. This dual washing step is performed twice to guarantee thorough removal of non-specific bindings.

The horseradish peroxidase (HRP)-linked detection antibody is accurately dispensed into 40 aliquot chambers using a zigzag aliquot structure. It is then transferred into the reaction chambers by centrifugal force, followed by a 15-minute incubation to facilitate complete binding between the HRP-linked detection antibody and the SARS-CoV-2 nucleocapsid antigen. Subsequently, centrifugation discharges the HRP-detection antibody into the waste chamber by another rotation.

To further eliminate weak and unspecific bindings of the HRP-linked detection antibody, an additional washing step employs washing buffer 2. This process mirrors the previous washing step, with precise dispensing into 40 aliquot chambers, introduction into the reaction chambers, and subsequent discharge into the waste chamber. These washing steps ensure the comprehensive removal of any weak and unspecific antibody bindings.

The final stage of the ELISA chip operation involves the injection of the 3,3',5,5'-tetramethylbenzidine dihydrochloride hydrate (TMB) substrate into a separate aliquot chamber, specifically 40 aliquot chamber #2. This separation is crucial to prevent contamination of the TMB substrate. Transferred to 40 reaction chambers, the TMB substrate undergoes a 5-minute incubation, allowing the colorimetric reaction between TMB and HRP to reach completion.

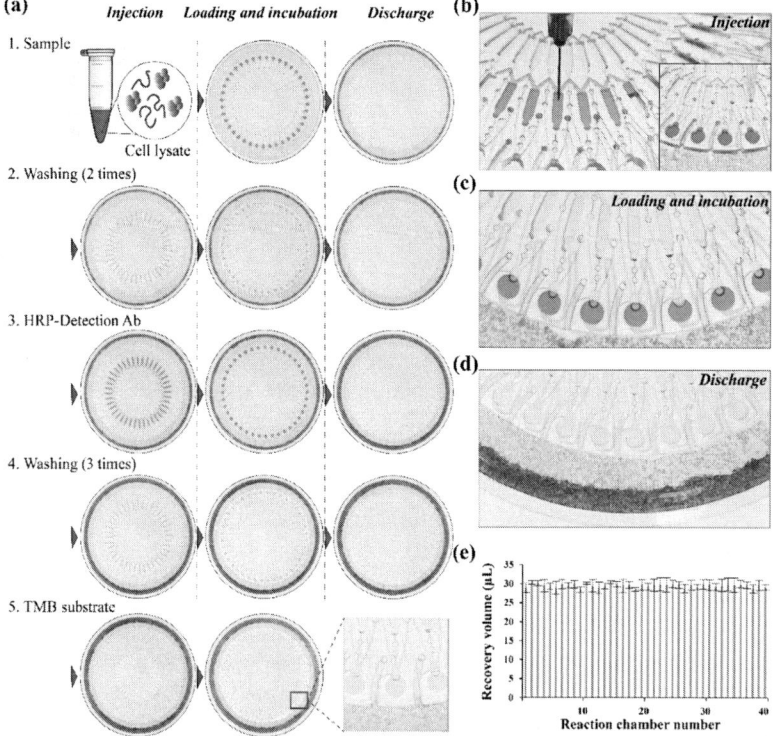

Fig. 5. Operation of the ELISA centrifugal microfluidic device for HTP SARS-CoV-2 detection. (a) On-chip operation processes of the microfluidic device. Forty cell lysate samples were manually loaded into 40 antibody-coated reaction chambers and discharged into the waste chamber via centrifugal force at 5000 RPM. Other reagents including the washing buffer 1 and 2, HRP-detection antibody, TMB substrate solution were divided automatically into the aliquot chambers with one-shot owing to the zigzag aliquot structure. Reagents were driven into the reaction chambers at 3000 RPM and discharged into the waste chamber at 5000 RPM. (b) The POCT ELISA analyzer could be programmed to load the designated solution such as the washing buffer 1 and 2, an HRP-detection antibody solution, and a TMB substrate solution into the aliquot chambers via a syringe. (c) At 3000 RPM (the first rotation), reagents were driven into the antibody-coated reaction chambers without overflowing into the waste chamber due to the siphon valve. (d) At 5000 RPM (the second rotation), reagents were discharged into the waste chamber and absorbed by the SAP. (e) Recovery volume of the reaction chamber showed the high accuracy for aliquoting (theoretically 30 μL each chamber). (Reprinted with permission from Ref. [29]. Copyright (2023) Elsevier.)

Ultimately, the portable UV-vis absorbance spectrometer measures the colorimetric reaction at a wavelength of 640 nm. The absorbance data obtained from each of the 40 reaction chambers is recorded and analyzed. This data reflects the intensity of the colorimetric signal, which is directly proportional to the concentration of the target in each chamber. By comparing these absorbance values to a standard curve or a set of reference samples with known concentrations, the precise concentration of the target molecule in the original

7. Application of COVID-19 detection

This book chapter highlights the HTP ELISA utilizing a centrifugal microfluidic device, taking COVID-19 as a pivotal model target. The choice of COVID-19 as a model is underpinned by its global significance, public health urgency, and the imperative to adapt diagnostics in response to genetic variability and the emergence of new variants.

Sensitivity and Specificity of HTP ELISA on the microfluidic chip

The study focused on the detection of SARS-CoV-2 nucleocapsid across a spectrum of concentrations, ranging from 1.6 ng/mL to 400 ng/mL. Notably, this innovative approach dramatically reduces the required sample volume, demanding only 30 μL, which stands in stark contrast to the 50-100 μL typically consumed by conventional ELISA protocols. The gradual reduction in blue color intensity corresponds to lower antigen concentrations, with deep blue observed at 400 ng/mL and a lighter shade at 4.1 ng/mL. At extremely low concentrations (1.6 ng/mL), the differentiation between the SARS-CoV-2 sample and negative

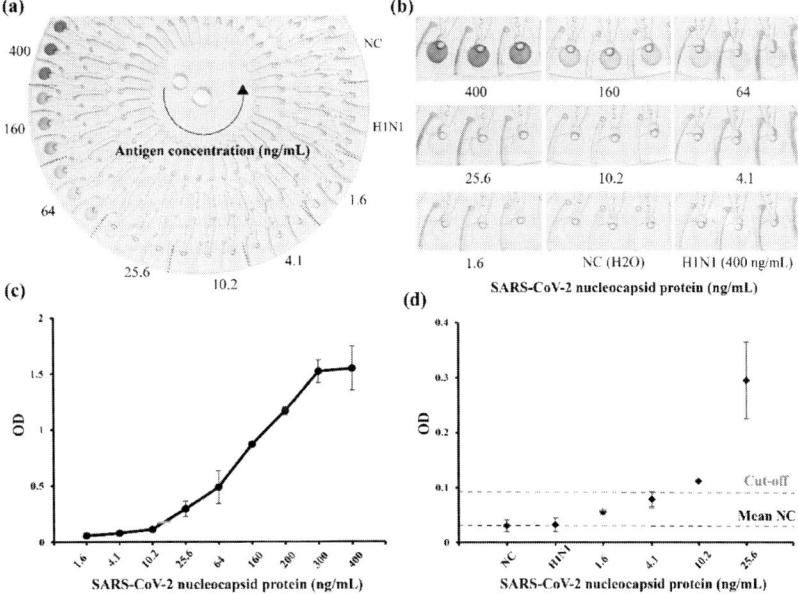

Fig. 6. HTP screening of SARS-CoV-2 nucleocapsid on the ELISA centrifugal microfluidic device. (a-b) Detection of SARS-CoV-2 antigens with different nucleocapsid concentration ranging from 400 ng/mL to 1.6 ng/mL. Instead of SARS-CoV-2 nucleocapsid antigens, H_2O and Influenza A H1N1 viral nucleocapsid (400 ng/mL) were used as the negative controls. (c) An ELISA standard curve in the range of 1.6–400 ng/mL of SARS-CoV-2 nucleocapsid. (d) LOD of the on-chip ELISA was 10.2 ng/mL based on the cut-off OD value of 0.093. (Reprinted with permission from Ref. [29]. Copyright (2023) Elsevier.)

controls (H1N1 and H₂O) becomes imperceptible (Fig. 6).

To gauge sensitivity, the limit of detection (LOD) was calculated using a predefined equation [31].

$$\text{Cut-off value} = a \times \overline{X} + f \times SD \quad (1)$$

where a and f are two multipliers and $\overline{X}$ and SD are the mean and the standard deviation of the negative control experiments, respectively.

To ensure the differentiation between the positive results and the negative controls, the cut-off value was set up by defining the multipliers for the negative control experiment as $f = 0$ and $a = 3$. Under these conditions, the LOD of the HTP ELISA centrifugal microdevice was determined to be approximately 10.2 ng/mL, positioning it among highly sensitive detectors capable of early-stage SARS-CoV-2 infection detection.

Beyond sensitivity, the technology offers a remarkable enhancement in time efficiency, completing ELISA procedures within a mere 75 minutes compared to the 4-6 hours typically required for conventional ELISA. Additionally, it significantly diminishes testing costs, rendering it a cost-effective solution in the landscape of COVID-19 diagnostics and other infectious diseases.

Another critical aspect of assessing the performance of the on-chip HTP ELISA is the rigorous evaluation of its specificity, particularly in distinguishing the target pathogen from other commonly encountered respiratory pathogens (Fig. 7). In this comprehensive evaluation, a panel of five frequently encountered respiratory pathogens, including Influenza A subtypes H1N1 and H3N2, Influenza B (Influ B), and respiratory syncytial viruses (RSV A and RSV B), was employed. To provide a robust benchmark for specificity assessment, commercial nucleocapids of SARS-CoV-2 served as a positive control (PC),

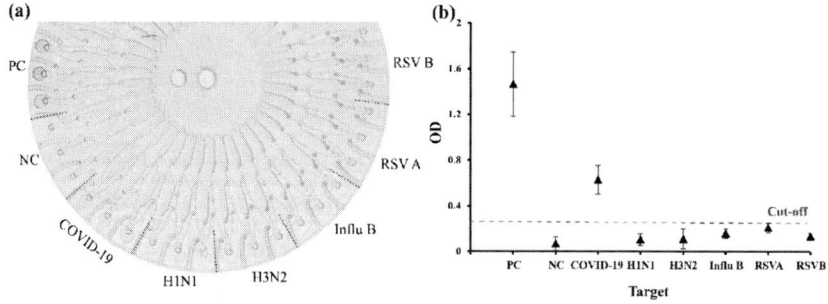

Fig. 7. Specificity test of the HTP ELISA centrifugal microfluidic device. (a) On-chip ELISA test with common respiratory infectious viruses including Influenza A – H1N1 and H3N2, Influenza B (Influ B), respiratory syncytial virus type A and B (RSV A, RSV B) together with NC, PC, and the COVID-19 samples. (b) OD values for the respiratory infectious viruses. Only the COVID-19 samples and PC showed higher than 0.27, while others revealed lower values. (Reprinted with permission from Ref. [29]. Copyright (2023) Elsevier.)

while H₂O functioned as a negative control (NC).

The results are strikingly indicative of the technology's selectivity. Clinical samples of other respiratory viruses elicited very weak signals, akin to the negative control, while the three COVID-19 samples, in tandem with the positive control, exhibited a distinct and vivid blue coloration. These findings are compelling, particularly when analyzed against the backdrop of a cut-off optical density (OD) value. Remarkably, only the COVID-19 samples and the positive control surpassed this cut-off threshold. These results unequivocally signify that the POCT ELISA analyzer possesses the exceptional capability to selectively and rapidly identify SARS-CoV-2 among a diverse array of respiratory infectious viruses, all while offering high speed, high-throughput, and complete automation.

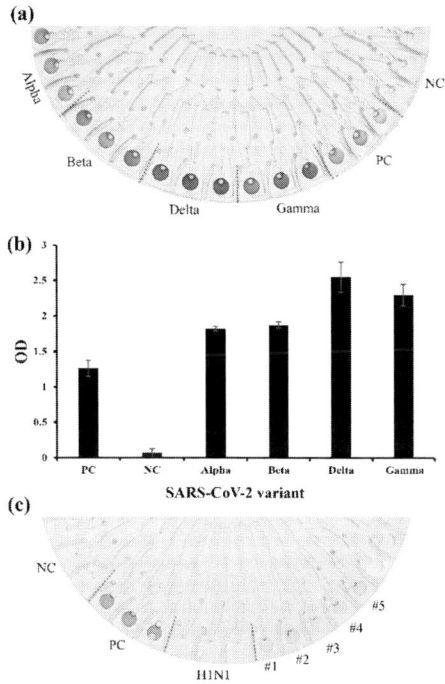

Fig. 8. COVID-19 variant and clinical sample tests on the HTP ELISA centrifugal microfluidic device. (a) On-chip ELISA test results with four common variants of SARS-CoV-2 (alpha, beta, delta, and gamma) and (b) The corresponding optical densities at 640 nm for each variant. (c) On-chip ELISA test results for five SARS-CoV-2 clinical samples (#1, #2, #3, #4, and #5). (Reprinted with permission from Ref. [29]. Copyright (2023) Elsevier.)

Variant detection and clinical testing

The dynamic nature of SARS-CoV-2, with the emergence of new variants, has underscored the critical importance of advanced diagnostic tools capable of effectively identifying these variants. In the pursuit of variant detection, the antibody pair was employed to detect multiple SARS-CoV-2 variants, including alpha, beta, gamma, and delta, utilizing the HTP ELISA analyzer (Fig. 8). The results underscore the analyzer's exceptional sensitivity and specificity in analyzing all four SARS-CoV-2 variants. The positive control (PC) sample, consisting of commercial nucleocapids of SARS-CoV-2, exhibited a robust blue coloration, with the variant samples displaying even deeper shades of blue. Notably, the delta and gamma variants exhibited the most intense blue coloration, indicating the analyzer's proficiency in diagnosing both the variants and the standard SARS-CoV-2.

To further underscore the practical utility of the POCT ELISA analyzer, clinical samples from SARS-CoV-2-infected patients were subjected to evaluation. All clinical samples (#1, #2, #3, #4, and #5) yielded positive results, manifesting significant blue coloration within the reaction chambers. In contrast, the two negative controls, represented by H_2O and an Influenza A H1N1 sample, exhibited minimal color intensity. These results decisively affirm the platform's capacity to perform precise ELISA diagnostics on real-world samples of SARS-CoV-2 in a high-throughput manner. This capability holds substantial promise in contributing to ongoing efforts to combat the evolving landscape of COVID-19 pandemics.

These results have far-reaching implications, not only for COVID-19 diagnostics but also for broader infectious disease surveillance. The platform's ability to rapidly and selectively identify SARS-CoV-2, combined with its high-throughput capabilities and automation, positions it as a formidable asset in public health settings. The prospect of efficiently and accurately detecting specific pathogens among a multitude of potential causes of respiratory illness is a significant advancement in disease management and control.

8. Conclusion

This chapter signifies a major advancement in the realm of serological testing, particularly through the innovative integration of ELISA with centrifugal microfluidic devices. This integration not only stands as a testament to the ingenuity within biomedical engineering but also marks a pivotal shift in diagnostic methodologies. The development of the centrifugal ELISA HTP chip embodies a leap forward, addressing several limitations inherent to traditional ELISA methods.

The essence of this innovation lies in its ability to enhance efficiency and automation. The centrifugal HTP chip streamlines the ELISA process, reducing the manual labor and time typically required. By automating and miniaturizing the assay processes, and enabling parallel processing of samples, this technology significantly boosts the precision and reproducibility of serological assays. Such advancements are crucial in contexts where high-throughput screening is essential, and time is of the essence. The system's adeptness in efficiently detecting various SARS-CoV-2 variants illustrates its potential in responding to emergent global health crises. This adaptability and efficiency in detecting rapidly evolving pathogens underscore the system's relevance in current and future epidemiological challenges. Looking ahead, the implications of this technological advancement extend far beyond immediate health crises. The integration of ELISA with centrifugal microfluidics is poised to revolutionize the landscape of high-throughput screening. It promises to make advanced diagnostic testing more accessible and feasible, even in resource-limited settings. This is particularly vital for disease surveillance, routine diagnostics, and epidemiological studies. Furthermore, this innovation opens up new pathways for research, potentially catalyzing further developments in biomedical and clinical laboratory sciences.

References

1. H. V. Nguyen, H. V. Nguyen, V. M. Phan, B. J. Park and T. S. Seo, *Chemical Engineering Journal*, 2023, 452, 139044.
2. B. H. Park, S. J. Oh, J. H. Jung, G. Choi, J. H. Seo, D. H. Kim, E. Y. Lee and T. S. Seo, *Biosens Bioelectron*, 2017, 91, 334–340.
3. Y. Rais, Z. Fu and A. P. Drabovich, *Clin Proteomics*, 2021, 18, 19.
4. T. Wang, M. Zhang, D. D. Dreher and Y. Zeng, *Lab Chip*, 2013, 13, 4190–4197.
5. R. Funari, H. Fukuyama and A. Q. Shen, *Biosens Bioelectron*, 2022, 208, 114193.
6. R. Huang, K. Zhang, G. Zhu, Z. Sun, S. He and W. Chen, *Sensors (Switzerland)*, 2018, 18, 3537.
7. H. Van Nguyen, V. M. Phan and T. S. Seo, *Sens Actuators B Chem*, 2022, 353, 131088.
8. J. H. Seo, B. H. Park, S. J. Oh, G. Choi, D. H. Kim, E. Y. Lee and T. S. Seo, *Sens Actuators B Chem*, 2017, 246, 146–153.
9. H. Van Nguyen, V. M. Phan and T. S. Seo, *Sens Actuators B Chem*, 2024, 399, 131088.
10. G. Choi, J. H. Jung, B. H. Park, S. J. Oh, J. H. Seo, J. S. Choi, D. H. Kim and T. S. Seo, *Lab Chip*, 2016, 16, 2309–2316.
11. H. Y. Heo, S. Chung, Y. T. Kim, D. H. Kim and T. S. Seo, *Biosens Bioelectron*, 2016, 78, 140–146.
12. B. K. Van Weemen and A. H. W. M. Schuurs, *FEBS Lett*, 1971, 15, 232–236.
13. E. Engvall and P. Perlmann, *Immunochemistry*, 1971, 8, 871–874.
14. R. M. Lequin, *Clin Chem*, 2005, 51, 2415–2418.
15. S. Aydin, *Peptides (N.Y.)*, 2015, 72, 4–15.
16. J. Kai, A. Puntambekar, N. Santiago, S. H. Lee, D. W. Sehy, V. Moore, J. Han and C. H. Ahn, *Lab Chip*, 2012, 12, 4257–4262.
17. T. Porstmann and S. T. Kiessig, *J Immunol Methods*, 1992, 150, 5–21.

18 J. E. Butler, L. Ni, R. Nessler, K. S. Joshi, M. Suter, B. Rosenberg, J. Chang, W. R. Brown and L. A. Cantarero, *J Immunol Methods*, 1992, 150, 77–90.

19 M. Tayyab, M. A. Sami, H. Raji, S. Mushnoori and M. Javanmard, *IEEE Sens J*, 2021, 21, 4007–4017.

20 M. Wu, S. Wu, G. Wang, W. Liu, L. T. Chu, T. Jiang, H. K. Kwong, H. L. Chow, I. W. S. Li and T.-H. Chen, *Sci Adv*, 2022, 8, eabn6064.

21 R.-Q. Zhang, S.-L. Liu, W. Zhao, W.-P. Zhang, X. Yu, Y. Li, A.-J. Li, D.-W. Pang and Z.-L. Zhang, *Anal Chem*, 2013, 85, 2645–2651.

22 N. Rezvani Jalal, P. Mehrbod, S. Shojaei, H. I. Labouta, P. Mokarram, A. Afkhami, T. Madrakian, M. J. Los, D. Schaafsma, M. Giersig, M. Ahmadi and S. Ghavami, *ACS Appl Nano Mater*, 2021, 4, 4307–4328.

23 M. Herrmann, T. Veres and M. Tabrizian, *Lab Chip*, 2006, 6, 555–560.

24 Y.-J. Yeh, T.-N. Le, W. W.-W. Hsiao, K.-L. Tung, K. K. Ostrikov and W.-H. Chiang, *Anal Chim Acta*, 2023, 1239, 340651.

25 W. Wei-Wen Hsiao, N. Sharma, T.-N. Le, Y.-Y. Cheng, C.-C. Lee, D.-T. Vo, Y. Y. Hui, H.-C. Chang and W.-H. Chiang, *Anal Chim Acta*, 2022, 1230, 340389.

26 A. Thiha and F. Ibrahim, *Sensors (Switzerland)*, 2015, 15, 11431–11441.

27 J. H. Lee, P. K. Bae, H. Kim, Y. J. Song, S. Y. Yi, J. Kwon, J.-S. Seo, J.-M. Lee, H.-S. Jo, S. M. Park, S. Chung and Y. B. Shin, *Biosens Bioelectron*, 2021, 191, 113406.

28 C.-T. Lin, S.-H. Kuo, P.-H. Lin, P.-H. Chiang, W.-H. Lin, C.-H. Chang, P.-H. Tsou and B.-R. Li, *Sens Actuators B Chem*, 2020, 316, 128003.

29 H. Van Nguyen, J. Yang, H. Van Nguyen, H. Poo and T. Seok Seo, *Chemical Engineering Journal*, 2023, 472, 144808.

30 J. Yang, V. M. Phan, C.-K. Heo, H. V. Nguyen, W.-H. Lim, E.-W. Cho, H. Poo and T. S. Seo, *Sens Actuators B Chem*, 2023, 380, 133331.

31 F. Lardeux, G. Torrico and C. Aliaga, *Mem Inst Oswaldo Cruz*, 2016, 111, 501–504.

Chapter 9
Centrifugal Microfluidic Device for High-throughput Nanoparticle Synthesis

1. Background

1.1. Limitations of traditional nanoparticle synthetic approaches

In the dynamic landscape of nanotechnology, nanoparticles (NPs) have emerged as pivotal components, finding applications in diverse domains such as catalysis, biosensing, energy conversion and storage, and diagnostics [1, 2]. The unique appeal of nanoparticles lies in their distinct physical, chemical, and biological characteristics, which significantly differ from their bulk material counterparts. These properties are primarily attributed to their nanoscale dimensions, typically ranging between 1 and 100 nanometers. At this scale, phenomena such as an increased surface area to volume ratio and dominant quantum effects come into play, opening a plethora of opportunities for tailoring these particles for specific applications across various fields, including medicine, electronics, energy, and materials science [3–9].

The traditional approach to nanoparticle synthesis has predominantly relied on flask-based batch reactors. This method, being a foundational technique in the development of nanotechnology, has facilitated the initial exploration and production of various types of nanoparticles [10–14]. However, despite its scalability and simplicity, this approach harbors several limitations that significantly impact the quality, efficiency, and environmental sustainability of nanoparticle production. One of the primary issues with flask-based batch reactors is their susceptibility to macro-environmental fluctuations [15]. For instance, in the synthesis of nanoparticles like gold, silver, or semiconductor quantum dots, variables such as temperature and concentration can greatly influence the outcome. Temperature fluctuations can alter the nucleation and growth rates of nanoparticles, leading to a broad spectrum of particle sizes and shapes [16]. This variation is critical as it directly affects the physical and chemical properties of the nanoparticles. Similarly, concentration inconsistencies, often due to incomplete mixing or localized concentration gradients within the reactor, can lead to uneven particle distributions. This is especially problematic in applications where uniformity in nanoparticle size is essential, such as in the production of quantum dots, where size uniformity directly impacts their optical properties. Another significant challenge with traditional synthesis methods is reproducibility. Batch-to-batch variations are a common concern, impacting the consistency of nanoparticle properties. In applications like the production of magnetic nanoparticles for medical imaging, any inconsistency in particle size or magnetic properties can lead to varying performance, which is undesirable. Moreover, the reliance on manual operations in these methods often leads to human errors, further exacerbating reproducibility issues [15, 17].

1.2. Microfluidics based nanoparticle synthesis

In response to these challenges, microfluidic technology has been increasingly utilized for the synthesis of high-quality nanomaterials. Microfluidic devices, with their design focused on manipulating fluids within microscale channels and reactors, offer several advantages. They provide a high surface-to-volume ratio and facilitate rapid heat and mass transfer. This technology has been explored through various formats, including continuous-flow-based single-phase microfluidics, droplet-based multiphase microfluidics, and microreactors. These systems exploit the characteristics of laminar flow and molecular inter-diffusion, or alternatively, convective flow with active or passive mixing structures. The ability to precisely control fluid transport and ensure uniform mixing of liquid precursors at a millisecond timescale leads to significantly improved reproducibility and homogeneity in nanoparticle production [18–28].

One of the most common types of microfluidic devices used in nanoparticle synthesis is the continuous flow microreactor. These devices, characterized by their channels through which reactants continuously flow and react, are the workhorses of microfluidic nanoparticle synthesis. They are particularly valued for their ability to maintain steady-state conditions, ensuring consistent production of nanoparticles. The control over parameters like flow rate and temperature in these systems is precise, allowing for the fine-tuning of nanoparticle characteristics [29–42]. This type of device is ideal for scaling up the production of nanoparticles, as it can operate continuously, making it suitable for industrial applications.

Droplet-based microfluidics presents a fascinating contrast. Here, the synthesis occurs within discrete droplets, which act as individual microreactors. This setup allows for the isolation of reactions in each droplet, minimizing cross-contamination and enabling high-throughput experimentation with different reaction conditions. The droplets can be precisely controlled in terms of size, allowing for uniform reaction environments. This approach is particularly useful for synthesizing nanoparticles that require highly controlled conditions or for applications where a high degree of uniformity in particle size and composition is crucial [41, 43–47].

Another innovative approach in the field is to utilize the static microreactors in a high-throughput manner [33, 39]. In particular, the use of centrifugal microfluidic devices has demonstrated their high potential for the automatic and parallel nanoparticle synthesis. These devices utilize the centrifugal force from a rotating disc to control fluid dynamics in microscale channels, allowing for the precise regulation of critical synthesis conditions [48, 49]. Such control is essential for the production of nanoparticles with specific desired characteristics.

The manipulation of variables like flow rates, reaction times, and mixing conditions within these devices plays a crucial role in defining the functionality and potential applications of the nanoparticles. Designed for efficiency, centrifugal microfluidic devices can perform multiple reactions either concurrently or in quick succession, significantly enhancing throughput compared to traditional batch processes. This feature is particularly valuable in both research and industrial production settings. Furthermore, the ability to integrate various synthesis steps, such as mixing, heating, and separation, into a single streamlined device, simplifies the synthesis process [50]. This not only reduces the complexity of the operations but also minimizes the risk of contamination, leading to a purer and more consistent nanoparticle product.

1.3. High-throughput nanoparticle synthesis on a microfluidic device

Despite the notable advantages of microfluidic systems, significant challenges persist, particularly in the high-throughput synthesis of nanoparticles using these devices. Synthesizing nanoparticles in a high-throughput microfluidic device presents a complex array of challenges, primarily stemming from the need to strike a balance between the precision control characteristic of microfluidic systems and the demands of large-scale production. This balance is crucial to ensure that the quality of nanoparticles produced on a larger scale mirror that achieved in smaller-scale, precision-controlled environments.

One of the primary challenges in this endeavor is scaling up the nanoparticle synthesis process without losing the precise control over particle size and distribution. Microfluidic devices are renowned for their ability to produce nanoparticles with high uniformity under controlled conditions at a small scale. However, replicating this level of precision on a larger scale necessitates uniform distribution of temperature, pressure, and reactant concentration, which poses significant technical difficulties. The challenge is not just in scaling up the process but doing so in a way that retains the meticulous control over the physicochemical parameters that define the quality of the nanoparticles. Besides, the design and fabrication of high-throughput microfluidic systems adds another layer of complexity. Such systems often require intricate and sophisticated designs that include larger or more complex channel networks and multiple reaction zones. This intricacy in design makes the fabrication process challenging, especially when it comes to maintaining the integrity and precision of microscale features over a larger area. Additionally, the materials used in constructing these devices must be selected with great care, ensuring chemical compatibility, durability, and cost-effectiveness.

Managing fluid dynamics in a high-throughput system also presents its own set of challenges, differing significantly from those encountered in traditional microfluidic devices. It is crucial to ensure uniform flow rates throughout the system while avoiding issues such as channel clogging or reactant precipitation in continuous flow microfluidic platforms. This often requires integrating complex pumping systems or developing novel methods to induce and control flow, like electrokinetic or acoustic techniques. Moreover, efficient heat and mass transfer are vital for the consistent synthesis of nanoparticles. Achieving and maintaining uniform temperature and efficient mixing on a larger scale is challenging, and any non-uniformity in these parameters can lead to variations in the size and properties of the nanoparticles.

Looking towards the future, the field of high-throughput nanoparticle synthesis is poised for further evolution. More integrated systems are expected to be developed, designed not only to synthesize nanoparticles but also to carry out their characterization and application testing in a streamlined and efficient manner. The advancement of centrifugal microfluidic devices is anticipated to address current challenges, further enhancing their utility in nanoparticle synthesis. These devices represent a significant step forward in the field, offering a novel and efficient approach to nanoparticles.

In this book chapter, we delve into how high-throughput centrifugal microfluidic platforms can be leveraged for the synthesis of nanoparticles, quantum dots, and bi-metallic nanoparticles. These examples serve as representative cases of advanced microfluidic design and control, highlighting the potential of these systems in overcoming traditional challenges and setting new benchmarks in nanoparticle synthesis. Through detailed exploration and analysis, we aim to shed light on the intricate processes, advantages, and future prospects of these innovative microfluidic devices in the realm of high-throughput nanoparticle production.

2. Design of a centrifugal HTP chip for nanoparticle synthesis

While common automated methods for HTP synthesis of NPs offer numerous advantages, they do come with certain limitations that should be considered. One notable limitation is the potential for bulkiness and complexity in the system setup. Liquid handling robots, microfluidic automation platforms, and other automated equipment can occupy significant laboratory space and require proper infrastructure, which may not be suitable for all research environments. Another important consideration is the cost associated with these automated systems. The initial investment in purchasing and setting up liquid handling robots or dedicated microfluidic automation platforms can be substantial. Additionally, ongoing maintenance, calibration, and software updates can contribute to the overall cost of ownership. For some research laboratories or smaller institutions with budget constraints, these expenses may pose a significant barrier to adoption.

To address these challenges, the development and design of centrifugal HTP microfluidic chips for nanoparticle synthesis represents a significant leap forward in the integration of engineering and nanotechnology. These chips are ingeniously tailored to facilitate rapid solution loading and enable precise, efficient synthesis of nanoparticles, meeting the increasing demands for high-throughput nanoparticle fabrication. At the core of their design lies a sophisticated microfluidic layout. This layout features an intricate network of microscale channels, chambers, and valves, meticulously engineered to control the flow and mixing of various reactants. This intricate microfluidic architecture plays a pivotal role in controlling the flow and mixing of various reactants, essential for the parallel synthesis of diverse nanoparticles.

2.1. A zigzag aliquot structure for efficient and rapid solution loading

The investigations conducted by Park et al. (2015) [51], Nguyen et al. (2020) [39], and Nguyen et al. (2023) [52] showcase an innovative method for HTP synthesis of NPs utilizing a singular microfluidic device. A shared pivotal characteristic among these devices is the integration of a zigzag aliquoting structure, an innovative microstructure designed for the automatic and rapid loading and aliquoting of reactants (Fig. 1). This structure significantly streamlines the process of nanoparticle synthesis, allowing for the quick and precise partitioning of reactants into the chip. The zigzag design optimizes the flow path of the reactants, ensuring even distribution and efficient mixing, crucial for the uniformity of nanoparticle synthesis. This aliquoting structure exemplifies the chip's ability to handle multiple reactants in a high-throughput manner, making it a cornerstone of the chip's functionality [53, 54].

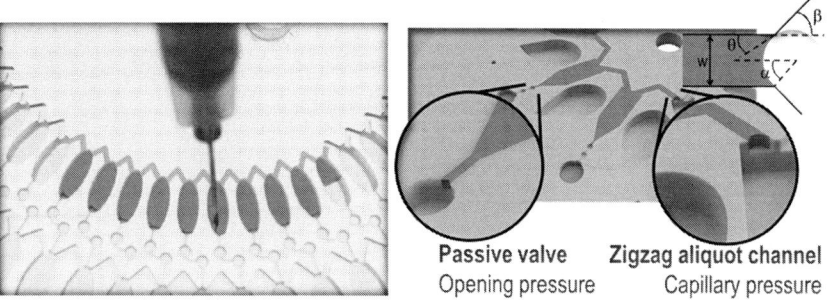

Fig. 1. The zigzag aliquoting structure enables rapid and automated aliquoting of multiple chambers in a single-shot injection by leveraging the pressure differential between the zigzag channel and the passive valves. (Reprinted with permission from Ref. [52]. Copyright (2023) Elsevier.)

The utilization of the zigzag aliquoting structure operates on the principle of rapidly and automatically aliquoting multiple chambers through a single-shot injection, primarily relying on the pressure differential between the zigzag channel and the passive valves. The opening pressure of these microstructures is determined through a specific mathematical formula that considers factors such as liquid surface tension, contact angles on different sides of the solution, and channel dimensions. The calculated capillary pressure in the zigzag aliquoting channel is calibrated to be lower than the opening pressure of the passive valves. This precision in calibration ensures that the solution remains confined within the aliquoting chamber, preventing any unintended leakage to other structures during the aliquoting process. Theoretically, the opening pressure of the passive valve (P) is computed using the following equation [55]:

$$P = -\gamma \left(\frac{\cos\theta_t + \cos\theta_b}{h} + \frac{\cos\theta_l + \cos\theta_r}{w} \right)$$

where γ is the surface tension of the solution, $θ_t$, $θ_b$, $θ_l$, $θ_r$ is the top, bottom, left, and right contact angle of the solution, h is the channel height, and w is the channel width.

The capillary pressure of the passive valve (Cp) was calculated using the following equation [56]:

$$C_P = \frac{Cγ \sin θ}{h \times w}$$

where C is the channel length from the reservoir to the passive valve, γ is the surface tension of the solution, θ is the contact angle, h is the channel height, and w is the channel width.

2.2. Regulating the release of the solution through centrifugal force

Moreover, the microreactor's solution loading process can be automated and carried out stepwise through RPM control. For efficient discharge of the loaded solution into the microreactors, it is essential for the centrifugal force to exceed the capillary pressure originating from the microchannel and the passive valve. Besides, the exertion of centrifugal force varies according to the distance from the axis, where centrifugal force varies with radial distance from the axis due to rotational dynamics. Points farther from the axis experience higher centrifugal forces than those closer to the axis. Consequently, by varying the RPM at distinct positions, we can systematically and automatically control the discharge of each solution into the reactor. For instance, a solution located at the outer edge of a centrifugal chip can be effortlessly released into the microreactor with relatively low centrifugal force, around 1000 RPM. In contrast, solutions nearer to the center of the chip may only be released at much higher forces, such as 4000-5000 RPM.

The opening RPM or burst RPM, marking the point at which the solution is expelled into the reactor, can be determined through a simple model. This model involves balancing the centrifugal pressure against the capillary pressure and is grounded in the Young-Laplace equation [56, 57].

$$\text{The opening RPM} = 60 \left(\frac{γ \sin θ}{π^2 ρ (R_2 - R_1) \times \frac{R_2 + R_1}{2} \times \frac{4hw}{C}} \right)^{1/2}$$

where γ is the surface tension of the fluid, ρ is its density, R_1 is the distance between the liquid plug in the channel and the center of rotation, R_2 is the distance between the passive valve and the center of rotation, θ is the contact angle of the reagents on the microchannels, C is the associated contact-line length, and h and w are the depth and width of the channel, respectively.

2.3. Design for a 30-unit centrifugal HTP chip

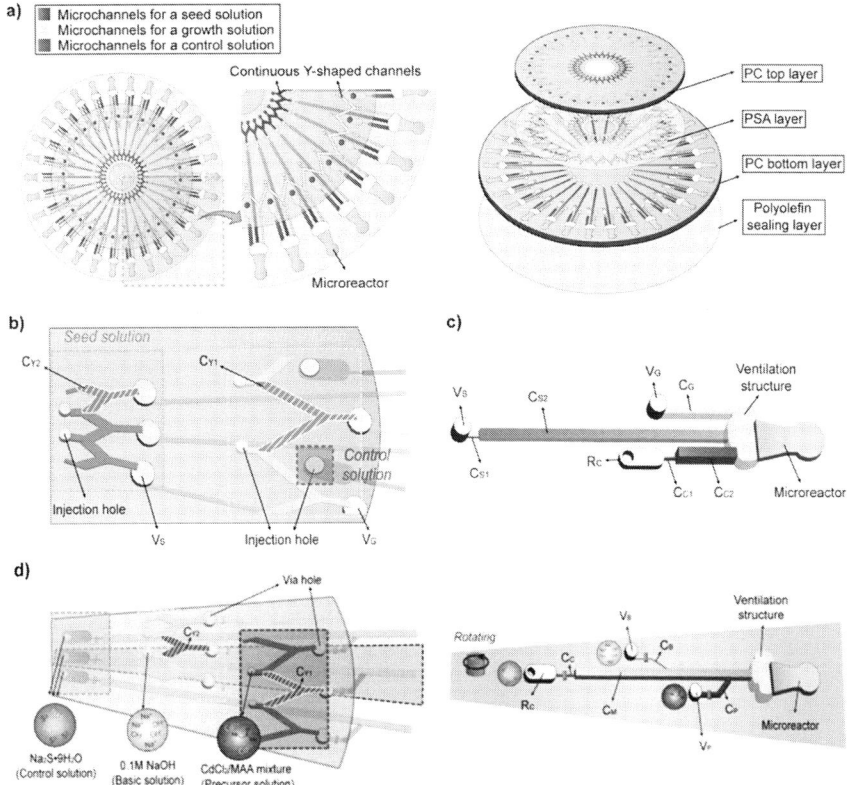

Fig. 2. a) The rotary microfluidic device for multiplex anisotropic Au NP synthesis. The diameters of the top PC and the bottom PC layer were 8 cm and 11.4 cm, and the diameter of a shaft hole was 3.2 cm. The blue, yellow, and red colors represent the seed, growth, and control solutions. b) Schematics of the top PC layer containing the continuous Y-shape microchannels for loading the seed (inner) and growth (outer) solution. C_{Y1} and C_{Y2} are the volumes of a single aliquoting growth and single aliquoting seed solution in the Y-shape microchannel. c) A functional unit in the bottom PC layer is composed of one control solution reservoir (R_C), two via holes (V_S and V_G, for connecting to the seed solution (C_{Y2}) and the growth solution (C_{Y1}), respectively), different dimensional microfluidic channels ($C_{S1, S2}$: seed solution channel, C_G: growth solution channel, $C_{C1, C2}$: control solution channel), a ventilation structure, and a microreactor. d) *The rotary microfluidic device for Cds nanocrystals synthesis.* A microfluidic subunit in the bottom PC layer is composed of one control solution reservoir (RC), two via holes (VP and VB which are connected to the precursor solution (CY1) and the basic solution (CY2), respectively), different dimensional microfluidic channels (CP: a precursor solution channel, CB: a basic solution channel, CC: a control solution channel), a main channel (CM), a ventilation structure, and a microreactor. (Reprinted with permission from Ref. [33]. Copyright (2015) Royal Society of Chemistry.)

The studies by Park et al. [33, 51] exemplify the application of centrifugal microfluidics for the HTP synthesis of NPs and quantum dots. These microfluidic chip designs share fundamental characteristics in their structure and operational principles. Both designs feature a two-layered architecture, consisting of a top polycarbonate (PC) layer and a PC bottom layer, sealed with two pressure sensitive adhesive (PSA) layers. Centrifugal force plays a pivotal role in both designs, facilitating precise reagent loading and controlled release into the microreactors. The incorporation of a zigzag aliquoting structure in both designs is instrumental in enabling one-shot injection and aliquoting of reagents, contributing to the efficiency of the HTP synthesis process. Additionally, ventilation structures are strategically integrated into both designs to prevent the backward flow of reagents.

Despite these shared elements, the two designs diverge in their specific applications and the nature of the synthesized NPs. The HTP chip from Park et al. [33], focusing on the HTP synthesis of anisotropic metallic NPs, particularly gold nanoparticles (Au NPs), features microreactor units with reservoirs for control solutions (ascorbic acid, AA), Y-shape microchannels for the gold seed (inner) and growth (outer) solution (Fig. 2A to C). By adjusting the concentration of AA as the controlling parameter, the chip can efficiently conduct 30 reactions at varying AA concentrations, providing a rapid and effective means to investigate the impact of AA on the morphology change of Au NPs.

On the other hand, the other design of Park et al. [51] is modified and tailored for the multiplex synthesis of quantum dots - CdS nanocrystals (NCs) (Fig. 2D). Its subunits include a chamber for control solution ($Na_2S \cdot 9H_2O$), Y-shape microchannels for basic solution and precursor solution ($CdCl_2$/MAA mixture). Similar to the first design, the volumes of the split solutions in the zigzag channels are calculated but differ based on the geometry of the Y-shaped structures. By tunning the concentration of sulfur in a control solution, a range of multi-color CdS nanocrystals is generated, spanning from green to red, contingent upon the molar ratios of Cd^{2+} to S^{2-}.

2.4. Design for a 60-unit centrifugal HTP chip

Nguyen et al.'s study underscores a significant leap forward in both design sophistication and operational efficiency [39]. The research represents a notable transition from a 30-unit chip to a cutting-edge 60-unit chip, specifically engineered for the synthesis of Pd@AuPt core-shell NPs (Fig. 3). The noteworthy innovation lies in the redesign of the HTP system, where the shift to a 60-unit configuration is accompanied by a host of intricate microstructures. Within the confines of a compact 6.5 cm radius chip, the researchers have ingeniously incorporated various microelements, such as chambers, zigzag channels, micro vents, and passive valves. This ingenuity in design not only maximizes spatial efficiency but also amplifies the chip's overall functionality, showcasing a remarkable fusion of microengineering and nanoparticle synthesis capabilities.

Radially, the chip features 60 inlet holes for loading Au^{3+} solutions with varying concentrations, a first zigzag microchannel for Pd-cube solution loading, a second zigzag microchannel for Pt^{2+} solution loading, and 60 reaction chambers. Microchambers are connected to the reaction chambers via passive valves and microchannels. Distinct distances of each microchambers from the spindle axis necessitate varying centrifugal forces for discharging solutions.

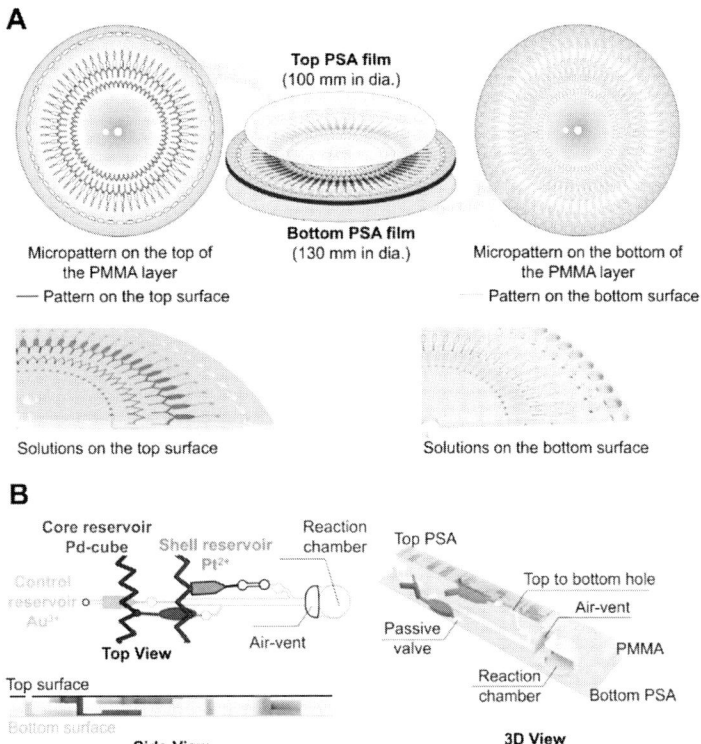

Fig. 3. (A) The schematics of the centrifugal microdevice consisting of three layers. The top side has two zigzag channels and the shell reservoirs for Pt^{2+}, and the bottom side has the core reservoirs for the Pd-cube solution as well as the control reservoirs for Au^{3+}. (B) An enlarged structure of one unit. From the inner to the outer, the reservoirs for the Au^{3+}, Pd-cube, Pt^{2+}, and the reaction chambers were patterned that were connected by the microchannels and passive valves. (Reprinted with permission from Ref. [39]. Copyright (2020) Royal Society of Chemistry.)

2.5. Design for a 60-unit centrifugal HTP chip with a serially diluting structure

Nguyen et al. presented an innovative HTP chip design geared towards the automated synthesis of various morphologies of AuNPs within a single device (Fig. 4) [52]. The microfluidic chip shares a fundamental similarity with its predecessor, featuring 60 parallel units. Specifically tailored for AuNPs synthesis, each chip unit is equipped with chambers designed for various stages of the reaction. These include dedicated chambers for Au seeds, a reducer (ascorbic acid, AA), water, and a growth solution comprising a mixture of Au^{3+}, surfactant, and Ag^+. The connection between these chambers is facilitated by a zigzag aliquoting channel, enabling the simultaneous injection of each reactant in a single shot. Despite these similarities, Nguyen et al.'s study represents an enhanced iteration of the earlier 60-units chip design. It is important to highlight that the original 60-units chip, devised for the synthesis of core-shell particles and featuring 60 distinct chambers for the injection of the control factor (control reservoirs for Au^{3+}) at varied concentrations, necessitated manual pipetting. The latest version incorporates an innovative component known as the serially diluting structure. This addition facilitates the automatic generation of a concentration gradient for the control factor. Consequently, the need for manual pipetting has been significantly reduced, transitioning from 60 instances to merely 2, marking a noteworthy improvement in efficiency and practicality.

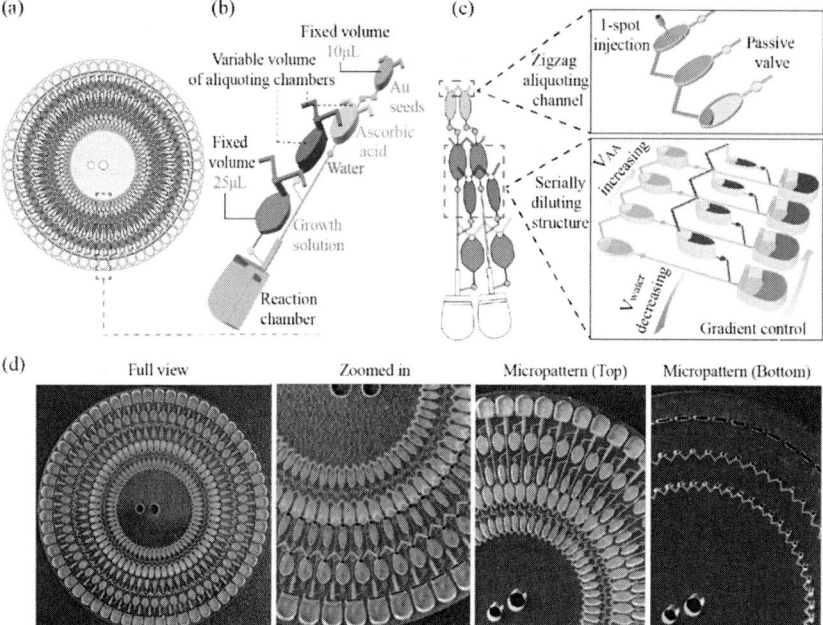

Fig. 4. Fabrication and design of the HTP centrifugal microdevice for the synthesis of Au NPs. (a) Schematics illustrating the centrifugal microfluidic chip having 60 reaction units. (b) Structure of a single unit, including 4 reagent chambers (for the Au seed solution, AA, water, and the growth solution). (c) The 60 units for each solution are connected by a zigzag aliquoting microchannel, which can be filled by a single injection shot because the capillary pressure exerted by a passive valve is higher than that by a zigzag aliquoting microchannel. The serially diluting microchannel was designed for generating concentration gradients of AA. (d) Digital images of the as-fabricated HTP centrifugal device. (Reprinted with permission from Ref. [52]. Copyright (2023) Elsevier.)

The serially diluting structure employs microfabrication to create chambers of varying depths, leading to different volumes within a chip. The key to achieving a series of concentrations lies in the controlled combination of different volumes of the AA solution with corresponding volumes of water, while ensuring that the total volume remains constant at 40 µL. The chamber depths are adjusted in a counterclockwise direction, with the depth of the AA solution gradually increasing and that of water decreasing. Following the mixing of each AA-contained chamber with its corresponding water-contained chamber, facilitated by centrifugal force, the process yields 60 distinct concentrations of the AA solution effortlessly. This method allows for a systematic and efficient tuning of the volume ratio between the reactant (AA) and the diluent (water), thereby creating a precise and reproducible concentration gradient on the chip.

3. Fabrication of a centrifugal HTP chip for nanoparticle synthesis

In the field of microfluidic-based nanoparticle synthesis, the material chosen for fabricating the devices is a critical factor that significantly impacts their performance and success. The careful selection of materials is essential to ensure the devices' compatibility with the synthesis processes, reliability, and the ability to produce nanoparticles with specific and desired characteristics. Researchers often turn to materials such as silicon and glass, which offer compatibility with various nanoparticle synthesis methods and the ability to withstand high temperatures and pressures. Silicon is renowned for its semiconductor fabrication compatibility, while glass, particularly borosilicate glass and fused silica, is favored for its chemical inertness and optical transparency, making it the go-to option when optical monitoring is essential.

For those seeking flexibility and ease of fabrication, polydimethylsiloxane (PDMS), a transparent elastomer with biocompatible properties, fits the bill. Additionally, polymers like PMMA and polyethylene offer cost-effective and readily fabricated alternatives. In some cases, researchers opt for microfluidic devices made from PC, known for its robustness, chemical resistance, and optical transparency. Polycarbonate-based microfluidic devices are suitable for various nanoparticle synthesis processes due to these advantageous properties. Even paper-based microfluidic devices, initially designed for diagnostics, have found their way into nanoparticle synthesis, especially in resource-limited settings, due to their simplicity and low cost. Researchers carefully weigh factors like chemical compatibility, thermal stability, and ease of fabrication to select materials that strike a harmonious balance for their specific nanoparticle synthesis processes and applications.

Park et al. introduced a chip fabrication process characterized by the integration of two main PC layers, emphasizing a dual-layer structure [51]. The microdevice is composed of a 1 mm thick top PC plate and a 2 mm thick bottom PC plate. Employing AutoCAD for meticulous design and a CNC machine for precise micropatterning on both PC layers, the fabrication process includes surface treatment of Y-shaped channels using a VISTEX solution in isopropanol [58]. Subsequently, a polyolefin sealing foil is affixed to the PC bottom layer, and the layers are systematically stacked with a pressing machine. This dual-layer configuration, with intricately patterned PC layers, showcases a methodical approach to microfluidic chip fabrication.

While the dual-layer configuration offers advantages, such as increased design complexity and the potential for diverse microfluidic functionalities, it also introduces significant challenges. Precise alignment between the two PC layers is a critical concern, as even slight misalignments during micropatterning and assembly can lead to disruptions in microfluidic channels and structures. Achieving high alignment precision requires advanced fabrication techniques, careful registration of patterns, consideration of material compatibility, and meticulous attention during the stacking and assembly process. Additionally, the inherent complexity of dual-layer fabrication can pose challenges in terms of cost and production time. Addressing these challenges is essential to harness the full potential of the dual-layer microfluidic chip design.

In another approach, Nguyen et al. [39] and Nguyen et al. [52] presented an innovative chip fabrication process that distinctly diverges from the dual-layer configuration. The chip fabrication centers around a singular main PMMA layer, strategically micropatterned on both sides to optimize microstructure placement within a confined area [59, 60]. Using Cut2D software for design and a CNC machine for precision micropatterning, the resulting microdevice comprises three layers: a PSA film, a double-side etched PMMA layer, and a second PSA film. The fabrication process incorporates a meticulous cleaning regimen, along with tailored hydrophobic surface treatments designed to ensure optimal fluid movement. These fabrication methods underscore the adaptability in microfluidic chip design, where material choice and optimization techniques cater to diverse synthesis processes and application requirements, particularly emphasizing the single-layer nature of the PMMA-based chips.

4. Chip operation

The operational dynamics of centrifugal microfluidic devices for nanoparticle synthesis, as demonstrated in the collective research, exhibit a harmonious blend of mechanical, fluidic, and nanotechnological principles, essential for HTP and precise nanoparticle production. This intricate process commences with the systematic introduction of various reagents into the microfluidic device, where the precision of each step is of utmost importance.

In the studies conducted by Park et al., the operation is defined by the one-shot, automated loading of reactants through a zigzag aliquoting structure (Fig. 5) [33]. The subsequent stepwise release and mixing of these reactants into a microreactor is facilitated by the utilization of centrifugal force. For example, in their study for the production of multiplex anisotropic metallic NPs, the operation commences with the manual pipetting of the control factor, AA, into 30 distinct chambers of the chip [33]. This is followed by the automated introduction of other reactants - Au seeds and growth solution - via the zigzag structure. After solutions are introduced into the microfluidic chip, they are poised for the releasing process, wherein the solutions are carefully aliquoted and then transferred to microreactors. The pivotal role in this process is played by rotational speed control, where centrifugal forces, adjusted to specific RPMs like 400, 600, and 1500, are meticulously calibrated. This precise modulation of RPMs ensures an exact and efficient distribution of the solutions into aliquots, leading to the microreactor. Such exactitude in control is crucial for achieving consistent quality and uniformity in the nanoparticles produced.

Park et al.'s subsequent exploration into multiplex CdS nanocrystal synthesis further showcases the advanced capabilities of these devices (Fig. 6) [51]. In this process, the precursor, basic, and control solutions are sequentially introduced into the microreactor, guided by a carefully programmed RPM control. The RPMs are strategically escalated from 600 to 1000 and finally to 1500, demonstrating the device's sophisticated handling of varied fluid dynamics. This progressive increase in RPMs ensures that each solution is accurately navigated through the microfluidic pathways, culminating in its precise placement within the microreactor. Such precision is testament to the device's ability to facilitate complex nanoparticle synthesis processes with remarkable accuracy and efficiency.

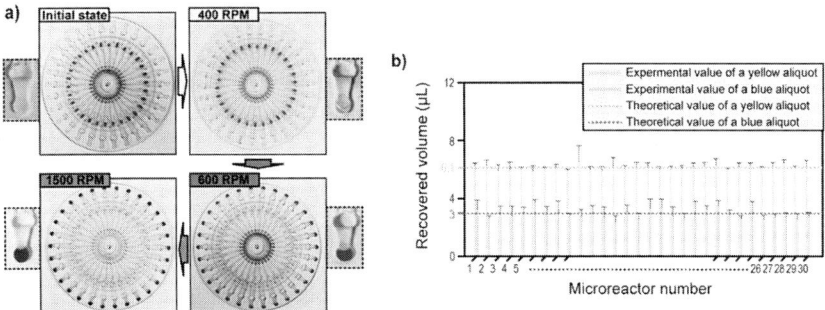

Fig. 5. a) Digital images of the rotary microdevice for stepwise reagent loading by RPM control. Through clockwise rotation starting from the top left, the yellow, the red, and the blue solutions were pumped out at 400 RPM, 600 RPM, and 1500 RPM in a sequential order. b) Comparison of the theoretical volume versus the experimentally recovered volume of the yellow and the blue solutions. (Reprinted with permission from Ref. [33]. Copyright (2015) Royal Society of Chemistry.)

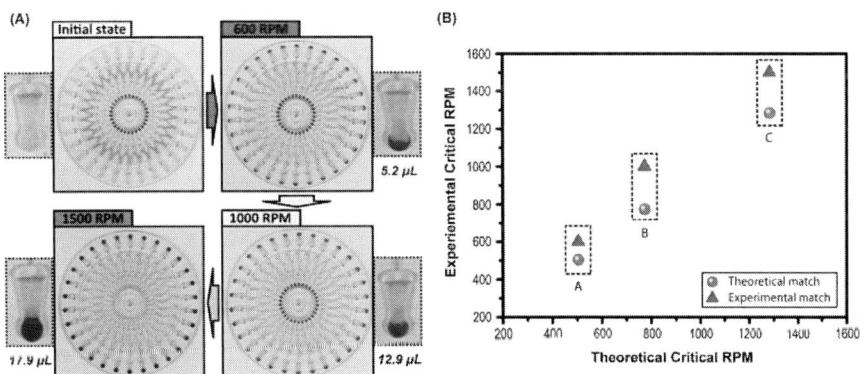

Fig. 6. (A) Digital images of the centrifugal microdevice for stepwise reagent loading by RPM control. Through clockwise rotation starting from the top left, the red, the yellow, and the blue solutions were pumped out at 400 RPM, 600 RPM, and 1500 RPM in a sequential order. (B) Theoretical versus experimental critical burst rotational speed. (Reprinted with permission from Ref. [51]. Copyright (2015) Elsevier.)

Similarly, in the research by Nguyen et al. [39], the centrifugal microfluidic approach is leveraged to handle diverse reagents necessary for complex nanoparticle synthesis (Fig. 7). Their study on Pd@AuPt core-shell nanoparticles involve the precise loading and aliquoting of Au^{3+}, Pd-cube, and Pt^{2+} solutions. Here, the solutions are introduced into 60 units of the chip through manual pipetting (for different concentrations of Au^{3+}) and the zigzag aliquoting channel (for Pd-cube and Pt^{2+} solution), with the centrifugal force required for discharging solutions into reactors being carefully calibrated at different RPM levels (-2000 RPM, 4000 RPM).

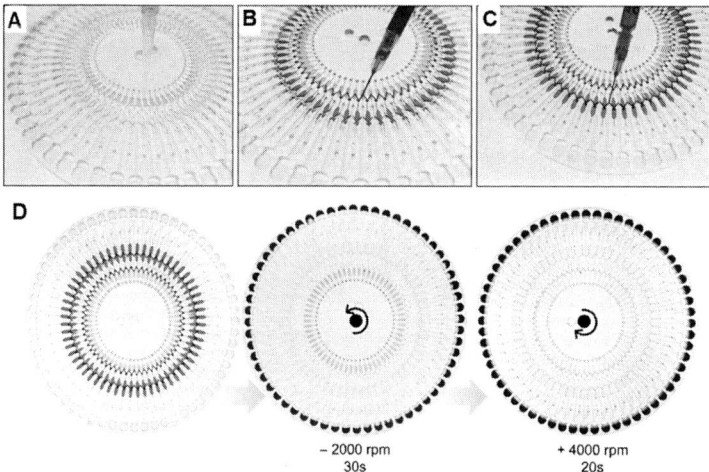

Fig. 7. Operation of the HTP chip for core-shell nanoparticle synthesis. The Au^{3+} solution (yellow) was loaded by a pipette, while the Pd-cube solution (blue) and the Pt^{2+} solution (red) were injected by a syringe. (A) Manual injection of 60 different concentration of the Au^{3+} solution into the control reservoirs. (B) Automatic filling in the zigzag microchannel and the core reservoirs with a single shot. (C) Automatic filling in the zigzag microchannel and the shell reservoirs with a single shot. (D) Digital images illustrating a stepwise elution of the Au^{3+}, Pd-nanocube precursor, and Pt^{2+} solution into the reactors by controlling RPM and the rotational direction. The blue (the Pd-nanocube precursor) and red (the Pt^{2+} solution) solution were released at –2000 rpm in a counter-clockwise direction, and the yellow solution (the Au^{3+} solution) was ejected at +4000 rpm in a clockwise direction. (Reprinted with permission from Ref. [39]. Copyright (2020) Royal Society of Chemistry.)

The 2023 study by Nguyen et al. for gold nanoparticle synthesis further exemplifies the sophistication of these devices (Fig. 8) [52]. It begins with the automatic injection of Au seed solution, AA, water, and growth solution into aliquoting chambers by the zigzag aliquoting structure. The device ingeniously combines capillary pressure with centrifugal force, employing burst RPM calculations for effective distribution and ejection of solutions into reaction chambers [39]. This method represents a significant advancement in nanoparticle synthesis, streamlining a process that traditionally requires extensive manual labor and time.

Throughout these studies, the use of a zigzag aliquoting structure for single-shot solution injection, alongside the application of centrifugal force for precise fluid movement, emerges as a central theme within microfluidic chips. Each study, while tailored to meet specific synthetic ambitions, adheres to these fundamental operational principles. These principles underscore the adaptability and efficiency of microfluidic devices across a spectrum of nanoparticle synthesis applications. The variations observed in the devices' rotational speeds, methods of fluid injection, and techniques for monitoring the synthesis process not only reflect the unique focus and objectives of each study but also illustrate the progressive sophistication and burgeoning potential of centrifugal microfluidic systems within the field of nanotechnology.

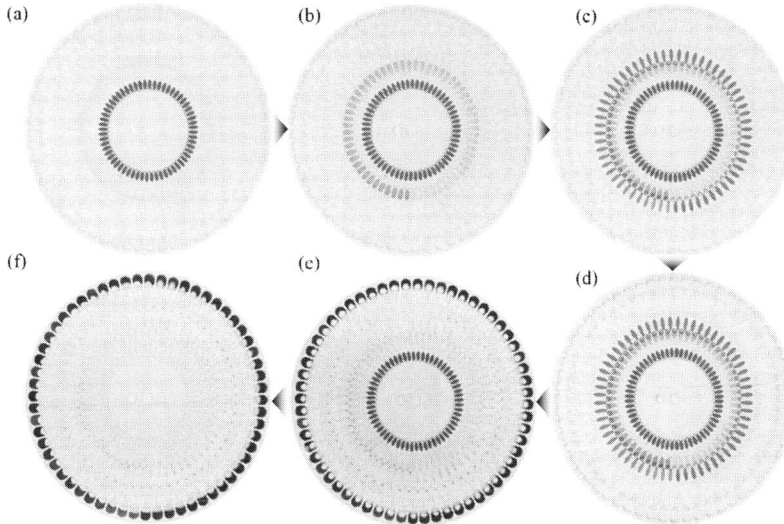

Fig. 8. Operation of the centrifugal microdevice for AuNPs synthesis. Solutions were injected into 60 reaction chambers each in the following order using a single shot per solution. (a) Au seed solution was injected first. (b) AA was injected thereafter. (c) Water was injected after AA. (d) Growth solution (25 μL per chamber) was injected last. (e) Rotation at 1000 RPM delivered AA, water, and the growth solution, while the Au seed solution remained in its chambers. (f) Rotation at 4000 RPM delivered the Au seed solution into the 60 reaction chambers. (Reprinted with permission from Ref. [52]. Copyright (2023) Elsevier.)

5. Application of nanoparticle synthesis

5.1. Quantum dots

In their innovative study, Park et al. focus on the synthesis of cadmium sulfide (CdS) quantum dots using a centrifugal microfluidic device, achieving a high level of precision and control over the synthesis process [51]. This approach marks a significant advancement in the field of nanocrystal synthesis, particularly for CdS quantum dots, known for their unique optical properties.

The synthesis process begins with the careful preparation of chemical reagents. The team uses a precursor solution of cadmium chloride ($CdCl_2$) mixed with mercaptoacetic acid (MAA), a basic solution of sodium hydroxide (NaOH), and a sulfur source in the form of sodium sulfide nonahydrate ($Na_2S·9H_2O$). These reagents are strategically loaded into the microfluidic device, which consists of multiple functional units, each featuring a reaction chamber and three reservoirs. The design of the device allows for an automated and sequential introduction of these reagents into the reaction chambers.

Central to the success of this synthesis is the manipulation of the molar ratio of Cd^{2+} to S^{2-}. Park et al. experiment with ratios ranging from 12:1 to 1:6, demonstrating the ability to fine-tune the optical properties of the resulting quantum dots. By adjusting these ratios, they successfully synthesize CdS nanocrystals that exhibit a spectrum of fluorescence colors, ranging from green to red. This variability in color is a direct consequence of changes in the particle size and composition of the quantum dots.

The mixing and reaction of these reagents are facilitated by the unique operation of the microfluidic device. The device employs a centrifugal force, controlled by adjusting the rotational speed, to move the reagents through the channels and into the reaction chambers. An innovative feature of this process is the use of alternating rotational directions, which ensures thorough mixing of the reagents, a critical factor in achieving uniformity in the nanocrystal synthesis.

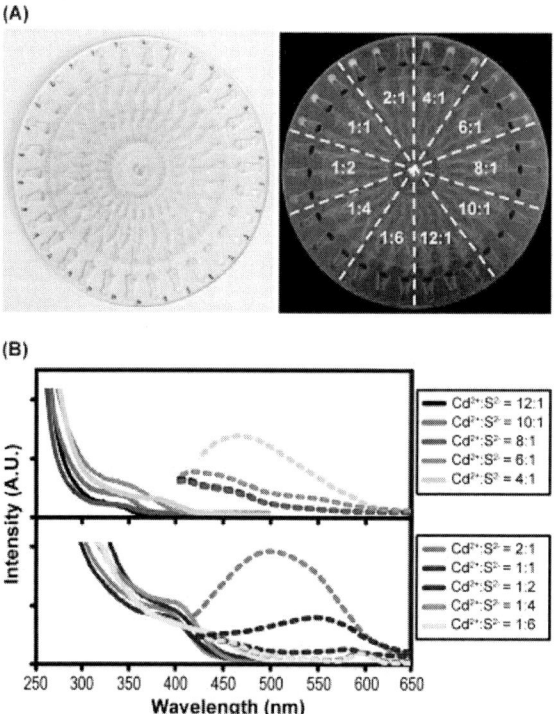

Fig. 9. (A) Digital image of the centrifugal microdevice after CdS nanocrystal synthesis (left). Upon UV irradiation, various fluorescence signals are emitted from the microreactors depending on the Cd^{2+}:S^{2-} molar ratio (right). (B) UV–vis absorption (solid lines) and fluorescence (dashed lines) spectra of CdS nanocrystals synthesized under different Cd^{2+}:S^{2-} molar ratios. Excitation wavelength was 365 nm for the ratio of 12:1–6:1, 380 nm for the ratio of 4:1–1:1, and 392 nm for the ratio of 1:2–1:6. (Reprinted with permission from Ref. [51]. Copyright (2015) Elsevier.)

Once the reaction is complete, the synthesized CdS nanocrystals are collected for analysis. Park et al. utilize various analytical techniques, including UV-Vis absorption and fluorescence emission spectrometry, to characterize the optical properties of the synthesized quantum dots. These analyses confirm that the synthesized nanocrystals exhibit the expected shifts in absorption and emission spectra, correlating with the changes in the $Cd^{2+}:S^{2-}$ molar ratio. The results underscore the device's capability to precisely control the size and composition of the CdS quantum dots, thus directly influencing their optical characteristics (Fig. 9).

The advancements in the synthesis of CdS nanocrystals through centrifugal microfluidic devices presents significant advantages over traditional off-chip methods, including improved efficiency, reduced reaction time, enhanced control over synthesis, scalability, and better environmental sustainability (Table 1).

Table 1. Comparison of the on-chip and off-chip multiplex CdS nanocrystal synthesis at thirty different conditions.

	On-chip	**Off-chip**
The number of reactors	1 centrifugal microdevice	30 glass vials
The number of pipetting steps	600 μL (each 20 μL)	75,000 μL (each 2500 μL)
Total reagent volumes	32 steps	90 steps
Time for synthesis	5 minutes	1 hour

One of the most notable differences is the efficiency in reactant consumption and waste reduction. The on-chip process, with its meticulously designed microfluidic channels, requires a fraction of the reactants used in off-chip methods. For instance, the total reaction volume in the on-chip process is just around 600 μL, a stark contrast to the 75,000 μL that might be needed in conventional methods. This not only makes the process more cost-effective but also aligns with environmentally sustainable practices by minimizing chemical waste. Additionally, the reaction time and throughput are greatly improved in the on-chip method. The unique design of the centrifugal microfluidic device allows for rapid mixing and heat transfer, significantly accelerating the reaction kinetics. This efficiency is evident in the reduced reaction time – about 5 minutes for the on-chip process, compared to approximately an hour with off-chip methods. The increased efficiency in reaction time directly translates to higher throughput, enabling the production of larger quantities of nanocrystals in a shorter period.

Moreover, the on-chip method offers enhanced precision and control over the synthesis process. The ability to finely tune the rotational speed of the device allows for exact control over mixing and reaction rates, which is crucial for the synthesis of quantum dots with specific optical properties. This level of control is a distinct advantage over off-chip methods, which often suffer from inconsistencies due to manual mixing and less precise temperature control. In terms of scalability and reproducibility, the on-chip method exhibits superior performance. The design of the microfluidic device allows for the parallel processing of multiple reactions under identical conditions, ensuring high reproducibility. This is a stark contrast to off-chip methods, where scaling up often leads to inconsistencies and batch-to-batch variations. Finally, the on-chip method's compact design and efficient use of resources make it highly versatile and adaptable for various applications, a significant advantage over the more rigid and extensive setups required in traditional off-chip methods.

5.2. Gold nanoparticles

In another experiment, Park et al. performed a comprehensive study on the synthesis of anisotropic gold nanoparticles (Au NPs) using a rotary microfluidic device [33]. This study is particularly focused on the effects of ascorbic acid (AA) concentration on the morphology of Au NPs, demonstrating a meticulous approach to controlling nanoparticle shape.

The experimental setup involved triplicate reactions with ten different conditions, each varying in AA concentration. The process began with a seed solution containing single crystalline Au NPs sized 13–17 nm and a growth solution of $HAuCl_4·3H_2O$ and CTAB mixture. The concentration of AA, a crucial control factor, was altered from 0.001 M to 0.2 M. Prior to introducing the reagents into the microdevice, it was exposed to UV ozone for 15 minutes to create oxygenous functional groups on the microreactor surfaces.

Park et al. observed a fascinating evolution in the morphology of the Au NPs as the AA concentration increased (Fig. 10A to C). At lower concentrations, hexagonal and triangular Au NPs were predominantly formed, indicative of an edge-biased growth from a spherical shape. As the concentration of AA rose, the Au NPs transitioned through intermediate shapes between triangular and tripodal, finally achieving a tripodal shape at the highest AA concentration of 0.2 M. This morphological transformation was driven by the need for excess AA to ensure the complete reduction of gold ions [61]. This shape evolution was visually evident through changes in the color of the Au NP solutions, shifting from pink to blue-purple proportional to the AA concentration. Complementarily, UV-Vis absorption spectroscopy revealed a red shift in the plasmonic absorption as the concentration of AA increased, signifying the elongation of the Au NPs [62]. For example, hexagonal-shaped Au NPs formed at 0.001 M AA showed a 535 nm absorption band, while the tripodal Au NPs at 0.2 M AA exhibited a peak at 550 nm with a broad shoulder peak around 700 nm.

High-resolution TEM analysis further confirmed these observations (Fig. 10D). Tripodal Au NPs produced with 0.2 M AA displayed three tips each extending to 49 nm, with a lattice plane d-spacing of ~0.142 nm, corresponding to the (220) plane of gold. Notably, these tripods formed without any stacking faults or twins, showcasing an epitaxial growth pattern attributed to the fast reduction of gold ions along <110> directions by AA in the presence of oxygenous groups.

In a similar study by Nguyen et al. [52], a novel approach for the synthesis of Au NPs using a HTP centrifugal microfluidic device is showcased. The study focuses on the impact of varying concentrations of AA on the morphology of the synthesized Au NPs, offering a comprehensive understanding of nanoparticle formation under different conditions. This method is particularly remarkable for its use of a serially diluting structure within the device, allowing for the synthesis of various Au NP morphologies on a single chip with 60 units, a significant increase from the previously discussed 30-unit system.

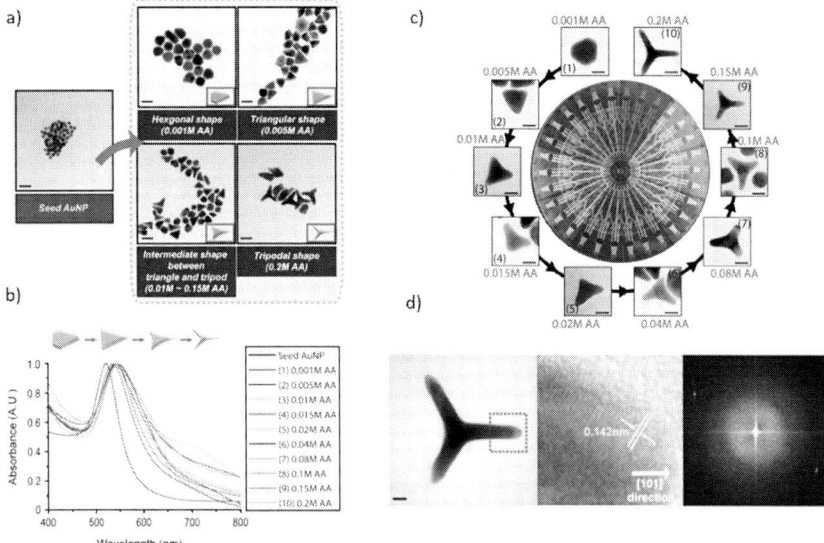

Fig. 10. a) Transmission electron microscope (TEM) images of the representative transformed Au NPs that were grown from the 13-17 nm seed Au NPs. Scale bar: 50 nm, b) The UV-Vis absorbance spectra of the Au NP solutions, which show a red-shift in proportion to the AA concentration. c) Representative TEM images showing the shape evolution of the Au NPs. Scale bar: 20 nm. d) TEM image of the well-developed tripodal Au NP which was synthesized at 0.2 M AA (left, scale bar: 10 nm), a lattice fringe image of the tripodal Au NP (middle), and fast Fourier transform pattern of the middle image (right). (Reprinted with permission from Ref. [33]. Copyright (2015) Royal Society of Chemistry.)

The microfluidic chip features a unique zigzag aliquoting microchannel design that facilitates the loading of each reactant in a single injection shot. In each unit, the chip is designed with specific volumes for the aliquoting chambers: 10 µL for the Au seed solution and 25 µL for the growth solution (was a mixture of HAuCl$_4$, AgNO$_3$, and Triton X-100). To establish a concentration gradient for AA, the research team implemented an innovative approach by the serially diluting structure. A key feature of the structure is its chamber design, where the depth allocated for the AA solution gradually increases in a counterclockwise direction, while the depth for water correspondingly decreases. This design ensures that the combined volume of AA and water in each unit sums up to 40 µL after the aliquoting process. By manipulating the volume ratio of AA and water, the team successfully established a gradient of AA concentrations from 0.4 mM to 3.6 mM across the device, allowing for a comprehensive exploration of the impact of AA concentration on Au NP morphology (Fig. 11).

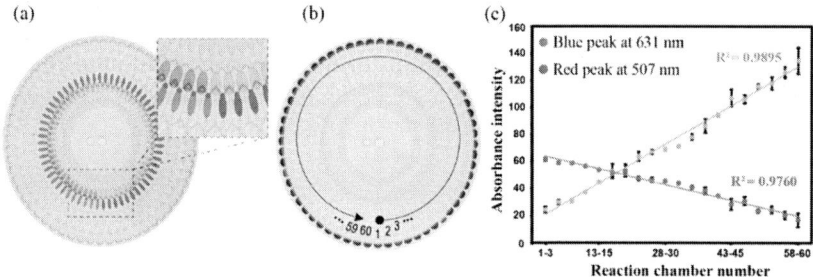

Fig. 11. Dilution structure for controlling the concentration of AA. (a) AA was introduced into 60 aliquoting chambers (blue color) and water was injected into the 60 corresponding aliquoting chambers (red color). The volumes of AA and water were varied by controlling the depth of the aliquoting chambers for each solution, while maintaining the sum of their volumes at 40 μL per unit (differences in color intensity caused by differences in depth). (b) A digital image showing the concentration gradient of AA after the centrifugation. The red color in the reaction chamber 1 was diluted with the blue color, resulting in the blue color in the reaction chamber 60. (c) Correlation curves for the color intensity versus the reaction chamber for the AA (blue line) and water (red lines) with high linearity. (Reprinted with permission from Ref. [52]. Copyright (2023) Elsevier.)

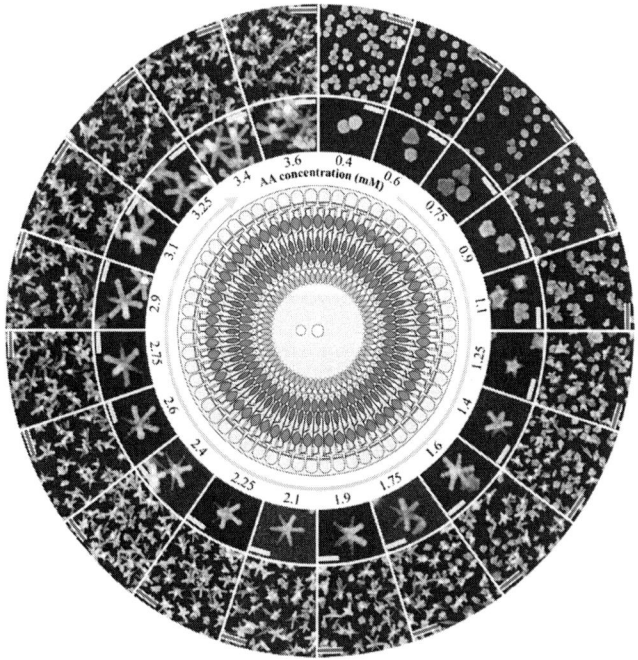

Fig. 12. SEM data illustrating the morphology change of Au NPs synthesized on a single centrifugal microfluidic device. Sixty separate reactions (20 conditions × 3 reaction chambers per condition) were carried out on a single chip in 30 minutes by serially-diluting the AA concentration from 0.4 mM to 3.6 mM. Scale: a yellow ruler = 50 nm and a red ruler = 150 nm. (Reprinted with permission from Ref. [52]. Copyright (2023) Elsevier.)

The morphological evolution of the Au NPs is closely monitored as the AA concentration varies. At the lower end of the concentration spectrum (0.4 to 1.1 mM), the nanoparticles transform from nanospheres to mixed shapes of spheres and triangles. As the concentration increases to between 0.9 and 1.1 mM, the nanoparticles begin to take on quadrangular, polyhedral, and small star shapes. A significant shift in morphology is observed when the AA concentration ranges from 1.25 to 1.9 mM, where the spikes of the Au nanostars start to grow more prominently, and the number of nanostars increases. In higher AA concentrations (2.1 to 2.75 mM), the density of the nanostars continues to increase, reaching a saturation level at 2.75 mM. Interestingly, at concentrations between 2.9 and 3.6 mM, the spike length of the nanostars decreases, and they tend to aggregate, highlighting the nuanced impact of AA concentration on nanoparticle morphology (Fig. 12).

Moreover, the team's exploration of scalability primarily focuses on the feasibility of translating the conditions and protocols used in the micro-scale on-chip synthesis to batch methods typically employed in larger volume productions (Fig. 13). By replicating the on-chip synthesis conditions in larger volumes, Nguyen et al. effectively demonstrate that the unique conditions optimized in the microfluidic environment can be successfully scaled up [52]. This scalability is a vital factor in bridging the gap between laboratory-scale research and commercial-level production, making the synthesized nanoparticles more accessible for practical applications.

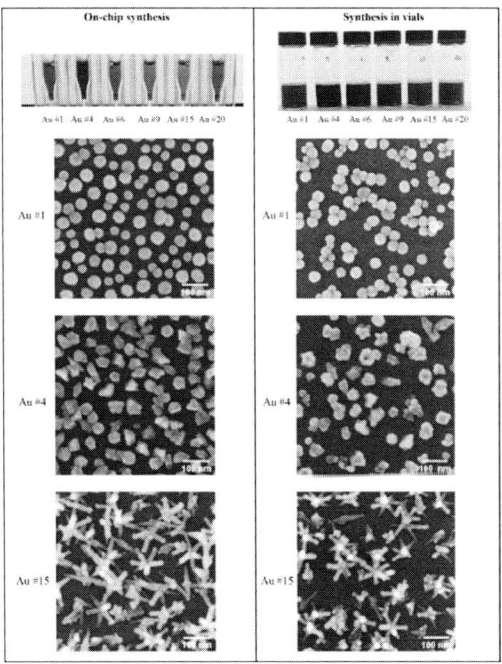

Fig. 13. SEM images of Au #1 (0.4 mM AA), Au #4 (0.9 mM AA), and Au #15 (2.75 mM AA) produced by (left panel) the on-chip (the total reaction volume of 75µL) and (right panel) a vial (the total reaction volume of 5 mL). The size and shape of the resultant Au NPs were quite similar to demonstrate the scale-up capability. (Reprinted with permission from Ref. [52]. Copyright (2023) Elsevier.)

5.3. Bimetallic catalysts

In their comprehensive study, Nguyen et al. explore the synthesis of Pd@AuPt core-shell NPs using a centrifugal microfluidic device, achieving remarkable results in the field of tri-metallic catalysts [39]. This research is significant for its systematic approach to synthesizing Pd@AuPt core-shell NPs under sixty different conditions, specifically varying the concentrations of Au^{3+} while maintaining constant concentrations of Pd nanocubes and Pt precursor solution.

The study's synthetic approach involves mixing an aqueous suspension containing Pd nanocubes and L-ascorbic acid with Pt and Au ion solutions in a specific order. This process results in the formation of core-shell NPs, where Pd forms the core, and a bimetallic shell of Au and Pt envelops it. On the centrifugal microdevice, the Au concentration constituting the bimetallic shell was adjusted across sixty different configurations, effectively altering the Au to Pt ratio from 0.028:1 to 12:1.

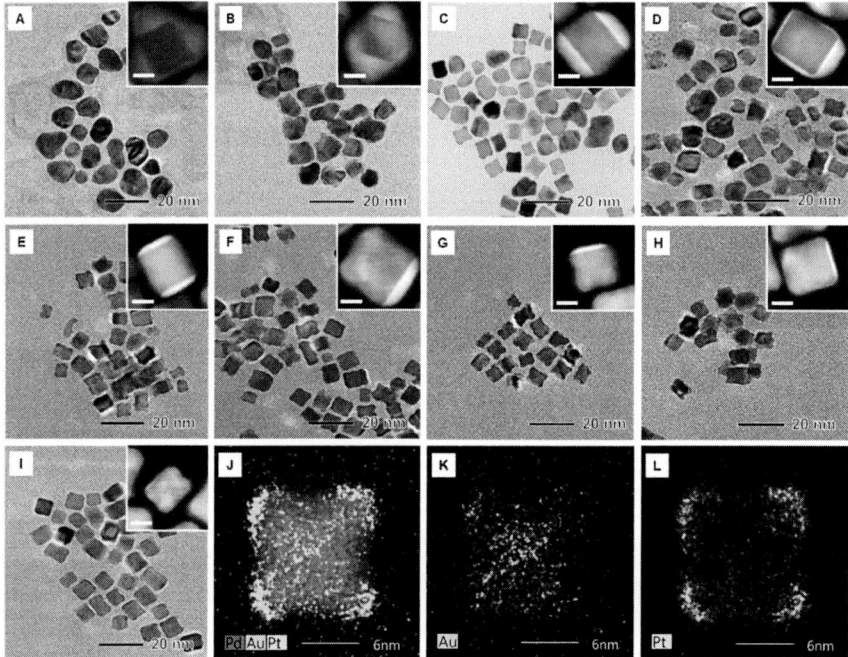

Fig. 14. (A-I) TEM images of Pd@AuPt core-shell NPs with different amount of Au while keeping the same concentration of Pd and Pt. Molar ratio of Au/Pt were (A) 12, (B) 8.4, (C) 5.14, (D) 2.86, (E) 1.48, (F) 0.74, (G) 0.36, (H) 0.18, and (I) 0.08, respectively. (J-L) HAADF-STEM images of Pd@AuPt core-shell NPs with an Au/Pt molar ratio of 0.36. (Reprinted with permission from Ref. [39]. Copyright (2020) Royal Society of Chemistry.)

The research provides a vivid depiction of the synthesized nanoparticles' morphology (Fig. 14). Transmission Electron Microscopy (TEM) images show that the Pd NPs have a cubic shape, while Pd@Pt NPs exhibit concave cubic morphology. Further TEM analyses reveal that the morphology of the Pd@AuPt core-shell NPs changes from quasi-spherical to cubic as the amount of Au precursor decreases. Simultaneously, the number of planes covered by the Au shell reduces approximately from three to one plane.

The study also employs Energy-Dispersive Spectroscopy (EDS) to provide detailed insights into the structural formation of the Pd@AuPt NPs. EDS images of Pd@AuPt NPs with an Au/Pt molar ratio of 0.36 show that the Pd core retains its cubic shape even after the formation of the core-shell structure, with a diameter of 10-12 nm. This observation suggests that galvanic replacement is not the primary reaction for the formation of the Pd@AuPt core-shell structure. The distribution of Au and Pt on the Pd cube is also revealed, showing that the Au shell predominantly covers the {100} plane of the Pd cube, while the Pt shell is mainly located on {111} corners and partially on {100} terraces.

After successfully producing a diverse range of Pd@AuPt NPs, Nguyen et al. conducted a detailed evaluation of the catalytic activity of these on-chip synthetic NPs for the direct synthesis of hydrogen peroxide (DSHP). A key finding of the research was the relationship between the production rate of hydrogen peroxide and the Au/Pt ratio in the catalysts. The study, illustrated in Fig. 15, revealed that the rate of H_2O_2 production increased gradually with the Au/Pt ratio, peaking at a ratio of 0.5. However, when the Au/Pt ratio exceeded 0.5, there was a noticeable decrease in the production rate. This observation identified the Au/Pt ratio of 0.5 as the optimal condition for enhancing the catalytic activity of the Pd@AuPt catalysts.

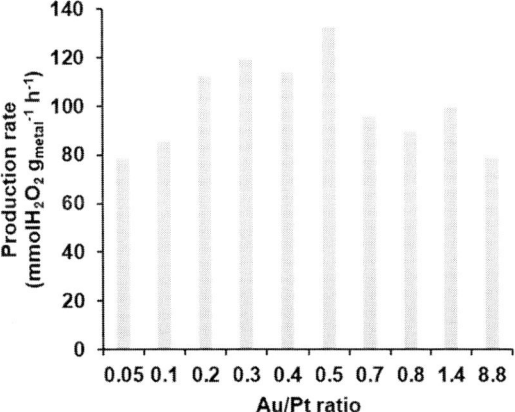

Fig. 15. The production rate of H_2O_2 depends on the Au/Pt ratio using a fast-screening platform. (Reprinted with permission from Ref. [39]. Copyright (2020) Royal Society of Chemistry.)

6. Conclusion

This chapter marks a notable advancement in the field of nanotechnology, with a particular focus on the HTP synthesis of nanoparticles using centrifugal microfluidic devices. The integration of nanotechnology and microfluidics has paved the way for significant improvements in the efficiency, control, and scalability of nanoparticle production.

The utilization of centrifugal microfluidic platforms for gold nanoparticle synthesis, quantum dots and bi-metallic nanoparticles, represents a leap forward in overcoming traditional challenges associated with nanoparticle fabrication. The design of these microfluidic devices, characterized by intricate microstructures like zigzag aliquoting channels and serially-diluting structures, allows for automatic solution loading, precise aliquoting and efficient HTP synthesis.

Moreover, the application of these devices extends beyond the synthesis of simple nanoparticles. Studies detailed in the chapter demonstrate the successful production of complex structures like anisotropic gold nanoparticles and Pd@AuPt core-shell nanoparticles, showcasing the versatility and adaptability of centrifugal microfluidic technology in nanoparticle synthesis. The ability to systematically vary conditions such as reactant concentrations within a single device underscores the precision and efficiency of this approach.

The advancements in the on-chip synthesis processes offer significant advantages over traditional methods. These include enhanced control over the synthesis conditions, reduced reaction times, improved scalability, and better environmental sustainability due to reduced chemical waste and efficient reactant consumption. In conclusion, the exploration of centrifugal microfluidic devices for nanoparticle synthesis in this chapter highlights a significant stride in nanotechnology. It not only demonstrates the potential of these devices in enhancing the efficiency and control of nanoparticle synthesis but also underscores their potential in revolutionizing approaches to material fabrication. As the field continues to evolve, these innovations are expected to open new avenues in various applications, ranging from medical diagnostics to environmental monitoring and nanomaterial synthesis.

References

1. M. Kim, J.-H. Lee and J.-M. Nam, *Advanced Science*, 2019, 6, 1900471.

2. N. Baig, I. Kammakakam, W. Falath and I. Kammakakam, *Mater Adv*, 2021, 2, 1821–1871.

3. C. M. Cobley, J. Chen, E. Chul Cho, L. V. Wang and Y. Xia, *Chem Soc Rev*, 2011, 40, 44–56.

4. M. Hu, J. Chen, Z.-Y. Li, L. Au, G. V. Hartland, X. Li, M. Marquez and Y. Xia, *Chem Soc Rev*, 2006, 35, 1084–1094.

5. X. Duan, Y. Huang, Y. Cui, J. Wang and C. M. Lieber, *Nature*, 2001, 409, 66–69.

6. H. Lee, S. E. Habas, S. Kweskin, D. Butcher, G. A. Somorjai and P. Yang, *Angewandte Chemie - International Edition*, 2006, 45, 7824–7828.

7   A. Mohanty, N. Garg and R. Jin, *Angewandte Chemie - International Edition*, 2010, 49, 4962–4966.

8   R. Narayanan and M. A. El-Sayed, *Journal of Physical Chemistry B*, 2005, 109, 12663–12676.

9   S. Cheong, J. D. Watt and R. D. Tilley, *Nanoscale*, 2010, 2, 2045–2053.

10 A. Gole and C. J. Murphy, *Chemistry of Materials*, 2004, 16, 3633–3640.

11 D. K. Smith and B. A. Korgel, *Langmuir*, 2008, 24, 644–649.

12 Y. Yin and A. P. Alivisatos, *Nature*, 2005, 437, 664–670.

13 C. J. Murphy, T. K. Sau, A. M. Gole, C. J. Orendorff, J. Gao, L. Gou, S. E. Hunyadi and T. Li, *Journal of Physical Chemistry B*, 2005, 109, 13857–13870.

14 M. R. Langille, M. L. Personick, J. Zhang and C. A. Mirkin, *J Am Chem Soc*, 2012, 134, 14542–14554.

15 A.-G. Niculescu, C. Chircov, A. C. Bîrcă and A. M. Grumezescu, *Nanomaterials*, 2021, 11, 864.

16 E. M. Chan, R. A. Mathies and A. P. Alivisatos, *Nano Lett*, 2003, 3, 199–201.

17 F. Dormont, M. Rouquette, C. Mahatsekake, F. Gobeaux, A. Peramo, R. Brusini, S. Calet, F. Testard, S. Lepetre-Mouelhi, D. Desmaële, M. Varna and P. Couvreur, *Journal of Controlled Release*, 2019, 307, 302–314.

18 J. Wagner and J. M. Köhler, *Nano Lett*, 2005, 5, 685–691.

19 S. Duraiswamy and S. A. Khan, *Small*, 2009, 5, 2828–2834.

20 I. Shestopalov, J. D. Tice and R. F. Ismagilov, *Lab Chip*, 2004, 4, 316–321.

21 Y. H. Kim, L. Zhang, T. Yu, M. Jin, D. Qin and Y. Xia, *Small*, 2013, 9, 3462–3467.

22 E. M. Chan, A. P. Alivisatos and R. A. Mathies, *J Am Chem Soc*, 2005, 127, 13854–13861.

23 S. Yao, Y. Shu, Y.-J. Yang, X. Yu, D.-W. Pang and Z.-L. Zhang, *Chemical Communications*, 2013, 49, 7114–7116.

24 J. Baek, P. M. Allen, M. G. Bawendi and K. F. Jensen, *Angewandte Chemie - International Edition*, 2011, 50, 627–630.

25 J. H. Jung, T. J. Park, S. Y. Lee and T. S. Seo, *Angewandte Chemie - International Edition*, 2012, 51, 5634–5637.

26 R. Kikkeri, P. Laurino, A. Odedra and P. H. Seeberger, *Angewandte Chemie - International Edition*, 2010, 49, 2054–2057.

27 D. J. Han, J. H. Jung, J. S. Choi, Y. T. Kim and T. S. Seo, *Lab Chip*, 2013, 13, 4006–4010.

28 M. Faustini, J. Kim, G.-Y. Jeong, J. Y. Kim, H. R. Moon, W.-S. Ahn and D.-P. Kim, *J Am Chem Soc*, 2013, 135, 14619–14626.

29 A. Abou-Hassan, R. Bazzi and V. Cabuil, *Angewandte Chemie - International Edition*, 2009, 48, 7180–7183.

30 A. Abou Hassan, O. Sandre, V. Cabuil and P. Tabeling, *Chemical Communications*, 2008, 1783–1785.

31 D. Lawanstiend, H. Gatemala, S. Nootchanat, S. Eakasit, K. Wongravee and M. Srisa-Art, *Sens Actuators B Chem*, 2018, 270, 466–474.

32 M. Thiele, A. Knauer, A. Csáki, D. Mallsch, T. Henkel, J. M. Köhler and W. Fritzsche, *Chem Eng Technol*, 2015, 38, 1131–1137.

33 B. H. Park, J. H. Lee, J. H. Jung, S. J. Oh, D. C. Lee and T. S. Seo, *RSC Adv*, 2015, 5, 1846–1851.

34 S. Bae, E. Jung, T. Yu and T. S. Seo, *Small*, 2018, 14, 1802851.

35 J. Wang and Y. Song, *Small*, 2017, 13, 1604084.

36 H. K. Bui, T. D. Dao and T. S. Seo, *Chemical Engineering Journal*, 2022, 429, 132516.

37 H. K. Bui, K. Y. Kim, H. Kim, J.-P. Ahn, T. Yu and T. S. Seo, *Particle and Particle Systems Characterization*, 2020, 38, 2000244.

38 D. Zhang, F. Wu, M. Peng, X. Wang, D. Xia and G. Guo, *J Am Chem Soc*, 2015, 137, 6263–6269.

39 H. V. Nguyen, K. Y. Kim, H. Nam, S. Y. Lee, T. Yu and T. S. Seo, *Lab Chip*, 2020, 20, 3293–3301.

40 M. Park, S. Kim, J. H. Jung and T. S. Seo, *Carbon Letters*, 2021, 31, 831–836.

41 H. Wang, K. Liu, K.-J. Chen, Y. Lu, S. Wang, W.-Y. Lin, F. Guo, K.-I. Kamei, Y.-C. Chen, M. Ohashi, C. K.-F. Shen and H.-R. Tseng, *ACS Nano*, 2010, 4, 6235–6243.

42 P. M. Valencia, E. M. Pridgen, M. Rhee, R. Langer, O. C. Farokhzad and R. Karnik, *ACS Nano*, 2013, 7, 10671–10680.

43 X.-H. Ji, N.-G. Zhang, W. Cheng, F. Guo, W. Liu, S.-S. Guo, Z.-K. He and X.-Z. Zhao, *J Mater Chem*, 2011, 21, 13380–13387.

44 C.-G. Yang, Z.-R. Xu, A. P. Lee and J.-H. Wang, *Lab Chip*, 2013, 13, 2815–2820.

45 J. S. Santana, K. M. Koczkur and S. E. Skrabalak, *Langmuir*, 2017, 33, 6054–6061.

46 T. Gu, C. Zheng, F. He, Y. Zhang, S. A. Khan and T. A. Hatton, *Lab Chip*, 2018, 18, 1330–1340.

47 G. Niu, L. Zhang, A. Ruditskiy, L. Wang and Y. Xia, *Nano Lett*, 2018, 18, 3879–3884.

48 G. Li, Q. Chen, J. Li, X. Hu and J. Zhao, *Anal Chem*, 2010, 82, 4362–

4369.

49 C. E. Nwankire, M. Czugala, R. Burger, K. J. Fraser, T. M. Connell, T. Glennon, B. E. Onwuliri, I. E. Nduaguibe, D. Diamond and J. Ducrée, *Biosens Bioelectron*, 2014, 56, 352–358.

50 R. Gorkin, J. Park, J. Siegrist, M. Amasia, B. S. Lee, J.-M. Park, J. Kim, H. Kim, M. Madou and Y.-K. Cho, *Lab Chip*, 2010, 10, 1758–1773.

51 B. H. Park, D. Kim, J. H. Jung, S. J. Oh, G. Choi, D. C. Lee and T. S. Seo, *Sens Actuators B Chem*, 2015, 209, 927–933.

52 H. Van Nguyen, H. Van Nguyen, V. M. Phan, B. J. Park and T. S. Seo, *Chemical Engineering Journal*, 2023, 452, 139044.

53 P. Andersson, G. Jesson, G. Kylberg, G. Ekstrand and G. Thorsén, *Anal Chem*, 2007, 79, 4022–4030.

54 J. Ducrée, S. Haeberle, S. Lutz, S. Pausch, F. Von Stetten and R. Zengerle, *Journal of Micromechanics and Microengineering*, 2007, 17, S103.

55 B. H. Park, J. H. Jung, H. Zhang, N. Y. Lee and T. S. Seo, *Lab Chip*, 2012, 12, 3875–3881.

56 S. Lai, S. Wang, J. Luo, L. J. Lee, S.-T. Yang and M. J. Madou, *Anal Chem*, 2004, 76, 1832–1837.

57 A. Olanrewaju, M. Beaugrand, M. Yafia and D. Juncker, *Lab Chip*, 2018, 18, 2323–2347.

58 M. Focke, F. Stumpf, G. Roth, R. Zengerle and F. Von Stetten, *Lab Chip*, 2010, 10, 3210–3212.

59 Q. Su, X. Ma, J. Dong, C. Jiang and W. Qian, *ACS Appl Mater Interfaces*, 2011, 3, 1873–1879.

60 Y. Wang, W. Qian, Y. Tan and S. Ding, *Biosens Bioelectron*, 2008, 23, 1166–1170.

61 L. Gou and C. J. Murphy, *Chemistry of Materials*, 2005, 17, 3668–3672.

62 X. Lu, M. Rycenga, S. E. Skrabalak, B. Wiley and Y. Xia, *Chemical synthesis of novel plasmonic nanoparticles*, 2009, vol. 60.

Chapter 10
Automatic Centrifugal Microfluidic Device

1. Background

The COVID-19 pandemic has highlighted the urgent need for efficient diagnostics of respiratory pathogens, which are highly transmissible and can cause both endemic and epidemic infections. Rapid and accurate identification of these viruses is crucial for effective disease control and management. As of January 2023, global confirmed COVID-19 cases exceeded 765 million. Additionally, seasonal influenza, caused primarily by Influenza A and B viruses, remains a significant health concern globally. Although rapid testing kits, such as those for SARS-CoV-2, have been invaluable, their accuracy, especially in the early stages of infection, is limited. The most reliable detection method remains the reverse transcription-polymerase chain reaction (RT-PCR) [1, 2], which, despite its accuracy, is labor-intensive and time-consuming. Integrated systems developed by various companies have attempted to enhance the accessibility and point-of-care capabilities of molecular diagnostics. However, these systems often have limitations, including high costs, single-sample analysis, lack of extraction steps, or complex pumping and valving systems.

Centrifugal microfluidic technology has emerged as a promising solution, capable of replacing external pump functions and simulating column extraction methods for nucleic acid purification. This technology allows for multiple units on a single chip [3–8]. However, previous centrifugal microfluidics devices have had limitations, such as the lack of full automatic integration or the necessity of multiple manual steps [8]. To advance the reach and point-of-care (POC) utility of molecular diagnostics, various companies have launched sophisticated automatic systems for pathogen detection. Notably, the BioFire FilmArray by bioMérieux, USA, approved by the FDA in 2018, has shown remarkable multiplexing abilities, capable of identifying multiple respiratory viruses and bacteria in a single assay [9]. Similarly, Roche Molecular Systems, Inc., USA, with their Cobas Liat system, obtained Emergency Use Authorization for a rapid POC nucleic acid test, distinguishing SARS-CoV-2, Influenza A, and B viruses via RT-PCR in about 20 minutes [10]. Abbott Diagnostics, USA, introduced the ID NOW, an isothermal nucleic acid amplification test using nasal swabs, delivering results in as little as 5 minutes for positive cases [11]. The GeneXpert system by Cepheid, USA, employs real-time RT-PCR for qualitative nucleic acid detection in respiratory samples, enhancing gene amplification efficiency through its innovative rotating valve system [12]. The adoption of microfluidics technology in these commercial products has spurred significant market growth, projected to reach USD 24,244.52 million by 2026. However, there are inherent limitations in these technologies, such as high costs (BioFire FilmArray), single-sample constraints (Cobas Liat), absence of extraction steps (ID NOW), and complex pumping systems (GeneXpert). In addition, the solution loading should be done by a manual pipetting.

In this chapter, we introduce a groundbreaking centrifugal POC system that automates the entire molecular diagnostic process, including solution loading, sample preparation, gene amplification, and data analysis. This microfluidics disc by Phan et al. (2023), with four separate units for analyzing seven targets each, allows for direct insertion of nasopharyngeal swabs into a solution loading cartridge [13]. It features a solution-loading cartridge for nasopharyngeal swabs, a zigzag-shaped aliquoting structure for rapid injection and division of reagents, and a robotic solution pipetting device for automated solution loading. Furthermore, a POC genetic analyzer equipped with a spindle motor, peltier heaters, and a fluorescence detector is constructed, enabling rapid and automatic diagnostics. Most importantly, the robot arm is integrated, enabling even the sample loading step to be performed automatically. The platform is designed for the simultaneous diagnosis of common respiratory viruses and has shown potential for high-speed, accurate, sensitive, and automatic molecular diagnosis based on real clinical sample detection and limit-of-detection tests.

1.1. Automation

The move towards automation in microfluidics has revolutionized the field, enhanced precision, and reduced manual intervention. This shift is pivotal in genetic analysis for infectious virus detection, demanding speed, and accuracy. Centrifugal microfluidic technology has evolved significantly, from manual operations to sophisticated automated systems over the past two decades. This evolution has enabled complex bioassays to be streamlined and has become crucial for point-of-care applications [6, 14]. The advent of fully automated systems has been essential for point-of-care testing, ensuring rapid, reliable diagnostics crucial for managing public health. During the COVID-19 pandemic, the importance of such systems was highlighted, automating the entire diagnostic process from RNA purification to result interpretation. Automation's integration into centrifugal microfluidic devices has led to the development of fully automatic platforms capable of handling multiple samples and targets in one run. This automation is made possible by control systems that precisely manage every operational aspect, broadening the potential applications of these devices.

The paper emphasizes the need for automation in centrifugal microfluidics, particularly for POC testing, where even the sample treatment step is automated. This approach is critical for enhancing the efficiency and reliability of diagnostic processes, especially in settings where quick and accurate results are paramount. To achieve full automation, the centrifugal POC genetic analyzer is integrated with a robotic solution pipetting device. This integration is a key advancement, enabling the automatic handling of solution loading and injection. The robotic device is modified with 3D-printed solution injecting components, capable of manipulating both a syringe and a pipette tip. The development of the novel centrifugal microfluidic disc, the POC genetic analyzer, and the robotic solution pipetting device collectively enables individuals without professional expertise to perform the entire molecular diagnostic process. This accessibility is crucial for broadening the scope of molecular diagnostics, allowing for its use in diverse settings, including remote or resource-limited areas. The automated system represents a significant advancement in the field of POC diagnostics. It combines the precision of molecular techniques with the ease of automation, making complex diagnostic procedures more accessible and user-friendly. This approach is particularly valuable in managing infectious diseases, where rapid and accurate diagnosis is essential for effective treatment and control.

This chapter delves into the design and development of an automatic centrifugal microfluidic device, detailing the components and technology that culminate in a fully autonomous system. It explores the novel integration of robotic arms and microfluidics, leading to innovations in sample processing and analysis. We will examine the components, integration, and operational challenges, culminating in the application of this technology for respiratory virus detection, a timely subject given the recent global health challenges. The focus will be on how this automated system could shape the future of diagnostics, making it accessible, efficient, and adaptable across various clinical scenarios.

1.2. Robot Arm

Robotic arms have become an indispensable tool in modern laboratories, particularly for their precision and consistency. In diagnostics, where accuracy is paramount, these robotic systems ensure that tasks such as liquid handling and sample sorting are performed with exceptional exactitude. This precision is important in molecular diagnostics, including PCR and microfluidic applications, where even minor discrepancies can significantly impact test outcomes. By reducing the risk of human error, robotic arms enhance the reliability and consistency of diagnostic results. One of the major advantages of employing robotic arms in laboratory settings is the significant increase in throughput. Capable of operating continuously without fatigue, these automated systems can handle multiple samples simultaneously, a feature in high-throughput screening and large-scale diagnostic testing. This efficiency is particularly beneficial in situations requiring rapid processing of a large number of samples, such as during epidemics or for large-scale research studies.

In the realm of microfluidics, robotic arms play a crucial role in automating the loading and manipulation of fluids on chips. Their integration into microfluidic systems is vital for developing point-of-care devices, aiming to simplify and expedite testing procedures. Automation in microfluidic devices is particularly relevant for onsite diagnostics and remote testing, where ease of operation and minimal manual intervention are key. The automation provided by robotic arms significantly reduces manual intervention in laboratory procedures. This aspect is crucial for maintaining sterile conditions, a vital component in various laboratory processes, especially those involving biological samples. By minimizing human handling, robotic arms help prevent cross-contamination, ensuring the integrity and purity of experimental and diagnostic results.

This chapter will examine the robotic solution pipetting device's design, its integration with the microfluidic disc, and how this combination propels the system towards complete automation. The discussion will highlight the design considerations, the operational programming of the robotic arm, and the challenges faced in harmonizing the robotic arm with the microfluidic system. We will also examine the broader implications of this technology, considering its potential to empower users without extensive technical training to conduct complex molecular diagnostics.

2. Design of components

2.1. A robotic solution pipetting device

The development of the robotic solution pipetting device was a key innovation in advancing the automation of centrifugal microfluidic systems for diagnostic purposes. In the study by Phan et al., the development of a robotic solution pipetting device is a key aspect in automating the process of molecular diagnostics. The robotic solution pipetting device, as conceptualized and implemented in Phan et al.'s study, represents a significant advancement in the automation of molecular diagnostic processes. This device was designed to enhance the precision and efficiency of solution handling within the microfluidic system, particularly in the context of respiratory virus detection from nasopharyngeal swab samples. The device plays a crucial role in the automated handling of solutions for molecular diagnostics. Upon inserting a nasopharyngeal swab and closing the lids of the cartridge, the robotic solution pipetting device is activated. It precisely loads the designated solutions from the solution racks onto the cartridge, which is securely attached to the microfluidic disc using strong double-sided adhesive tape. This setup prevents solution leakage and ensures the accurate dispensation of reagents necessary for the diagnostic process.

A commercial robotic arm from WLKATA Robotics forms the core of the device. This robotic arm serves as the traversing component, with its design and moving axis meticulously engineered for optimal performance in solution handling (Fig. 1A). The configuration of the robotic arm allows for precise movements, essential for the accurate transfer of solutions onto the microfluidic disc. The solution injecting component of the robotic device was specifically designed using Fusion 360 software. It was then fabricated using a Raise 3D E2 printer, utilizing polylactic acid as the material. This custom design ensures that the component is perfectly suited to the specific requirements of the microfluidic device, particularly in terms of solution handling and dispensation. To further refine the device's precision and control, two servo motors were employed (Fig. 1B). These motors are responsible for the elevation and gear maneuvering of the robotic arm, enabling it to precisely control the dispensing of solutions into the microfluidic system.

The robotic solution pipetting device, as part of the automatic centrifugal microfluidic system, includes a unique assembly that enables it to manipulate

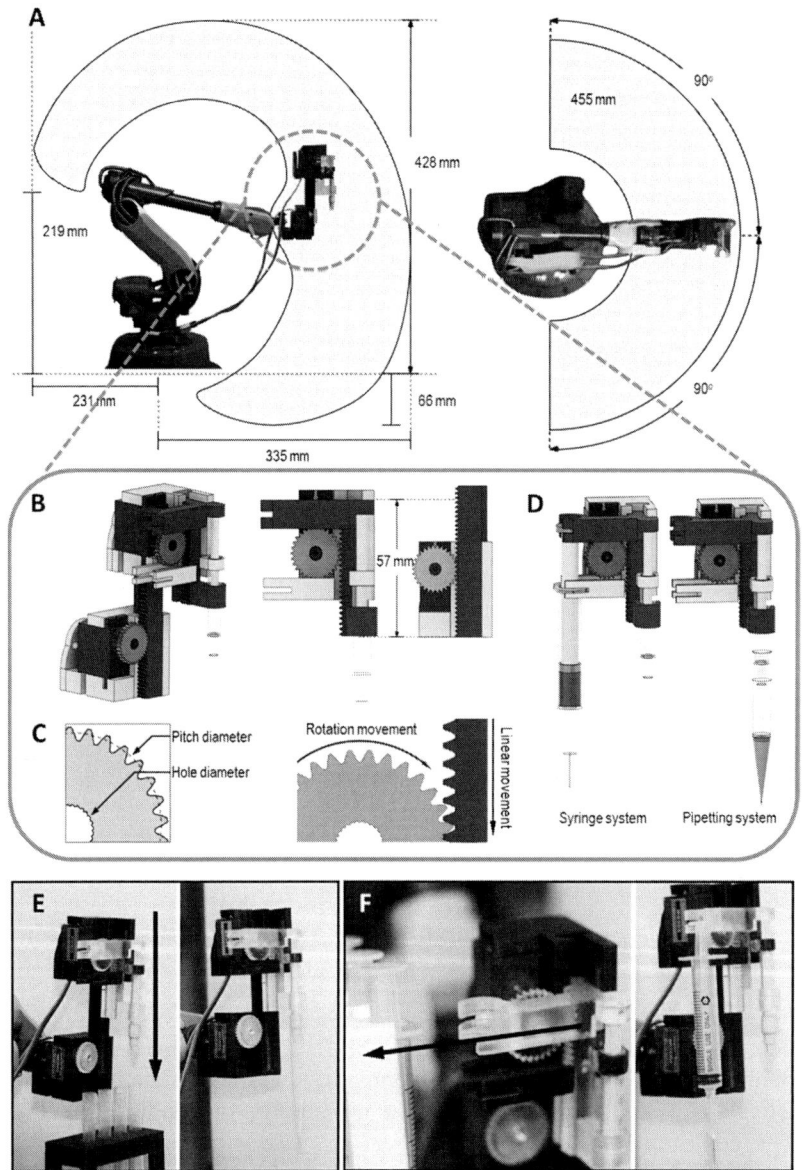

Fig. 1. The design of the robotic arm for the automatic solution injection. (A) Schematics for the movement range of the robotic arm. (B) Schematics for the 3D-printed solution injecting structure. (C) The rack and pinion mechanism. (D) The syringe system and the pipetting system. Digital images of (E) the automatic pipetting system and (F) the automatic syringe system. (Reprinted with permission from Ref. [13]. Copyright (2023) Elsevier.)

both syringes and pipette tips using a single motor. This capability is crucial for the automation of the molecular diagnostic process. At the heart of this device is a rack and pinion mechanism, which is instrumental in converting rotational movement into linear movement (Fig. 1C and D). This conversion mechanism results in notably smooth motion, essential for precise and controlled pipetting actions. Such smooth motion is critical in ensuring accuracy and reliability in the dispensing and transferring of liquids in the microfluidic system. The distance moved by the mechanism is calculated using the following formula:

$$\text{Distance to move} = \frac{\pi \times \text{Angle to move} \times \text{Pitch diameter}}{360}$$

A key feature of the device is the integration of a 1 mL Kovax-syringe, which is used for pipetting functions. This syringe, with a maximum input volume of 600 µL, is modified to fit a universal 1000 µL tip. Such customization allows for versatile use of the device in various diagnostic applications. The following formula allows for precise control over the movement of the pipette or syringe, ensuring accurate liquid handling.

$$\text{Volume} = \text{Distance to move} \times \pi \times (\text{Syringe radius})^2$$

To facilitate the removal of used pipette tips and enable subsequent injections, the device includes a ring frame with a smaller diameter than the pipette tip. This frame can move in sync with the dispensing motion, pushing the used tip into a waste container, thus automating the process of pipette tip disposal. The entire process of pipetting, including the movement and disposal of tips, is controlled through in-house coding, using Node-RED, a visual programming tool. This programming capability allows for the customization and automation of the pipetting process, adapting it to various diagnostic requirements.

The development of this robotic solution pipetting device marks a significant stride in automating molecular diagnostics, particularly for respiratory virus detection. By automating the process of solution loading and dispensing, the device minimizes the need for manual intervention, thereby enhancing the efficiency, accuracy, and reliability of the diagnostic process.

2.2. A solution loading cartridge

In the development of the automatic centrifugal microfluidic device by Phan et al., the solution loading cartridge plays a pivotal role. This component is crucial for the precise and secure loading of reagents necessary for molecular diagnostics. The cartridge is constructed using polymethyl methacrylate (PMMA), a material chosen for its durability and optical clarity. The main body of the cartridge is made from a 7 mm thick PMMA sheet, providing a robust structure to hold the solutions securely. The solution loading cartridge is designed to include four sample chambers, each with a capacity of 300 µL (blue color in Fig. 2C). These chambers are specifically intended to accommodate nasopharyngeal swabs, which serve as the primary sample type in the device. The physical design and dimensions of the chambers ensure that the swabs can be securely and effectively placed for processing. A key feature of the cartridge

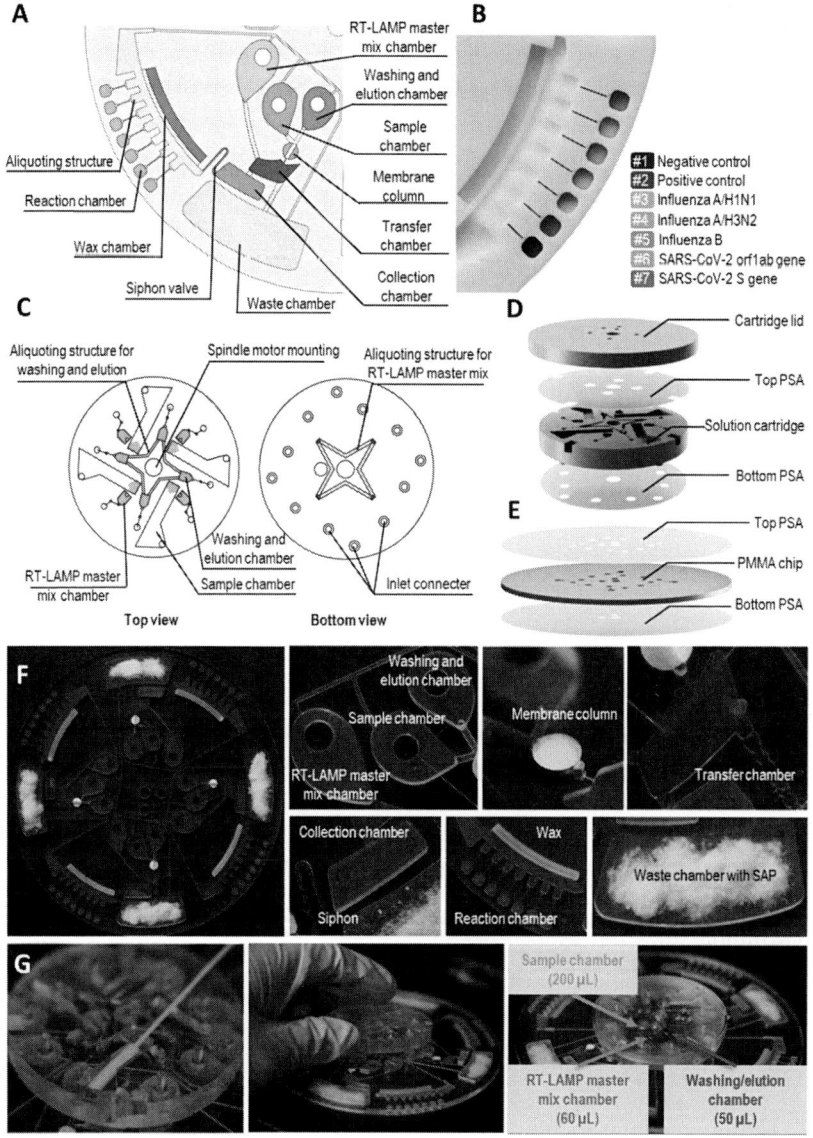

Fig. 2. The proposed centrifugal microfluidic disc and the solution loading cartridge. (A) Schematics of one unit of the disc. (B) Position order of the primers on the 7 reaction chambers. (C) Schematics for the solution loading cartridge having a zigzag aliquoting structures. Assembly of (D) the solution loading cartridge and (E) the microfluidic disc. (F) Digital images for the 4-unit centrifugal microfluidic disc. (G) Digital images for the insertion of a nasopharyngeal swab and the attachment of the cartridge on a disc. (Reprinted with permission from Ref. [13]. Copyright (2023) Elsevier.)

is its zigzag aliquoting structure. This innovative design facilitates the efficient and precise injection of various solutions, including a washing buffer, an elution buffer, and the RT-LAMP master mix (Fig. 2C). The structure is laid out in such a way that both the washing and elution buffers share the same aliquoting microchannel, while the RT-LAMP master mix is stored in a separate structure to prevent contamination. The zigzag structure is specifically engineered to fully fill the four chambers with a single injection shot, highlighting its efficiency in solution dispensing (green and orange colors in Fig. 2C). The dimensions of the structure between adjacent chambers are meticulously calculated at 1 mm in width and depth, ensuring proper distribution of solutions within the cartridge. The design also includes a channel connecting each chamber to the inlet connector, with dimensions of 0.2 mm in width and 0.15 mm depth. This feature is essential for directing the flow of solutions from the inlet to the respective chambers, ensuring accurate and controlled distribution of reagents. Fluid flow through a narrow channel experiences greater resistance than that through a wider channel, a phenomenon referred to as hydrodynamic resistance [15]. This resistance can be calculated using the Hagen-Poiseuille equation:

$$\Delta P = \frac{8\mu LQ}{(\pi r^4)}$$

where ΔP represents the pressure drop across the channel, μ is the dynamic viscosity of the fluid, L is the length of the channel, Q is the volumetric flow rate of the fluid, and r is the radius of the channel. A higher pressure drops results in greater hydrodynamic resistance. Consequently, under similar conditions, the channel's dimensions influence the fluid flow direction. To further reduce hydrodynamic resistance in the microchannel, the zigzag aliquoting structure was coated with a hydrophobic substance, Novec™ 1700. After curing at 80 °C for 30 minutes, the cartridge was sealed by two PSA films (Fig. 2D). This sealing mechanism is used for preventing any contamination, evaporation, or leakage of the solutions. It maintains the purity and concentration of the reagents, which is essential for the accuracy and reliability of the diagnostic tests performed using the microfluidic device. The cover of the cartridge, fabricated from a 2 mm thick PMMA sheet, ensures a tight seal while allowing for visual inspection of the contents. The choice of PMMA not only ensures structural integrity but also facilitates the handling and observation of the solutions within the cartridge.

The solution loading cartridge, as conceptualized and implemented by Phan et al., represents an important element in the automated centrifugal microfluidic system. Its thoughtful design and fabrication underscore the precision and reliability in molecular diagnostics, particularly in the context of rapid and accurate detection of respiratory infectious viruses.

2.3. Design and fabrication of a centrifugal microfluidic disc

The centrifugal microfluidic disc in Phan et al.'s study was intricately designed and fabricated to facilitate the detection of various respiratory viruses. The disc was designed using Cut2D software, which allowed for precise layout and structuring of its components. It was fabricated using a CNC machine, ensuring high accuracy in its construction. The disc featured four units, each equipped with a silica filter column for RNA purification and seven reaction chambers for target amplification (Fig. 2B). This multi-unit design allowed for the simultaneous processing of multiple samples, enhancing diagnostic throughput and efficiency. A siphon valve within the disc served as an on-off valve, critical for fluid control. High centrifugal force keeps the solution secured in the collection chamber, while low centrifugal force permits flow through the thin channel, allowing for precise fluid management. PMMA, which is compatible with RT-LAMP amplification solutions, was used for the disc's fabrication. The PMMA sheets were 3 mm thick with a diameter of 130 mm, providing a sturdy base for the microfluidic structures.

After CNC etching, the disc underwent several post-processing steps, including sonication in Deconex 11 Universal cleaner and rinsing with DNase-free water and ethanol for cleaning. It was then sterilized using a UV lamp in a clean bench. The siphon valve was coated with a hydrophilic mixture of 6% Vistex 111-50 reagent in 33% isopropanol, while other surfaces were coated with a hydrophobic substance. This coating ensured proper fluid movement and control within the disc. The disc was cured at 80 °C for 30 minutes to stabilize the coated layers. Super absorbent polymer (SAP) was weighed and divided into the four waste chambers to prevent backflow, ensuring directional fluid flow (Fig. 2F).

The chamber #1 was a negative control without any primer set, while a primer set targeting the human 18S ribosomal RNA was loaded into the chamber #2 as a positive control (Fig. 2B). Primer sets for Influenza A (H1N1, H3N2) and Influenza B were loaded into chambers #3, #4, and #5, respectively. For Influenza A (H1N1 and H3N2), two primer sets were mixed in a 1:1 ratio. Primer sets for SARS-CoV-2 orf1ab and S genes were loaded into chambers #6 and #7, enabling the detection of seven targets simultaneously. The disc was then sealed with two PSA films (Fig. 2E).

The design and fabrication of this centrifugal microfluidic disc highlights the intricate engineering and attention to detail necessary for creating a functional and efficient diagnostic tool for respiratory viruses. The disc's multiple units, coupled with the precise control mechanisms and careful selection of materials, demonstrate its potential in enhancing the automation, accuracy, and efficiency of diagnostic processes.

3. Integration of a robotic solution pipetting device with a centrifugal chip

The integration strategy for the centrifugal microfluidic system orchestrates a harmonious interaction between the robotic arm, the solution loading cartridge, and the microfluidic chip, designed to ensure a seamless workflow for molecular diagnostics. The integration of the robotic solution pipetting device with the centrifugal microfluidic disc in the study by Phan et al. represents a significant advancement in automating the molecular diagnostic process. In Phan et al.'s study, a key focus was automating the sample treatment step in centrifugal microfluidics, particularly for POC testing. Traditionally, while centrifugal microfluidic technology allowed for automated operations like disc rotation, bioassay execution, and optical sensing, the injection of reagent solutions onto the disc remained a manual process [4, 8]. Automating this step was crucial to achieve full automation and enhance the functionality and efficiency of the system.

The solution to this challenge was the integration of a commercial robotic arm with 3D-printed solution injecting components. This modification enabled the robotic arm to manipulate solution loading and injection using both a syringe and a pipette tip. The entire sample treatment process, including picking up the pipette tip or syringe, suctioning the designated solution, transporting the solution-loaded pipette/syringe to the disc, and controlling the height for accurate solution dispensing, was programmed through in-house coding. This programmability allowed for precise control over each step, ensuring the accurate and efficient delivery of solutions to the microfluidic disc.

In tandem with the development of the robotic solution pipetting device, Phan et al. designed a novel centrifugal microfluidic disc. This disc was specially created to work seamlessly with the robotic device, facilitating the automated handling of molecular diagnostic processes. The integration of the robotic solution pipetting device with the centrifugal microfluidic disc led to the creation of an advanced POC system. This system could perform all necessary molecular diagnostic processes, including solution loading, sample pretreatment, gene amplification, and data display, in a fully automated manner.

The authors believe that this full integration marks a significant advancement in centrifugal microfluidics, realizing a truly automatic platform. This development is particularly important for facilitating rapid, user-friendly, and efficient molecular diagnostics, especially in point-of-care settings. This innovative integration demonstrates how automation can revolutionize molecular diagnostics, making complex processes more accessible and less reliant on professional expertise. It highlights the potential of combining robotic technology with microfluidic systems to enhance the capabilities and reach of diagnostic tools.

4. Construction of a portable genetic analyzer

The construction of the portable POC genetic analyzer embodies a significant advancement in the field of molecular diagnostics. This sophisticated device integrates the key stages of genetic analysis into a compact, user-friendly system designed for rapid deployment in various clinical settings. The POC genetic analyzer's design is centered around portability and ease of use without compromising the analytical capabilities. The centrifugal POC system integrates a spinning motor, essential for the centrifugal distribution of reagents, with precision temperature control facilitated by peltier heaters (Fig. 3A). This combination ensures that the conditions for the RT-LAMP are optimal throughout the assay. Apart from the spinning motor and peltier heaters, the team incorporates a miniaturized fluorescence detector, and a touch screen for operational control and data display. Central to the analyzer's functionality are the peltier heaters. These heaters are important for controlling the reaction temperature and melting the wax in the microfluidic disc. Precise temperature control is vital for the success of the RT-LAMP reactions conducted within the

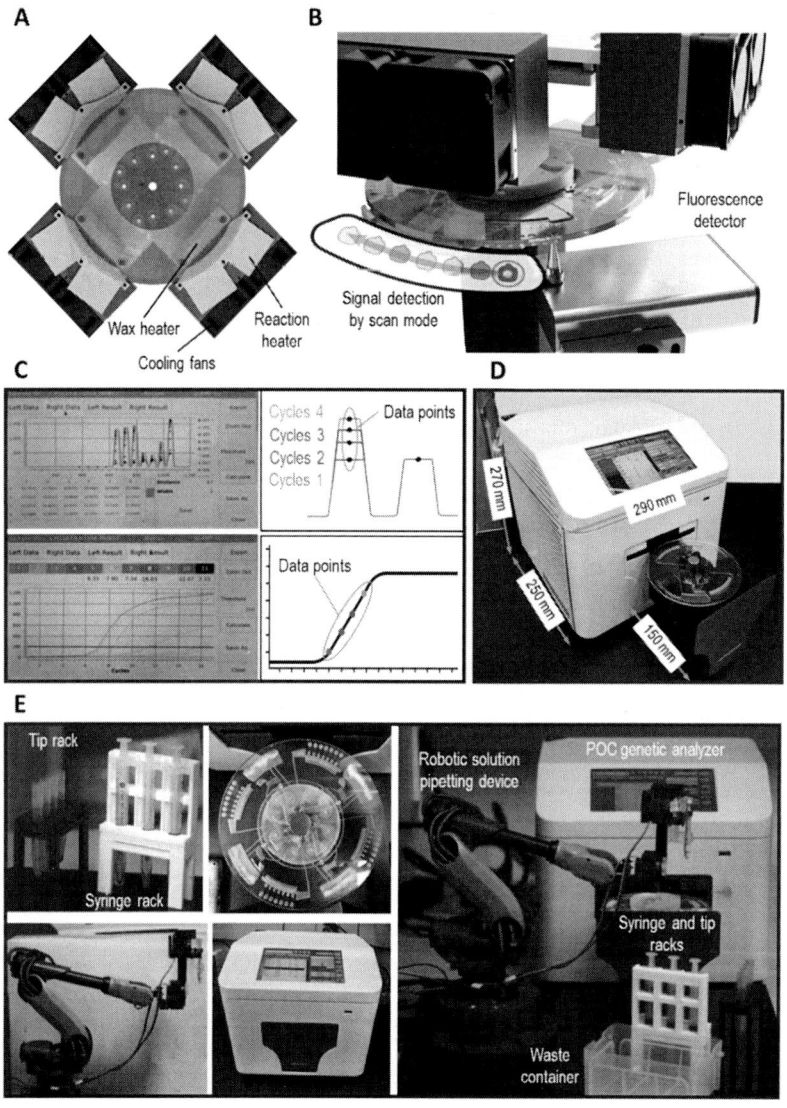

Fig. 3. The design of the POC genetic analyzer. (A) Schematics of the peltier heaters and the cooling fans. (B) The scan mode of the fluorescence detector to measure the fluorescence signal of the reaction chambers. (C) Amplification results in the LED screen: 1500 data point display (upper panel) and the amplification curve (bottom panel). (D) Dimension of the POC genetic analyzer. (E) Digital image of the fully integrated centrifugal molecular diagnostic platform. (Reprinted with permission from Ref. [13]. Copyright (2023) Elsevier.)

device. Each peltier heater in the analyzer is attached to a heatsink and a cooling fan. This assembly is designed to facilitate rapid temperature regulation, ensuring that the device can quickly adjust to the optimal temperatures required for different stages of the diagnostic process. The spinning motor in the analyzer is responsible for rotating the centrifugal microfluidic disc. This rotation is essential for moving liquids through the microfluidic pathways, enabling the automated processing of samples and reagents.

The fluorescence detector is a critical component for detecting the amplified genetic material from the samples. It scans the reaction chambers in the microfluidic disc and measures the fluorescence signals, which are indicative of the presence or absence of target pathogens (Fig. 3B). The optics having an excitation wavelength of 470 nm and an emission wavelength of 520 nm were suitable for detecting the intercalating dye of SYTO 9. Real-time amplification data were displayed on a touch screen, presenting up to 1500 data points in one measurement to scan the seven reaction chambers (Upper panel in Fig. 3C). For the analysis of amplification data, the POC genetic analyzer employs a sophisticated 5-parameter sigmoidal (5PL) model fitting. This advanced statistical method is instrumental in generating non-linear or logistic regression curves that accurately represent the amplification process based on the fluorescent signals. The 5PL model fitting is particularly adept at handling the variable sigmoidal shapes of biological assays, providing a robust and sensitive analysis that enhances the reliability of the diagnostic results [16]. By applying this model, the analyzer can precisely quantify the concentration of nucleic acids present in the sample, even at low levels, making it extremely effective for early detection of pathogens. The model's parameters are fine-tuned to match the dynamics of the RT-LAMP reaction, ensuring that each amplification curve reflects the actual biological process with high fidelity.

A standout feature of the POC genetic analyzer is its ability to function with minimal user intervention. The analyzer's internal mechanisms are designed to perform automatic release of solutions, gene amplification, and signal detection. A user only needs to insert the nasopharyngeal swab into the cartridge; from there, the system manages the entire diagnostic process, culminating in the display of the results. The analyzer features an ergonomic interface with a touchscreen that simplifies the operational procedure. This design choice makes the system accessible to operators with varying levels of expertise and is particularly advantageous in high-pressure situations where rapid diagnosis is paramount.

The touch screen interface provides an interactive and user-friendly means of controlling the analyzer. It displays experimental parameters, real-time data, and diagnostic results, making the information easily accessible to the operator. The POC genetic analyzer's design highlights the integration of various technological elements to create a fully automated, efficient, and user-friendly platform for molecular diagnostics. Its compact size, 290 mm [width] × 250 mm [length] × 270 mm [height] (Fig. 3D), combined with advanced heating, cooling, and detection systems, makes it ideal for POC applications, especially in rapid testing scenarios such as those required for respiratory virus detection.

5. Chip operation

The operation of the centrifugal microfluidic disc capitalizes on a symphony of forces arising from rotational motion, namely centrifugal force, Euler force,

Coriolis force, and capillary force, to meticulously control the flow of solutions within the diagnostic device [3, 17]. At the start, the robotic arm picks up the AVL lysis buffer and injects it into the nasopharyngeal swab-incorporated cartridge. This step is repeated for each of the four sample chambers in the cartridge, with the robotic arm discarding the used tip after each injection and rotating the centrifugal disc by 90° for subsequent injections. Following a 10-minute incubation for viral lysis, the centrifugal microfluidic disc is spun counter-clockwise at +5000 rpm for 60 seconds. This action, driven by the Coriolis force, allows the capture of viral RNAs on GF/F glass filter papers, while other residues are directed to the waste chamber. The robotic arm then picks up the AW2 washing buffer with a syringe and injects it into the washing buffer chamber. The washing buffer fills four chambers through the zigzag aliquoting structure. After spinning the disc at +5000 rpm for 60 seconds, the buffer passes through the filter paper, washing away undesired residues. This process is repeated for the AVE elution buffer and the RT-LAMP master mix. After the elution buffer and RT-LAMP master mix chambers are filled, the disc is rotated for 60 seconds at -5000 rpm to drive both solutions into the collection chamber. Here, the elution solution delivers purified RNAs to the chamber. The mixture is retained due to the dominant centrifugal force over capillary force. The mixture of eluted RNAs and RT-LAMP master mix is thoroughly mixed using Euler force by rotating 180° at ±1000 rpm. This step ensures proper homogenization of the mixture before the RT-LAMP reaction. The mixture then flows into the aliquoting structure and is directed to the 10 µL reaction chambers. The wax is melted at 80 °C for 3 minutes to seal the chambers and prevent backflow and cross-contamination. The RT-LAMP reaction then proceeds at 64 °C, with amplification signals measured every 3 minutes.

This operation sequence, as demonstrated in Fig. 4, reflects the sophisticated integration of robotic automation with centrifugal microfluidics. It illustrates a significant advancement in automating molecular diagnostics, particularly in terms of efficiency, precision, and ease of use.

6. Respiratory infectious virus detection from nasopharyngeal swab samples

The performance evaluation of the centrifugal molecular diagnostic platform was a comprehensive process, assessing various critical aspects to ensure its effectiveness in detecting respiratory viruses. One of the first tests conducted was to evaluate the potential for cross-contamination between reaction chambers during the RT-LAMP reaction. This is a crucial aspect, as cross-contamination could lead to false positives and compromise the accuracy of the diagnostics. The testing protocol involved placing a primer set for the positive control in chamber #1, while the primer set for Influenza A H3N2 was alternately coated in chambers #2, #4, and #6. The results showed fluorescence signals only in the chambers where the specific primers were placed, and gel-electrophoresis confirmed these findings (Fig. 5A). The absence of fluorescence signals in the other chambers indicated no cross-contamination, validating the integrity and reliability of the reaction chambers.

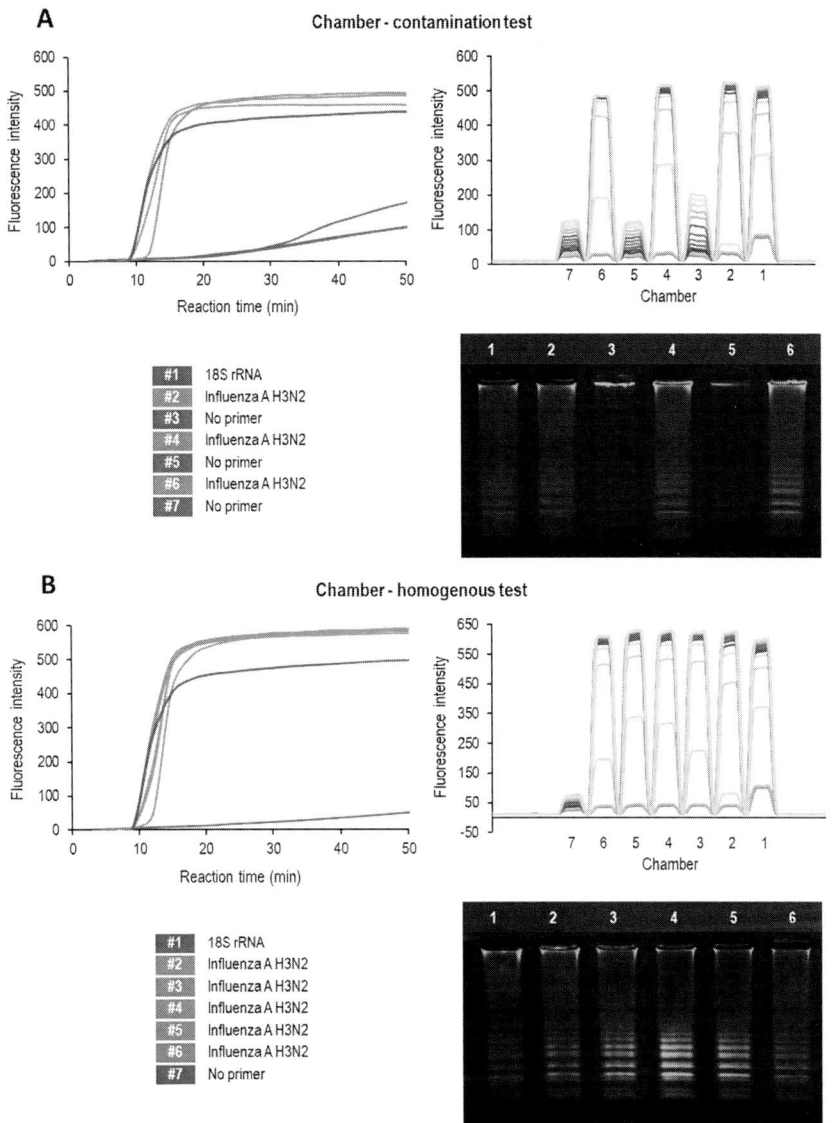

Fig. 5. (A) On-chip cross-contamination test. (B) On-chip homogeneity test for the aliquoting performance. (Reprinted with permission from Ref. [13]. Copyright (2023) Elsevier.)

The accuracy of aliquoting, or the division of the sample and reagents into the reaction chambers, is important for consistent and reproducible results. To test this, a primer set for 18S rRNA was placed in chamber #1 as a positive control, while chambers #2 to #6 received another primer set. The amplification curves from these chambers were analyzed for uniformity. Remarkably, all chambers showed homogeneous amplification curves, with a consistent threshold time of 11 minutes across the chambers (Fig. 5B). This uniformity in reaction times across multiple chambers demonstrated the precision of the aliquoting process in distributing the washing, elution, and RT-LAMP master mix solutions.

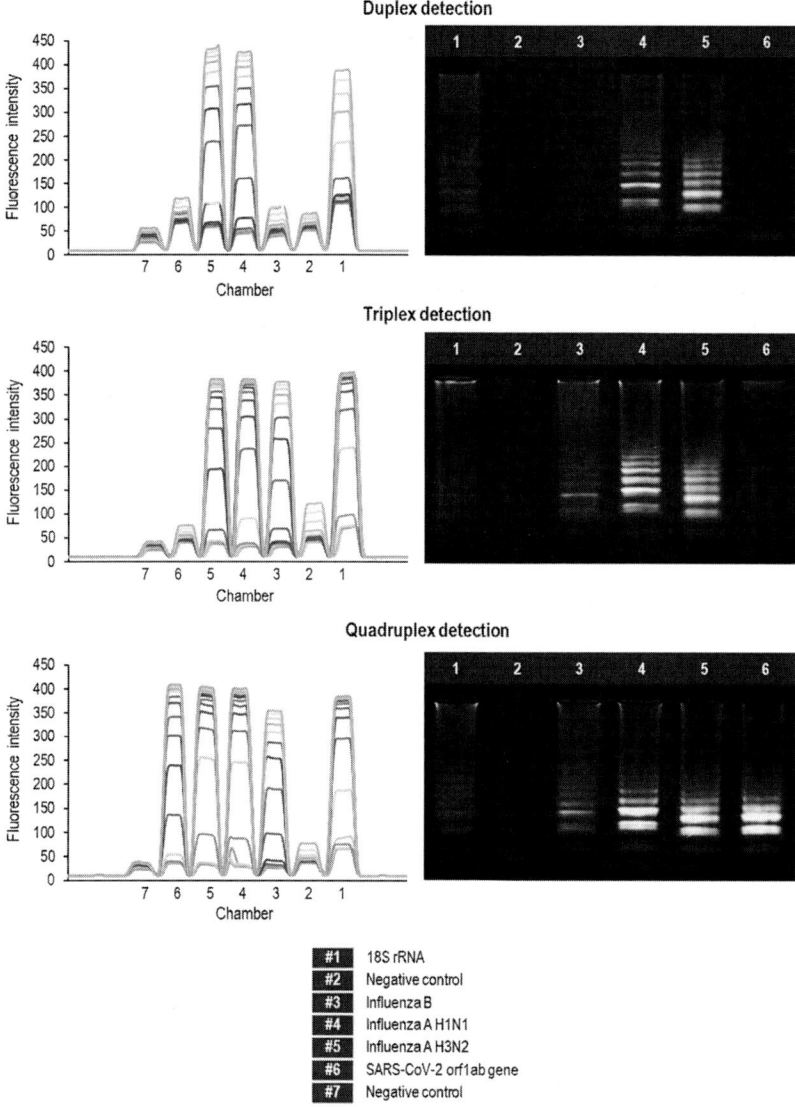

Fig. 6. Evaluation of the capability of the simultaneous multiple virus detection using the proposed POC molecular diagnostic platform. (Reprinted with permission from Ref. [13]. Copyright (2023) Elsevier.)

A critical aspect of the platform's performance was its ability to simultaneously detect multiple respiratory viruses in a single test. This capability was evaluated through duplex, triplex, and quadruplex tests (Fig. 6). In the duplex test, equal amounts of Influenza A H1N1 and H3N2 samples were mixed in a 1:1 volume ratio and tested. The results showed specific fluorescence signals in the corresponding chambers for both H1N1 and H3N2, in addition to the positive control. The triplex test involved a 1:1:1 mixture of Influenza A H1N1, H3N2, and Influenza B, with the platform successfully identifying each virus in its respective chamber. The quadruplex test extended this to include SARS-CoV-2, with the system effectively detecting all four viruses. These tests highlighted the platform's exceptional multiplexing capabilities, a significant advantage for comprehensive viral diagnostics.

To validate the platform's practical application, 19 clinical samples were tested. These samples included four cases each of Influenza A subtypes H1N1 and H3N2, seven cases of Influenza B, and four cases of SARS-CoV-2. The total process time from sample to result was 1 hour and 30 minutes, fully automated without manual intervention. The diagnostic performance was impressive, with all 19 samples showing amplification before 45 minutes (Fig. 7). The 18S rRNA gene control consistently amplified before 20 minutes in all trials, ensuring no false negatives. The influenza subtypes showed average threshold times of 25.32 minutes for H1N1 and 21.96 minutes for H3N2. Influenza B samples also showed robust amplification with an average threshold time of 27.13 minutes. For SARS-CoV-2, both the S gene and orf1ab gene were monitored, showing early detection in most samples.

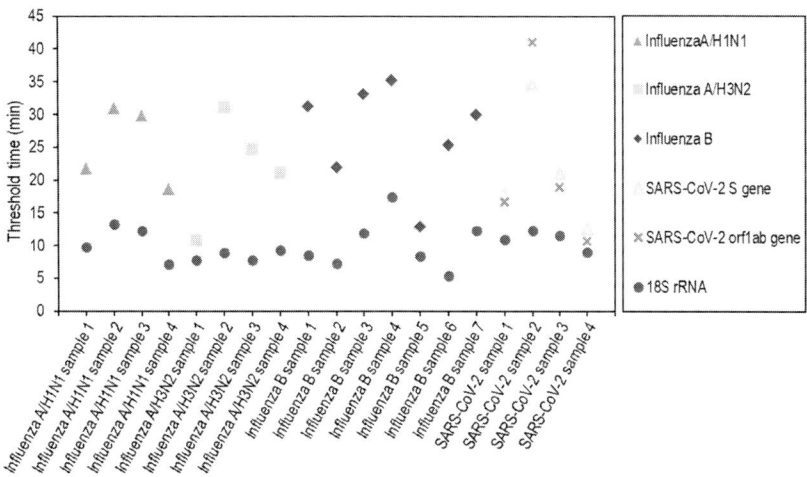

Fig. 7. Results of the clinical sample test with the centrifugal molecular diagnostic platform for detecting 4 respiratory viruses. (Reprinted with permission from Ref. [13]. Copyright (2023) Elsevier.)

The sensitivity of the platform was further assessed through an LOD test using commercially available samples of Influenza A subtypes H1N1, H3N2, and Influenza B. These samples were serially diluted to various concentrations and tested, with the results indicating an LOD of 10^1 pfu/mL for influenza viruses (Fig. 8). The amplification curves and threshold times were consistent with these findings, demonstrating the platform's high sensitivity suitable for respiratory virus diagnostics and are comparable with other references related to respiratory virus diagnostic [18–20].

These comprehensive tests underscore the robustness, accuracy, and efficiency of the centrifugal molecular diagnostic system. The platform's ability to prevent cross-contamination, achieve uniform aliquoting, and simultaneously detect multiple viruses, along with its proven efficacy with clinical samples and high sensitivity, highlights its potential as a rapid, accurate, and automated solution for common respiratory virus detection.

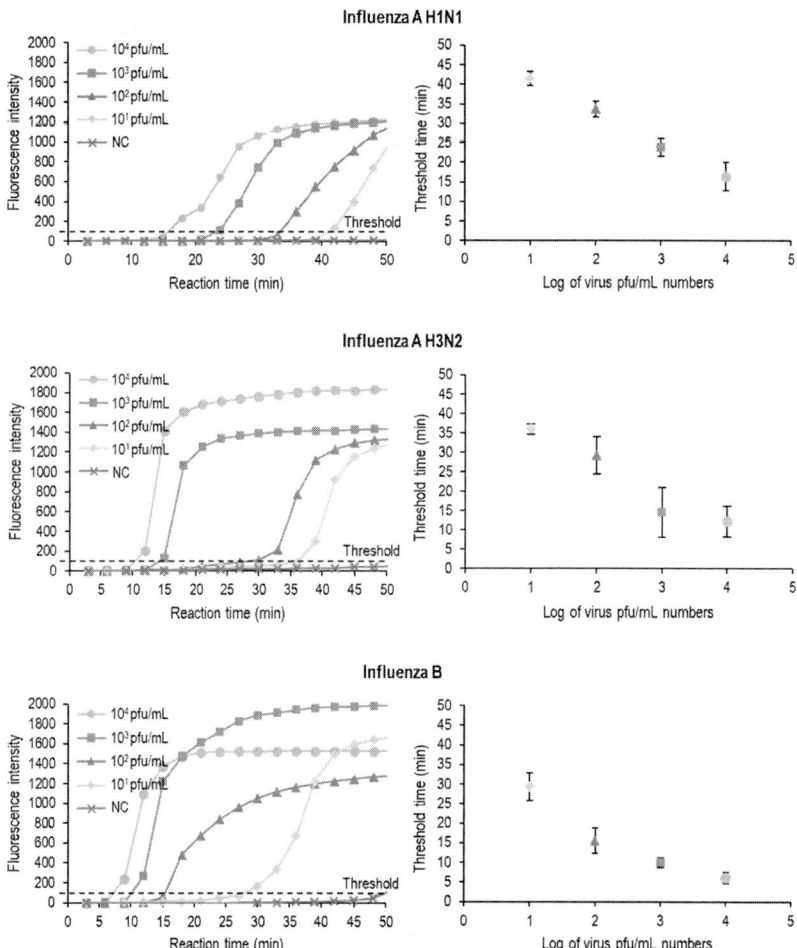

Fig. 8. LOD test for influenza A H1N1, influenza A H3N2 and influenza B on the proposed automatic POC genetic analyzer. (Reprinted with permission from Ref. [13]. Copyright (2023) Elsevier.)

7. Conclusion

This chapter has explored the intricate design, operation, and diagnostic capabilities of automatic centrifugal microfluidic devices, with a particular focus on their application in the detection of respiratory infectious viruses. The journey from conceptual design to clinical validation illustrates the transformative potential of these devices in the realm of point-of-care diagnostics. The presented centrifugal microfluidic platform represents a significant advancement in molecular diagnostics, leveraging the principles of automation, precision engineering, and user-friendly interface design. It successfully integrates a robotic arm for solution handling, a sophisticated cartridge for sample preparation, and a microfluidic chip for nucleic acid purification and amplification. This integration culminates in a portable genetic analyzer capable of real-time, sensitive detection of various viral pathogens. The device's operation, hinging on the coordination of rotational forces and precise temperature control, demonstrates an elegant solution to the challenges of miniaturization and automation. The design of RT-LAMP primers and the chip's architecture ensure broad strain coverage and efficient biochemical processing, which are critical for accurate diagnostics. Clinical validation results underscore the platform's high sensitivity and specificity, with LODs comparable to or exceeding those of conventional laboratory methods. Such performance paves the way for its deployment in diverse settings, from traditional labs to remote locations lacking sophisticated infrastructure. As we look forward, the scalability and adaptability of these devices suggest a promising avenue for rapid response to emerging infectious diseases. The potential for integration with digital health records and telemedicine platforms could further enhance their impact, leading to improved disease surveillance and management. In summary, the automatic centrifugal microfluidic device stands as a testament to the ingenuity of modern biomedical engineering, offering a glimpse into a future where rapid, accurate, and accessible diagnostics are the norm, not the exception. The implications for global health, particularly in the management of infectious diseases, are profound and hold the promise of a more responsive and resilient healthcare ecosystem.

References

1. W. Feng, A. M. Newbigging, C. Le, B. Pang, H. Peng, Y. Cao, J. Wu, G. Abbas, J. Song, D.-B. Wang, H. Zhang and X. C. Le, *Anal Chem*, 2020, 92, 10196–10209.

2. S. A. Bustin, *J Mol Endocrinol*, 2000, 25, 169–193.

3. H. V. Nguyen, V. D. Nguyen, H. Q. Nguyen, T. H. T. Chau, E. Y. Lee and T. S. Seo, *Biosens Bioelectron*, 2019, 141, 111466.

4. V. D. Nguyen, H. Van Nguyen, J. W. Seo, S. H. Lee and T. S. Seo, *Biosens Bioelectron*, 2022, 199, 113877.

5. H. Van Nguyen and T. S. Seo, *Biosens Bioelectron*, 2021, 181, 113161.

6. H. Van Nguyen, V. M. Phan and T. S. Seo, *Sens Actuators B Chem*, 2022, 353, 131088.

7. O. Strohmeier, M. Keller, F. Schwemmer, S. Zehnle, D. Mark, F. Von Stetten, R. Zengerle and N. Paust, *Chem Soc Rev*, 2015, 44, 6187–6229.

8. Y. Yao, X. Chen, X. Zhang, Q. Liu, J. Zhu, W. Zhao, S. Liu and G. Sui, *ACS Sens*, 2020, 5, 1354–1362.

9. D. M. Webber, M. A. Wallace, C.-A. D. Burnham and N. W. Anderson,

J Clin Microbiol, 2020, 58, e00343.

10. G. Hansen, J. Marino, Z.-X. Wang, K. G. Beavis, J. Rodrigo, K. Labog, L. F. Westblade, R. Jin, N. Love, K. Ding, J. Sickler and N. K. Tran, *J Clin Microbiol*, 2021, 59, e02811.

11. W. Stokes, A. A. Venner, E. Buss, G. Tipples and B. M. Berenger, *Clinical Microbiology and Infection*, 2023, 29, 247–252.

12. T. Phan, A. Mays, M. McCullough and A. Wells, *Journal of Clinical Virology Plus*, 2022, 2, 100067.

13. V. M. Phan, H. Q. Nguyen, K. H. Bui, H. Van Nguyen and T. S. Seo, *Sens Actuators B Chem*, 2023, 394, 134362.

14. H. Van Nguyen, V. M. Phan and T. S. Seo, *Sens Actuators B Chem*, 2024, 399, 134771.

15. J. X. J. Zhang and K. Hoshino, *Molecular Sensors and Nanodevices*, 2014, 103–168.

16. H. Q. Nguyen, V. D. Nguyen, H. Van Nguyen and T. S. Seo, *Sci Rep*, 2020, 10, 15123.

17. S. Li, C. Wan, B. Wang, D. Chen, W. Zeng, X. Hong, L. Li, Z. Pang, W. Du, X. Feng, Y. Li and B.-F. Liu, *Anal Chem*, 2023, 95, 6145–6155.

18. R. Bukasov, D. Dossym and O. Filchakova, *Anal Methods*, 2021, 13, 34–55.

19. B. V. Ribeiro, T. A. R. Cordeiro, G. R. Oliveira e Freitas, L. F. Ferreira and D. L. Franco, *Talanta Open*, 2020, 2, 100007.

20. E. A. Tarim, B. Karakuzu, C. Oksuz, O. Sarigil, M. Kizilkaya, M. K. A. A. Al-Ruweidi, H. C. Yalcin, E. Ozcivici and H. C. Tekin, *Emergent Mater*, 2021, 4, 143–168.

Author

Dr. Tae Seok Seo is a Professor of Chemical Engineering at Kyung Hee University, South Korea. He earned his bachelor's degree from Seoul National University, Korea (1996), and his master's degree from KAIST, Korea (1998). He completed his doctoral degree at Columbia University, USA (2004), where he developed a next-generation sequencing technology. Subsequently, he moved to UC Berkeley, USA as a post-doctoral researcher (2007) and contributed to point-of-care genetic analysis projects based on the lab-on-a-chip technique.

Dr. Seo held a faculty position at KAIST from 2007 and has been at Kyung Hee University since 2016. Well-regarded for his research in microfluidics, molecular diagnostics, biosensors, flow chemistry, and nanobiotechnology, Prof. Tae Seok Seo has authored over 150 scientific papers and holds more than 60 patents, in addition to facilitating 8 technology transfers to venture companies. His notable recognitions include Excellent Result 100 in National Research, KAIST Technology Innovation Award, and the Best Industrial Technical Minister Award, etc. He lives in Suwon with his wife and four children. He enjoys golfing, meditating, and participating in Catholic religious activities.

Editor:

Prof. Tae Seok Seo

Kyung Hee University

Yongin-si, South Korea

Cover illustration:

The graphic was provided by Dr. Hau Van Nguyen and Mr. Vu Minh Phan

© 2024 EduContentsHuepia Books

All rights reserved including those of translation into other languages. No part of this book may be reprinted or reproduced in any form – by photoprinting, microfilm, or any other means – nor transmitted or translated into machine language without written permission from the author.

Registered names, trademarks, etc. used in this book, even when not specifically marked as such, are not to be considered unprotected by law.

First edition on February 24th, 2024

Identifiers: ISBN 978-89-6356-451-7